Environmental Science

Series Editors: R. Allan • U. Förstner • W. Salomons

Springer
Berlin
Heidelberg
New York
Hong Kong
London
Milan
Paris
Tokyo

Dmitry G. Matishov
Gennady G. Matishov

Radioecology in Northern European Seas

With 163 Figures

Springer

Dr. Dmitry G. Matishov
Murmansk Marine Biological Institute
Vladimirskaya St. 17
183010 Murmansk,
Russia
Email: dima_matishov@mmbi. info

Prof. Gennady G. Matishov
Murmansk Marine Biological Institute
Vladimirskaya St. 17
183010 Murmansk,
Russia
Email: matishov@mmbi.info

ISSN 1431-6250
ISBN 3-540-20197-1 Springer-Verlag Berlin Heidelberg New York

Cataloging-in-Publication Data applied for

A catalog record for this book is available from the Library of Congress.

Bibliographic information published by Die Deutsche Bibliothek
Die Deutsche Bibliothek lists this publication in the Deutsche Nationalbibliografie; detailed bibliographic data is available in the Internet at <http://dnb.ddb.de>.

Springer-Verlag Berlin Heidelberg New York
a part of Springer Science+Business Media GmbH

springeronline.com

Printed in Germany

Camera ready by the author
Cover design: Struve & Partner, Heidelberg
Printed on acid-free paper 30/2132/AO 5 4 3 2 1 0

Foreword

This volume represents a major synthesis of literature on the oceanographic and biological conditions influencing the transfer and accumulation of radionuclides throughout the Northern European seas. The volume provides an integrated view of artificial radionuclide transformation and trophic dynamics; inputs of global, regional and local sources to the radioactivity inventory of the Polar Ocean, its marginal seas, and North Atlantic ecosystems; and the bio-oceanographic processes and phenomena that contribute to the assimilation capacity of marine ecosystems for radioactive pollutants. Through their extensive activities in this remarkable region the authors have achieved a unique perspective on radioecology which they encapsulate in this volume.

The presented information was derived through several decades of investigations carried out by many dedicated scientists from Murmansk Marine Biological Institute. The institute itself is located in the Kola region, an area of significance from a radiological perspective, as it contains the world's highest concentration of nuclear reactors, active and derelict. The book presents their contributions, augmented and complimented by others, to the understanding of radionuclide migration and accumulation in the environment and biota of the Arctic and Sub-Arctic seas. Much of the data has never appeared in western scientific literature and thus provides a window into literature that otherwise is not readily available. This synthesis is rich in data about anthropogenic radionuclide sources, transportation, fluxes and budgets. From their unique multi-disciplinary perspective the authors present a new methodical and theoretical framework describing radionuclide bioaccumulation by marine invertebrate and vertebrate animals, with special emphasis on marine food webs leading to humans.

The original translation of this work was completed by Sergey L. Dzhenyuk. The final translation, conducted by myself, has been a large project requiring considerable effort. Through this project, I have gained a greater awareness and appreciation of historical and contemporary perspectives and issues in Russian radioecology. I am pleased to have had the opportunity to work with the authors in preparing this book on the radioecology of the northern European seas.

Dr. JoLynn Carroll
Akvaplan-niva
Polar Environmental Center
Tromsø, Norway

Contents

Introduction

The phenomenon of artificial radioactivity in the environment has existed since the beginning of nuclear testing in 1945. About 1500 nuclear explosions have been carried out worldwide; more then 90 % of them have been carried out by the United States and Russia in the Northern Hemisphere. Nuclear fallout resulting from atmospheric and underwater tests is the largest source of anthropogenic radioactivity found in the world's oceans. Nuclear testing is followed by atomic power station accidents and wastewater discharges from European and Siberian nuclear fuel processing industries (Polikarpov 1964a, b; 1971; Shvedov end Patin 1968; Kuznetsov 1971; Polikarpov end Aarcrog 1993; Braginsky et al. 1994; Izrael et al. 1994; Kuznetsov et al. 1994; Matishov et al. 1994a,b,c; Strand et al. 1997). More than 150 nuclear-related accidents have occurred since the beginning of the nuclear era: the Chernobyl accident being the greatest environmental disaster based on both the amount of radioactivity released and areal extent of contaminated land. The characteristic radioactive signature of Chernobyl fallout can be identified on continents and sea, in land and marine organisms.

Radioactive elements enter the Ocean and seas attached to particles suspended in seawater or in a dissolved form. Today anthropogenic radionuclides are found in seawater in either form from the Ocean surface to its bottom. Artificial radionuclides propagate through the oceans by the major current systems, spreading many thousands of kilometers from sources.

In the northern seas and the Arctic, dumping and burial of liquid and solid radioactive waste was carried out during the second half of the 20th century. During this time period, nuclear submarines and icebreakers equipped in total by 180 nuclear reactors cruised the waters of the Barents and Kara Seas (Nilsen et al. 1996). Anthropogenic radionuclides entered the sea becoming a part of the ocean biogeochemical cycle and entering all components of the marine ecosystem.

The accidental sinking of the atomic submarines «Komsomolets» (07.04.1989) and «Kursk» (12.08.2000) and possible radioactive leaks from the atomic reactors were cause for concern by the potentially affected populations. The potential radioactive contamination of marketable fish species harvested from the Barents Sea has a psychological influence on fishery economics in northern countries.

In the 1990s the radioactive contamination of Arctic Seas was the focal point of numerous international research programs: the International Arctic Monitoring and Assessment Program (AMAP); the United States Arctic Nuclear Waste Assessment Program (ANWAP); the Atomic Energy Agency's International Arctic Seas Assessment Programme (IASAP). The Arctic Environmental Atlas was published as part of ANWAP (Crane end Galasso, 1999). It is a consolidated work containing information on radioactivity concentrations in seawater and bottom sediments. At the same time, a synthesis of knowledge on the interactions between biogeochemical cycling in unconsolidated sediments and the organisms living there has not been similarly produced.

The phenomenon of radioactive contamination has affected the global environment and is anyway perceived negatively by humankind. Radiation influences organisms both internally and externally as a result of the concentration of radioactivity in response to irradiation from external sources. Radioactive substances affect marine biological resources and humans. Current understanding is that ionizing radiation leads to morpho-physiological and functional changes in exposed organisms (Polikarpov 1964 a,b; 1971: Ilyenko 1974). Therefore the issues of radiation safety, environmental standardization and extrapolation of effects from high to low radiation doses on marine biota are of paramount importance in the field of marine radioecology.

However, the idea of radioactive contamination of different marine shelf flora and fauna species has received only fragmentary attention. There are only a few works on ^{137}Cs and ^{90}Sr accumulation by marine mammals (Roos et al. Gordon et al. 1999). As a result, there was only limited knowledge of radionuclide migration and transformation for the marine food web: from phyto- and zooplankton to benthos, fishes, birds, seals and whales. Therefore many of the questions arising in marine radioecology and radiochemistry are related to problems in oceanography.

The influence of global levels of radioactive contamination on individual organisms, communities and populations in the Arctic seas was generally not addressed during the earliest studies on radioactivity in these areas. The ecosystems of the bays and inlets, where atomic submarines are based or nuclear tests were carried out (Kola Bay, Zapadnaya Litsa, Sayda, Chernaya), remained unexplored for many years.

Regulatory agencies must make decisions on environmental protection issues, especially for local waterways with relatively high levels of artificial isotopes: ^{137}Cs, ^{134}Cs, ^{134}I, ^{129}I, ^{90}Sr, ^{238}Pu, ^{239,240}Pu, ^{241}Am, ^{60}Co. Contemporary information on the extent of radioactive pollution in marine ecosystems is vital for making informed regulatory decisions. This knowledge is no less important for the system of environmental safety designed for offshore oil and gas extraction, atomic energy installations, or emergency situations on nuclear-powered vessels. All these measures certainly proceed from the study of ecosystems affected by low-level radiation and the determination of protection strategies to ensure the conservation of biodiversity in northern seas.

This work is based on the original radioecological studies carried out by the Murmansk Marine Biological Institute in 1990-2000 (Fig. 1). The topic was elaborated within a framework of research programs and plans of the Russian Academy of Sciences, Russian Fund of Fundamental Researches, projects: «Comparative analysis of radionuclide transfer and accumulation processes in the Barents-Kara and Azov-Black Sea regions», «Reconstruction of radioactive contamination processes of marine ecosystems of European Russia», «Assessment of values of artificial radionuclide migration in the food chains of marine ecosystems» and others. A significant part of the work was carried out through international grants from the IAEA, NATO, TACIS etc., for example, «Contemporary level of radioactive contamination and risk assessment in the Barents Sea coastal waters» (University of Umeo, Sweden).

The aim of this work is to develop a methodical and theoretical framework based on the information contained within our extensive ocean biogeographical database. The following subjects are addressed: radionuclide migration and accumulation in the environment and biota of the Arctic and Sub-Arctic seas; artificial radionuclide transformation and trophic dynamics; input of global, regional and local sources to the radioactive contamination inventory of the Polar Ocean and North Atlantic ecosystems; bio-oceanographic processes and phenomena that contribute to the self-purification capacity of marine Arctic and Sub-Arctic ecosystems for radioactive pollutants.

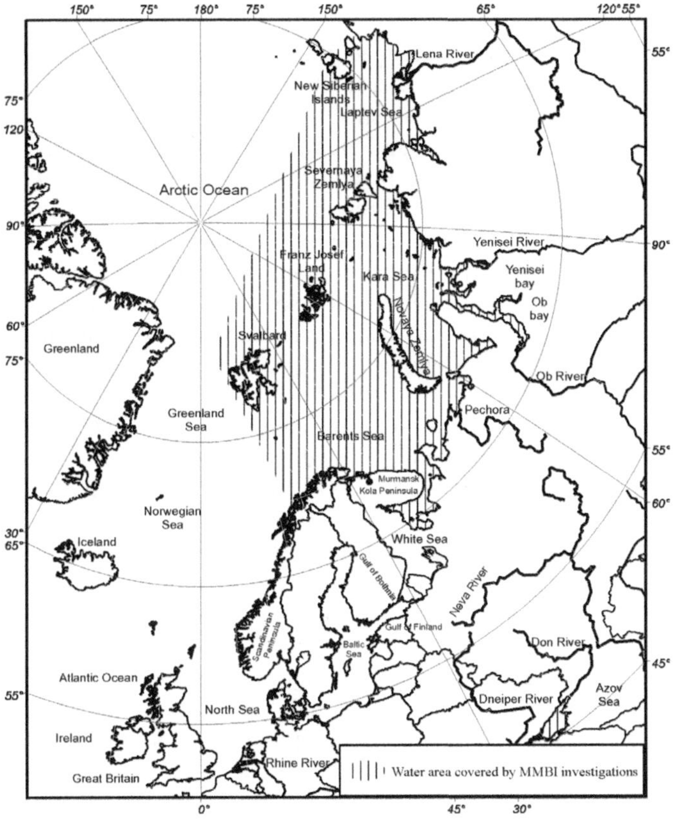

Fig. 1. Area of radioecological investigation

Main tasks are the following:

To ascertain general oceanographic and biological conditions promoting the transfer and accumulation of radionuclides in marine sediment and biota from different geographical regions;

To reconstruct the dynamics (1950-1990s) of radioactive contamination of the seas taking into account discharges from regional sources;

To determine the role of local sources of radioactive pollutants in the bays, inlets and shelf areas of northern seas;

To investigate the composition, quantity and general features of radionuclide accumulation in bottom sediments of northern seas.

To study the significance of pelagic (plankton) and benthic (macrophyte algae, benthos) filter-feeders in the primary distribution of radioactive substances;

To examine radionuclide bioaccumulation by marine invertebrate and vertebrate animals, paying special attention to marine food webs leading to humans: crustaceans and mollusks, fishes, birds, seals, whales, humans;

To formulate a conceptual model for the assimilation of artificial radionuclides in marine ecosystems;

To elaborate the system of radioecological monitoring using lithological and biological indicators, and also the forecast of influence of weak radiation doses on marketable species of marine organisms.

The study of anthropogenic radionuclides in marine ecosystems follows both from scientific objectives and the practical demands of societies. This work is in part the result of the goal-oriented research supported by scientific programs of the Russian Academy of Sciences, Ministry of Science and Technical Politics, Russian Fund of Fundamental Researches and other organizations. The analytical and theoretical results, achieved in the course of this work, have been used for the elaboration of the Stockman gas-condensate deposit Environmental Impact Assessment for the «Rosshelf» Company.

The authors are grateful for the valuable advice and assistance provided by colleagues during the preparation of this study: Prof., Dr of Chemistry, Yu.V. Kuznetsov, A.A. Namyatov, Prof. Ye. Shzhypa (Poland), Dr. J. Carroll (Norway), Dr. K. Rissanen (Finland), Dr. H. Nies (Germany), Dr. J. Smith (Canada).

Dr. of Biology I.A. Shparkovsky, Dr. of Biology S.F. Timofeev, Dr. of Geography V.V. Denisov, Dr. of Geology G.A. Tarasov, Dr. of Biology N.V. Lebedeva, Cand. of Biology P.R. Makarevich, Cand. of Biology A.D. Chinarina, Cand. of Biology G.M. Voskobojnikov and others contributed to various parts of the book and provided a number of valuable suggestions. Many specialists helped the authors in research and assisted them during different stages of data collection and processing, laboratory experiments and analytical works: D.V. Moiseev, Cand. of Biology A.A. Kondakov, Cand. of Geography G.V. Iljin, Cand. of Biology E.G. Berestovsky, S.V. Zyryanov, I.V. Bronnikov, M.S. Gromov, P.A. Lyubin, D. I. Alexandrov and others. The authors express their sincere gratitude for all of the valuable contributions provided throughout the course of this effort.

List of figures

Introduction

Chapter 1

Chapter 2

Chapter 3

Chapter 4

Chapter 5

Chapter 6

Chapter 7

Chapter 8

1 History and methods

1.1 History of investigations

The first investigations of radioactivity in the marine environment were carried out by Russian and foreign institutes in the 1960-1980s. This time period coincided with the period of nuclear tests on the island of Novaya Zemlya and with the large-scale release of radioactivity by Western European nuclear facilities. Radioecological observations in the Barents and Kara Seas were conducted by PINRO, Murmansk Hydrometeorological Service, Typhoon enterprise (Obninsk), the Khlopin Radium Institute the Kurchatov Research Centre. German and British institutes began monitoring artificial radionuclide levels in water and bottom sediments of the Norwegian-Greenland Basin and the western part of the Barents Sea in order to trace the transport of radioactivity from Western European nuclear facilities (Kautsky 1980, 1986; Vakulovsky et al. 1985; Nies and Nielsen 1996; Nies et al. 1997, 1998) (Fig. 1.1). The Norwegian Institute of Marine Research investigated the contamination of marketable fish species in the zone of direct impact of the Novaya Zemlya nuclear tests.

Important results, characterizing the contamination of the fjord ecosystems of the eastern bays of Novaya Zemlya were obtained by a Russian-Norwegian expedition on the Victor Bujnutsky vessel (Strand and Nielsen 1995). Numerous containers, barges, and reactor compartments containing radioactive substances were located and investigated in the bays Tsivolki, Stepovogo and Abrosimova using side-view sonar, video camera and nuclear detectors. Similar works were carried out in the Novaya Zemlya bays and on the adjoining shelf by the Russian Institutes VNIIOceangeologia and Sevmorgeologia, together with the Bedford Oceanographic Institute (Ivanov et al. 1995 a, b; Ivanov 1999). Additional information was collected on the distribution of radionuclides in other bays of Novaya Zemlya including Chernaya and Pechorskaya Bays. In 1994 the Murmansk Hydrometeorological Service conducted a detailed investigation of radionuclide accumulation in bottom sediments in the area adjoining Atomflot, the nuclear ship repair and transport enterprise in Murmansk (Namyatov 1995, 1998).

In 1990-1995 a series of surveys were carried out in the Barents and Norwegian Seas by the British Ministry for Agriculture, Fisheries and Food (MAFF), the German Hydrographic Institute, the Norwegian Fishery Directorate, the Norwegian Institute of Marine Research and others (Dahlgaard 1993; Sickel et al. 1995; Foyn and Swaren 1995; Solberg et al. 1995). These investigations resulted in a large amount of environmental data on the radioactivity content of seawater, bottom sediments, fish and other organisms (Fig. 1.2).

In 1990-1996 a number of investigations were carried out in the Norwegian Sea area at the site of the sunken atomic submarine Komsomolets (Blindheim et al.

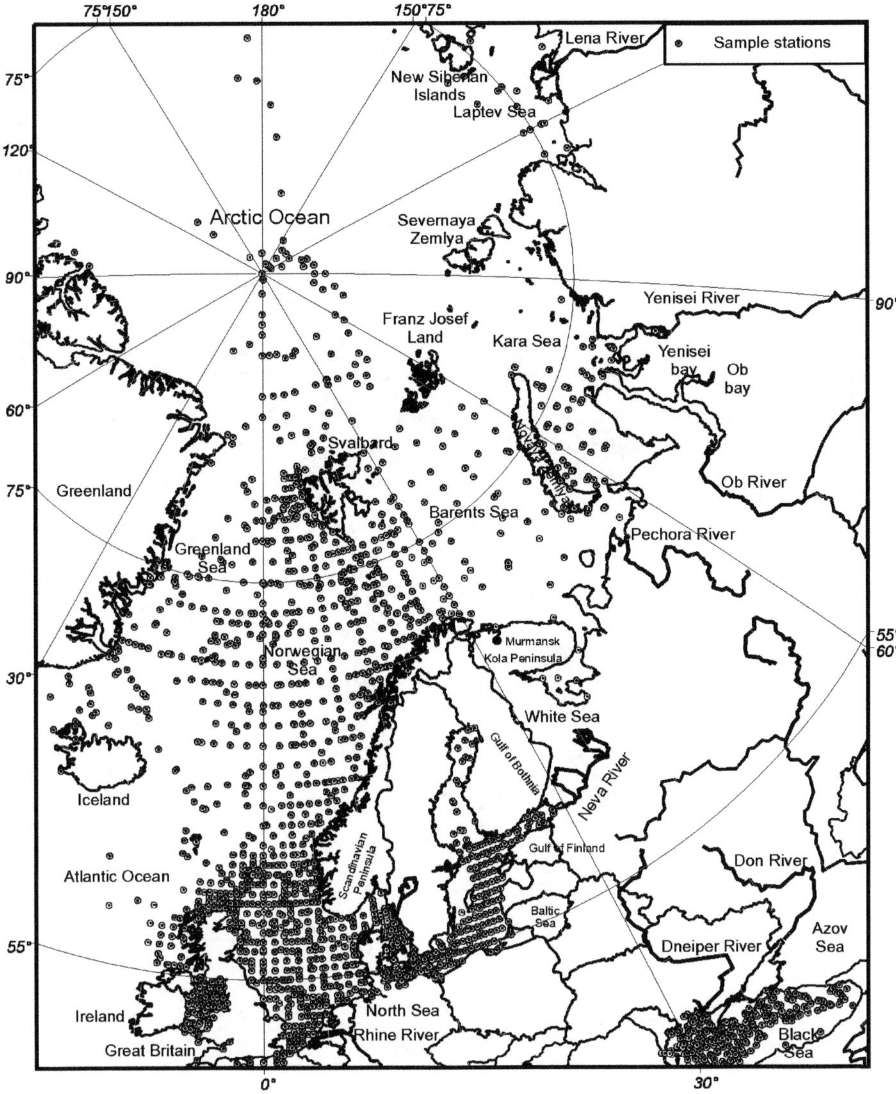

Fig. 1.1. Sampling locations for ^{137}Cs in the surface layer of the European seas and Arctic Basin (1970-1990s)

1994; Strand et al. 1994; Kuznetsov et al. 1996, 1997 a; Strand 1998; Lisovsky 1999). Participating institutes included the Norwegian Institute of Marine Research together with the North Atlantic Treaty Organization (NATO), and the Institute of Oceanology of the Russian Academy of Sciences (IO RAS) together with the Russian Ministry of Emergency Situations and the Hydrographic Service

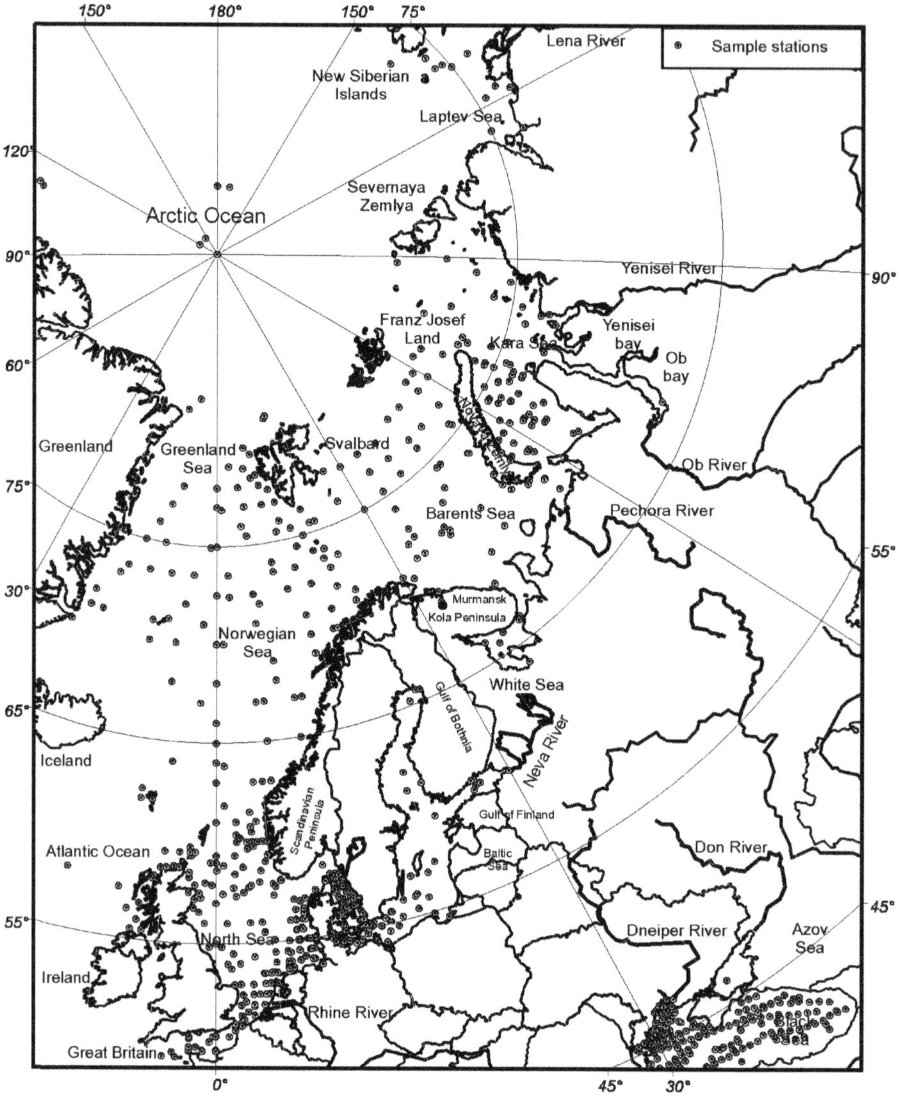

Fig. 1.2. Sampling locations for ^{90}Sr in the European seas and Arctic Basin (1950-1990s)

of the Russian Northern Fleet. The general conclusion of these investigations was that radioactive contamination of sediments and bottom fauna at the site of the sunken submarine is insignificant.

A number of other investigations were aimed at assessing the impact of artificial radionuclide discharges to the Kara and Laptev Seas via the Ob and Yenisey Rivers. Investigations were carried out in 1992-1998 by the Typhoon enterprise, Texas University, IO RAS, Geochemical Institute of RAS, Khlopin Radium Institute, and GEOMAR (Germany) (Vakulovsky et al. 1994; Sapozhnikov and Plishkin 1994; Kuznetsov et al. 1994; Champ et al. 1994; Brooks et al. 1995; Panteleev et al. 1995; Trapeznikov et al. 1997; Chumichev 1997; Nikitin et al. 1997). The role of the Siberian nuclear industries in the contamination of both living and non-living components of the marine environment became apparent during these investigations. Previous investigations in the area were focused almost entirely on investigating radioactivity levels in the non-living components of the marine ecosystem, e.g. seawater and bottom sediment (Fig. 1.3, 1.4).

In 1990-2000 the scientific staff of Murmansk Marine Biological Institute (MMBI) dedicated themselves to investigating marine radioactivity in numerous Arctic and Sub-Arctic marine ecosystems using a multidisciplinary oceanographic investigative approach. Several cruises were carried out in cooperation with M. Sklodovska-Kurie University (Lublin, Poland), the Norwegian Polar Institute, the Norwegian company Akvaplan-niva, the Finnish Institute of Marine Research, the Finnish Institute of Radiation and Nuclear Safety, Alfred Wegener Polar Institute, the Marine Research Institute (Germany), and others. For example, MMBI carried out investigations of the plutonium content in bottom sediments, fish and algae in the White Sea and adjacent areas (Rissanen et al. 1997 a, b).

The cooperation of specialists from different countries and institutes has greatly enriched methodological and scientific aspects of the MMBI investigations. As a result of these endevours, advances have been achieved in the understanding of environmental radioactivity and a number of fundamental contributions have been made in the field of radioecology (Matishov et al. 1992 a, b, c, 1993 d, 1994 a, b, c, 1995 a, b, 1996 e, f, 1997 a, b, c, d, e, f, 1998 b, c, 1999 a, b, c, d, e, i, h, 2000 a, b, c; Matishov and Denisenko 1996; Matishov G. and Matishov D. 1997; Matishov D. 1992 b, 1993 b, 1994, 1997, 1998; Matishov D. 1993; Matishov D. and Rissanen 1993; Matishov D. et al. 1993, 1995 a, b, 1996, 1997; Rissanen et al. 1995; Matishov D. and Namyatov 1997; Matishov D. and Matishov G. 1998).

Radioactivity that is present in Arctic and sub-Arctic marine ecosystems as a result of human interventions is the focus of this book. The work is based on the analysis of an extensive collection of radioactivity measurements on marine organisms representing diverse taxonomic groups combined with supporting environmental data. Anthropogenic radionuclides (^{137}Cs, ^{90}Sr, ^{60}Co, and $^{239,\,240}$Pu) in the Barents-Kara region is emphasized in this work. Particular attention is given to the investigation of the bays of the Kola Peninsula and Novaya Zemlya, as well as the estuaries of the Yenisey, Ob, and Pechora Rivers. Attention is drawn to this region due to the existence of several important regional and local radionuclide emission sources: the radioactive waste burial grounds on the continental shelf, the bases for Russia's nuclear fleet, and the former nuclear test site at Chernaya Bay, Novaya Zemlya (Matishov et al. 1993 c). Conditions in these areas are compared with other areas where sources of radioactivity are present such as the Baltic Sea. It is equally important to study the spectrum of artificial radionuclides on shelves and coastlines where local sources of contamination are absent. In this connection we include a discussion of the radioecological situation in the Azov Sea based on studies conducted from 1995-1997 (Matishov et al. 1999 a, b, 2000 c).

For the first time, the levels of ^{137}Cs, ^{90}Sr, $^{239,240}Pu$ accumulation in different species and populations of marine organisms is shown, as well as the

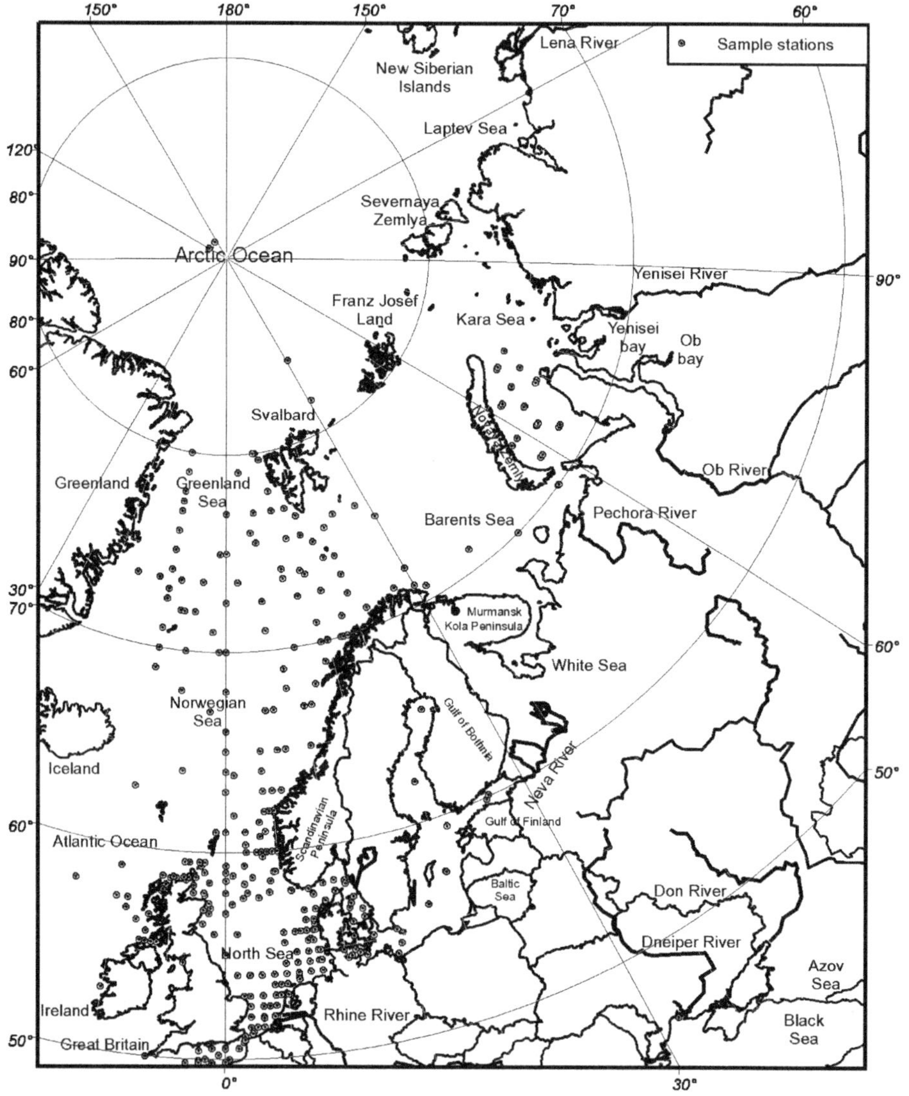

Fig. 1.3. Sampling locations for $^{239,240}Pu$ in the European seas and Arctic Basin (1950-1990s)

characteristics of radionuclide migration across trophic chains from macrophytes andplankton to zoobenthos, fish, birds, seals and whales. This work is more than a compilation and review of a unique radioactivity database. A synthesis of the data is carried out based on the principals of three scientific fields- radioactivity,

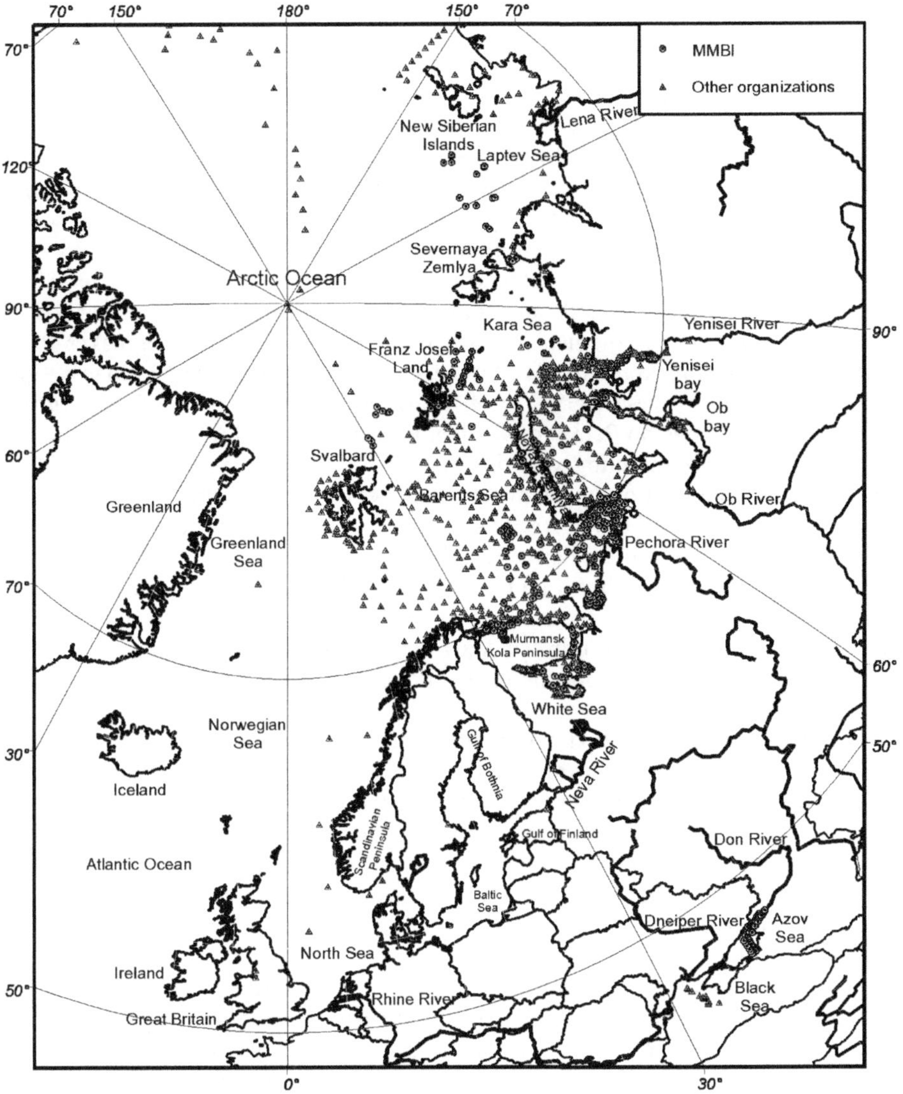

Fig. 1.4. Sampling of radionuclides in bottom sediments of the European seas and Arctic Basin (1950-1990s)

ecology, and oceanography. This work culminates in a new scientific perspective of how living organisms, in conducting their normal life-sustaining activities, sequester and transport anthropogenic radionuclides and thereby contribute to the assimilation capacity of marine ecosystems impacted by this form of environmental pollution.

1.2 Methods

The results of the Murmansk Marine Biological Institute expeditions conducted between 1990 and 2000 serve as the foundation for this work. These expeditions brought together MMBI scientists working in all fields of oceanography, as well as ecotoxicology. A special oceanographic and radioecological survey was carried out onboard the RV Dalnie Zelentsy in September 2000 in the area of the sunken atomic submarine Kursk on the Murmansk Bank (Fig. 1.5).

In addition to the expeditions carried out on the MMBI research vessel RV Dalnie Zelentsy and research and transport vessel Pomor, we have carried out oceanographic and radiological investigations on the German icebreaker Polarstern, atomic icebreakers Arktika, Soviet Union, Tajmyr of the Murmansk Shipping Company, the vessel GS-440 of the Hydrographic Service of the Northern Navy, RV Academician Petrov of Geohemical Institute of RAS and many others.

Our strategy of sampling was directed on the elucidation of the artificial radionuclide contamination levels of all components of the marine environment and the ecological pyramid of the Arctic seas. Such an approach is necessary to (1) determine the migration of radionuclides ^{137}Cs, ^{134}Cs, ^{60}Co, ^{90}Sr, $^{239, 240}$Pu over the food chain from macrophytes and plankton to sea mammals, (2) reconstruct the pathways of radionuclide transfer, and (3) identity radionuclide sources. MMBI studies were also carried out in the Azov Sea consisting of the sampling of bottom sediments and benthic organisms for the determination of ^{137}Cs, ^{134}Cs, ^{60}Co, ^{90}Sr, $^{239, 240}$Pu.

The new approach applied in our radioecological investigations involved the efforts of MMBI experts in taxonomy (algologists, benthologists, ichthiologists, ornithologists) to determine species and classify flora and fauna specimens into groups. As a result, γ-spectrometric and radiochemical analyses have been carried out on more than 300 species of marine biota.

Samples were collected using routine methods and equipment in accordance with the methods adopted by the International Atomic Energy Agency (IAEA 1996).

Sea water samples were collected in 100 liter volumes (Fig. 1.6) and filtered through the sorbent Cu_2, Fe $(CN)_6$ to extract cesium according to the methods of the German Federal Maritime and Hydrographic Agency (GFMHA 1997). Zooplankton samples were collected using towed plankton nets in the pelagic layer to a depth of 100 meters. Zoobenthos were sampled with a Van-Veen grab, and Ocean-50 box corer.

Bottom sediments were sampled by Van-Veen grab, Ocean-50 box corer, dredge, gravity corer Gemini and a 4 meter long gravity corer. (Fig. 1.7, 1.8). If the sediments in the corer were not mixed, the surface layer (0-2 cm) was removed using a plastic tube of 104 mm diameter. The sediment column retrieved by the

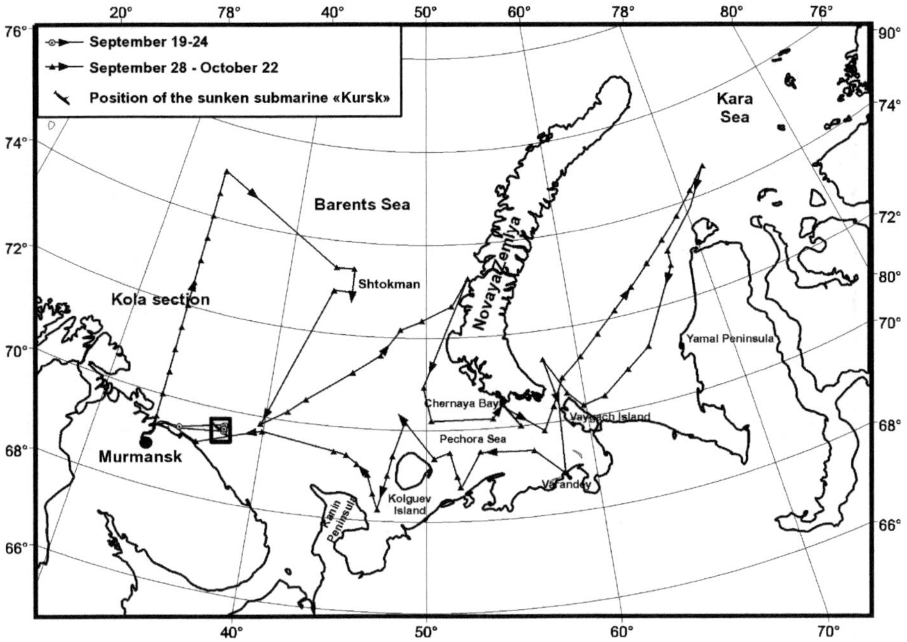

Fig. 1.5. MMBI expeditions on RV «Dalnie Zelentsy» in 2000

plastic tube was subsequently sliced into 10 mm layers. In some cases the first 10 cm of the sediment column was sliced into 2 cm interval. Below 2 cm sediments were sliced into 5 cm intervals.

Red and Phaeophyceae algae (*Laminaria Saccarina, Alaria sp., Fucus vesiculosus, Fucus inflatus, Ascophyllum nodosum* and others) were sampled in the littoral zone of the White, Barents, Pechora and Kara Seas (Fig. 1.8). Many samples of macrophytes were collected in the bays adjoining the different atomic bases. In addition, lichens of species *Cetraria nivalis*, *Cladonia mitis*, *Cladonia Rangiferina*, *Cladonia alpestris*, were collected on the Kola Peninsula coast and on the south-eastern islands of the Barents Sea and Novaya Zemlya.

Benthic organisms were collected in representative geomorphological zones of the Barents-Kara Sea continental shelf. The benthic collections were sorted by species, and the most abundant organisms were taken for radiochemical analyses. Specimens from more than ten groups were analyzed: Annelida, Bryozoa, Gastropoda, Echinodermata, Decapoda, Amphipoda, Ophiuroidea, Asteroidea and others.

Sea birds were taken for analyses in August 1996 on the eastern coast of the Kola Peninsula and in the adjoining bays Ivanoskaya and Drozdovka: two sea gulls (*Larus marinus*), two common gulls (*Larus carnus*), two great skuas (*Catharacta skua*), one goosander (*Mergus merganser*), one stint (*Tringa erythropus*) and six Temmink's stints (*Calidris minuta*). Sea birds were taken for analyses in

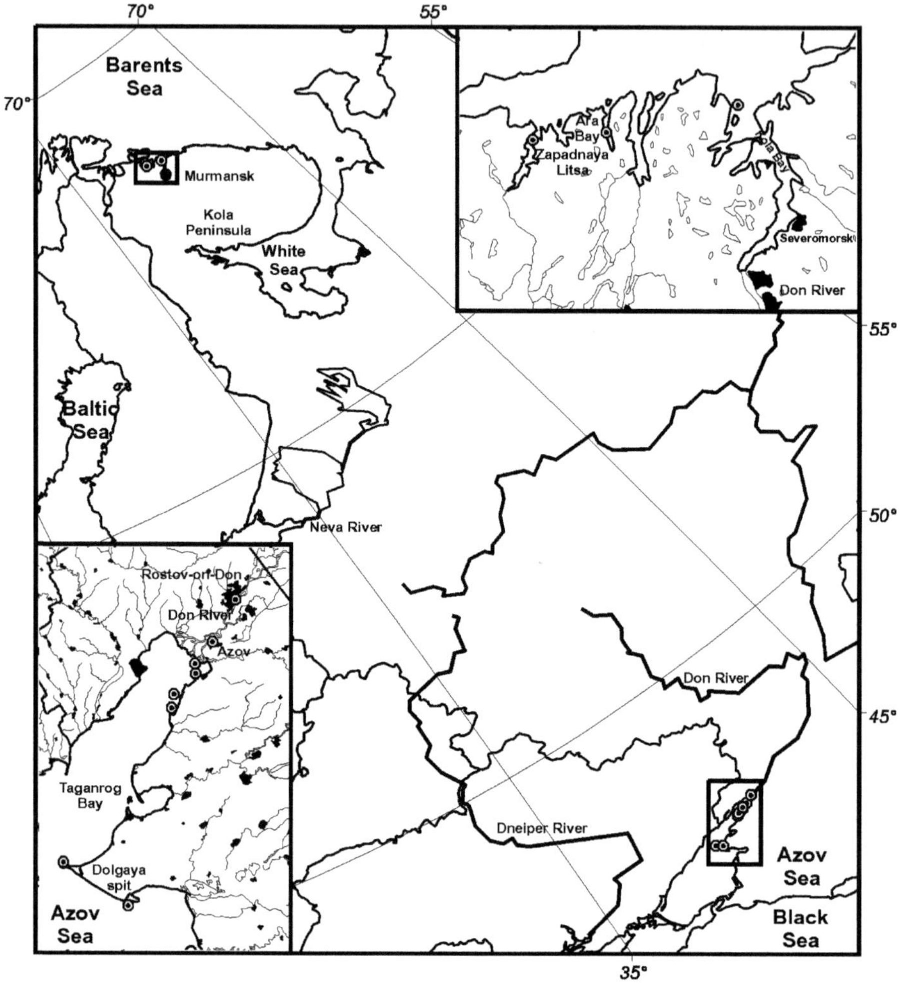

Fig. 1.6. ^{137}Cs investigations in the European seas during 1990s

February 1996 in the Dalnie Zelentsy area (Fig. 1.10) and adjoining Yarnyshnaya Bay: four eiders (*Somateria mollissima*), one black guillemot (*Cepphus grylle*) and four purple sandpipers (*Calidris maritima*).

The preliminary processing of algae, benthos, fish, bird, and seal samples varied depending upon the organism type. For example, fish bones were separated from muscle and the liver and other internal organs were collected separately. Muscles and bones were dried (105 °C) and homogenized before analyses. Approximately 60 g of dried muscle and 40 g of bone material were processed for each analysis. Large bivalve and Gastropoda molluscs were separated from shells before drying. Starfish were cut and the piloric appendages, stomachs and bowels

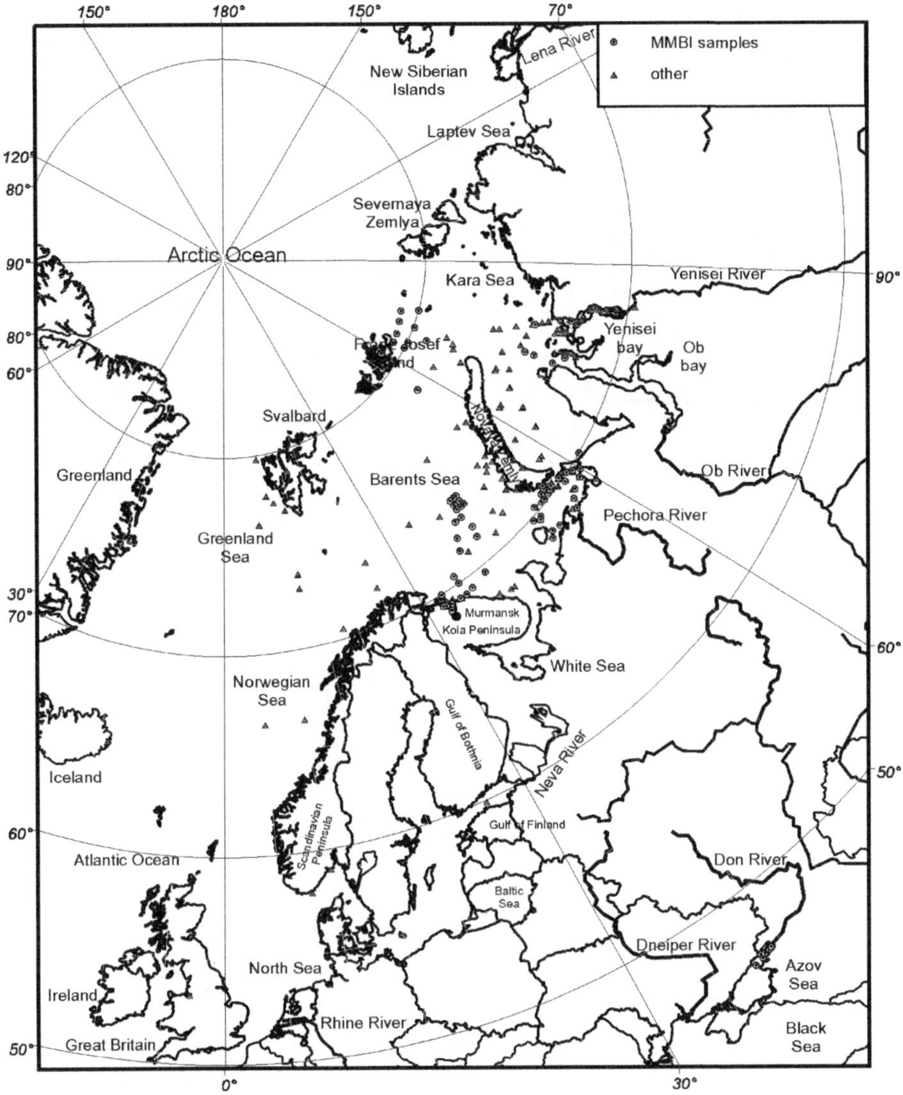

Fig. 1.7. Bottom sediment samples for the analysis of $^{239, 240}Pu$ in the European seas and Arctic Basin (1950-1990s)

were separated from the skin, dried and homogenized. Only the cod liver with high fat content was analyzed wet. Samples were placed in Marinella containers of 550 ml or plastic vessels of 35 or 100 ml for analysis by high-purity gamma spectrometry analysis. The minimum counting time on the detector was 1000 min.

The following laboratories provided γ-spectrometric and radiochemical analyses: the Chair of radiochemistry and colloidal chemistry of M. Sclodovska-Kurie University, (Lublin, Poland); Northern Branch of the Finnish Centre of Radiation

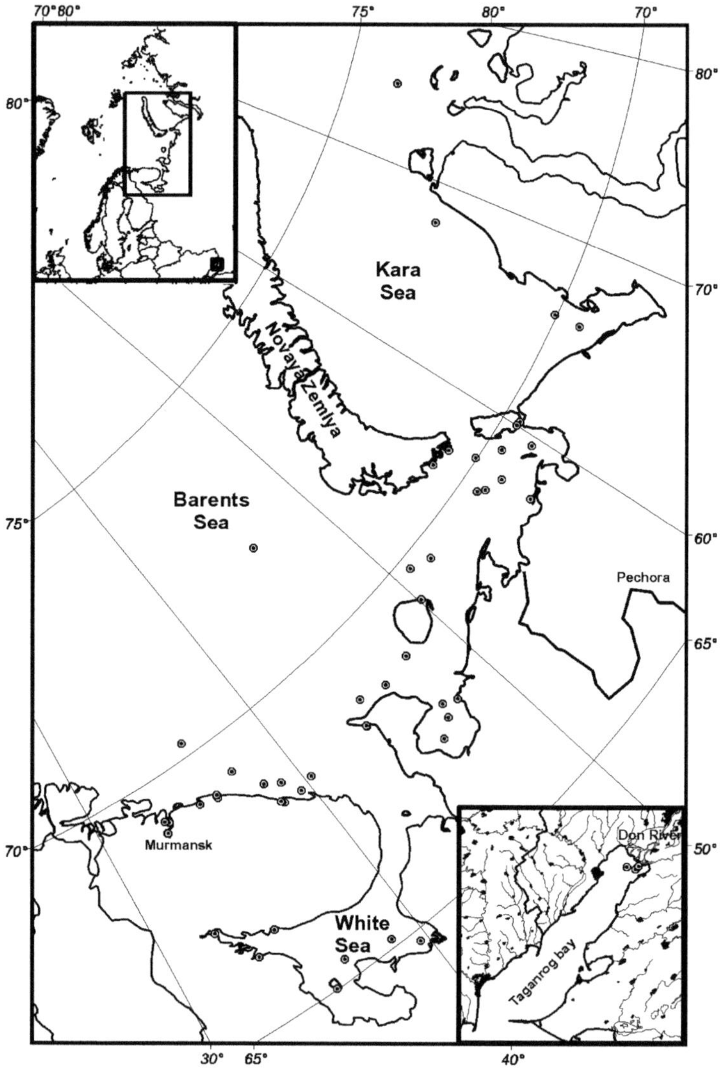

Fig. 1.8. Benthos and macrophyte sampling stations (MMBI unpublished)

and Nuclear Safety (STUK, Finland); Laboratory of Radiochemistry of the Bedford Oceanographic Institute (Canada); Department of Marine Radioactivity of German Hydrographic Institute (Hamburg); Typhoon enterprise (Obninsk, Russia); V.G. Khlopin Radium Institute (S.-Petersburg) and Murmansk Hydrometeorological Service.

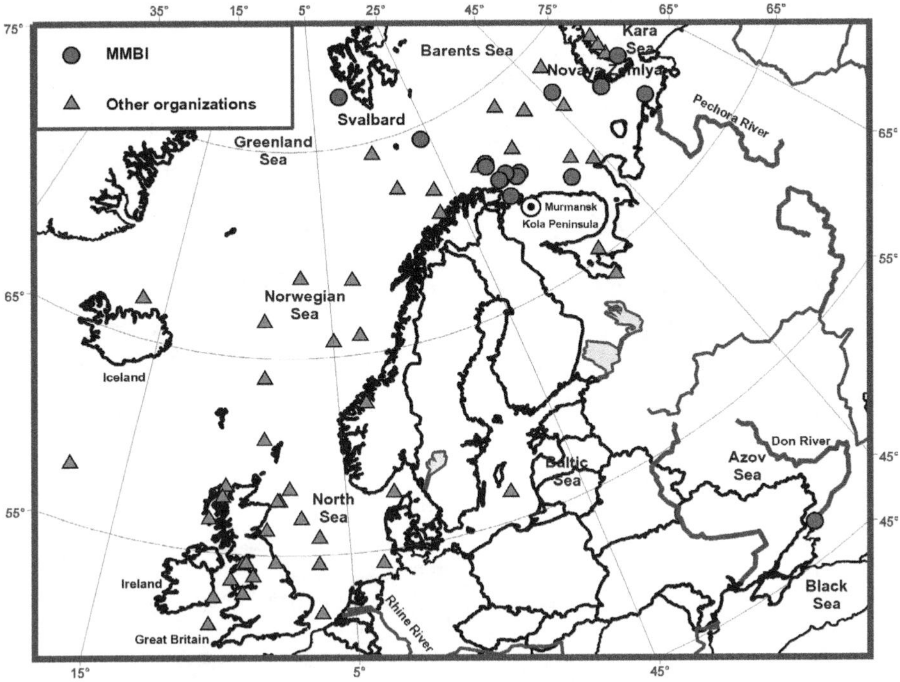

Fig. 1.9. Samples taken for the investigation of radioactivity in marine fish

Analyses were primarily carried out to determine the content of ^{137}Cs, ^{60}Co and other γ-radiating isotopes in environmental material and biota. The determination of γ-emitting radionuclides is performed using semi-conductor γ-spectrometers, containing Ge(Li) detectors of high resolution according to internationally established methods (IAEA 1996).

For the determination of ^{238}Pu and $^{239,240}Pu$, weighed sediment, water, or marine biota samples are heated to temperatures not higher than 350-400 °C. The radionuclide ^{242}Pu is used as a tracer (IAEA 1996). The radiochemical determination of $^{239,240}Pu$ is then carried out on a sample (not more than 10-20 g) by the addition of acids, followed by plutonium separation using ion-exchange resins and plutonium electrolytic deposition onto stainless steel discs. Sample quantification is carried out by alpha particle spectrometry at the V.G. Khlopin Radium (Matishov et al. 1999 c, e) and at the Finnish Centre of Radiation and Nuclear Safety (Rissanen et al. 1995, 1997 a). ^{90}Sr determinations are carried out by the conventional 'fumic nitric acid' method after the addition of ^{85}Sr isotope as tracer. The β-activity of the prepared ^{90}Sr sample is measured on a scintillation counter NaI(TI).

Thus, during 1990-2000 more than 2000 analyses on the content of ^{137}Cs, ^{60}Co, ^{90}Sr, and $^{239,240}Pu$ in environmental and biological samples were carried out (Table 1). As a result, new, more comprehensive, information is available on the content

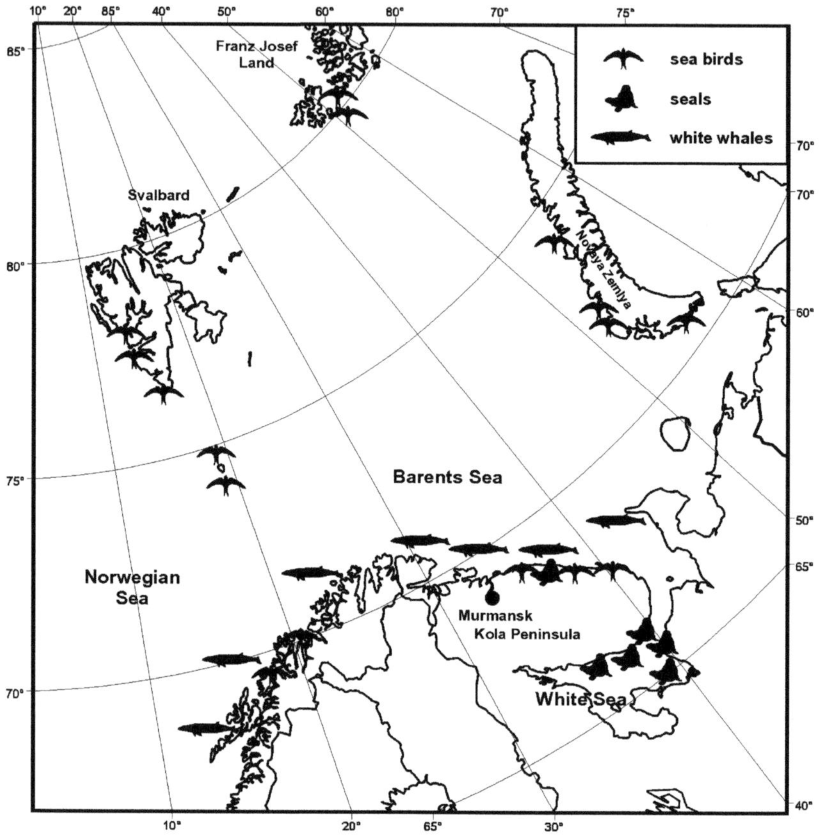

Fig. 1.10. Sampling locations for sea birds, seals, and whales

and distribution of the above-mentioned radionuclides in different compartments of Russia's Arctic and sub-Arctic marine ecosystems. This radioecological database provides the foundation for presenting monitoring methodology and critical tasks for marine radioecology in the future - the monitoring of $^{239,240}Pu$ will be a critical task. Plutonium is a main irradiator of marine biological resources and will require monitoring to assess its impact on ecosystems as a whole.

Systematizing and analyzing the large amount of radioecological, radiochemical, oceanographic and biological data on the Arctic Seas required the use of traditional cartographic methods as part of the research (Matishov et al. 1994 a; Ionov et al. 1996). In the course of this work, a series of special radioecological biogeographic maps were created to present information on the entire spectrum of artificial radionuclides in both living and non-living components of the environ ment. The cartographic basis for these maps was adopted from the Numerical Atlas GEBCO 97.

Table 1.1. Analyses (^{137}Cs, ^{60}Co, ^{90}Sr, $^{239,240}Pu$) performed on samples from different components of the marine ecosystem (MMBI data)

Ecosystem component	Azov Sea			White Sea				Barents Sea				Kara Sea			Laptev Sea	Total			
	^{137}Cs	^{90}Sr	$^{239,240}Pu$	^{137}Cs	^{90}Sr	^{660}Co	$^{239,240}Pu$	^{137}Cs	^{90}Sr	^{60}Co	$^{239,240}Pu$	^{137}Cs	^{60}Co	$^{239,240}Pu$	^{137}Cs	^{137}Cs	^{90}Sr	^{60}Co	$^{239,240}Pu$
Water	7	5	-	-	-	-	-	3	-	-	-	-	-	-	-	10	5	-	-
Sediment	121	47	21	51	4	1	18	440	21	60	96	95	8	22	19	726	72	99	157
Macro-phytes	3	3	3	22	-	-	-	53	7	6	22	-	-	-	-	78	10	6	25
Plankton	1	1	1	-	-	-	-	-	-	-	-	-	-	-	-	1	1	-	1
Benthos	-	-	-	19	-	-	-	150	-	4	-	39	-	-	-	208	-	4	-
Fish	5	3	-	1	-	-	-	28	-	-	14	-	-	-	-	34	3	-	14
Birds	8	-	-	-	-	-	-	30	-	-	-	-	-	-	-	38	-	-	-
Seals	-	-	-	133	5	-	5	-	-	-	-	-	-	-	-	133	5	-	5
Total																1228	96	109	202
Total (all radionuclides)																1635			

2 Sources and pathways of artificial radionuclides in northern seas

Artificial radioactivity has been present in marine ecosystems for more than a half-century. Present concentrations of radionuclides in all compartments of the marine ecosystem are low. Moreover since the beginning of the 1980s, decreasing concentrations of artificial radionuclides has been the trend for northern seas, considering both physical and ecological pathways of radionuclide transport. The content, ratio and behaviour of isotopes, as well as their rates of accumulation in shelf and deep ocean sediments, depend on the source of radioactive contamination. At present only the long-living isotopes, such as ^{90}Sr, ^{137}Cs, $^{239,240}Pu$, ^{241}Am are ecologically dangerous.

The relative importance of the major sources in the general balance of radioactive contamination of the seas has not changed since the beginning of the 1990s. The present content of radioactive pollutants in the polar seas is the result of radioactive fallout from nuclear tests, the Chernobyl accident and also radioactive wastes carried by sea currents from Siberian rivers and European nuclear processing facilities. At the same time, the input of regional and local sources of radionuclides to the overall balance of radioactivity in Arctic marine ecosystems is not as well known. Here we discuss some major aspects concerning local and regional contamination sources. We do not consider the dumping of radioactive waste in the Arctic marginal seas as this has not influenced the offshore radioactivity level (Gerlach 1981; Matishov et al. 1994 a, b, 1997 b, c, 1998 b, 1999 a, 2000 b; Matishov D. et al. 1995 a; Dovgusha and Tikhonov 1995; Foyn and Svaren 1995).

Our analysis of the content and distribution patterns of radionuclides in the Arctic seas has lead to the development of a classification scheme separating radioactive environmental contamination into five types or levels, depending on the sources (Fig. 2.1). To some extent, the impact of contamination is apparent on different radioecological scales (global, regional and local), but in addition, the impact reflects post-depositional migration events and interactions between primary (direct) and secondary sources.

2.1 Primary sources of radioactive pollutants

In 1940-1950s the dumping of radioactive wastes to the Ocean and seas was routinely conducted by nuclear states. The generally accepted expert opinion is that today, the radiation situation in the Arctic Ocean and the Arctic as a whole is the result of several main sources: global fallout from past nuclear tests and the Chernobyl accident; and discharges of liquid radioactive waste from western European reprocessing facilities in Sellafield (Great Britain), Cap de la Hague (France) and others (Gromov and Spitsyn 1975; Vakulovsky et al. 1985, 1988, 1994 a ; Kautsky

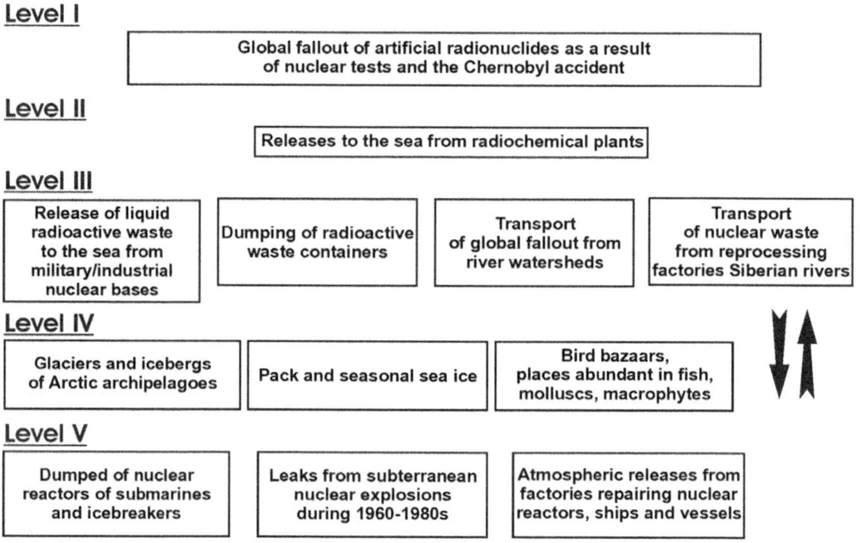

Fig. 2.1. Formation of radioactive contamination in the northern seas

1988; Izrael and Tsyban 1989; Izrael et al. 1994; Matishov et al. 1994 a, c; Strand et al. 1994; Vartanov and Hollister 1997; Strand 1998). These phenomena have resulted in the influx of a broad spectrum of radionuclides into the Ocean environment.

2.1.1 Nuclear explosions

The first type of anthropogenic contamination in northern seas is from nuclear explosions. Nuclear explosions in the atmosphere and in the hydrosphere were carried out in 1950-1960s in the area of the Novaya Zemlya archipelago (Fig. 2.2). Atmospheric nuclear tests were carried out around the globe from 1945 to 1990s and the Chernobyl accident occurred in 1986. As a result of nuclear testing, 910 x 10^{15} Bq ^{137}Cs, 600 x 10^{15} Bq ^{90}Sr, and 240 x 10^{18} Bq ^{3}H were released into the atmosphere. The atmospheric nuclear tests conducted on Novaya Zemlya resulted in the release of 410 x 10^{15} ^{137}Cs, 270 x 10^{15} ^{90}Sr and 240 x 10^{18} Bq ^{3}H (UNSCEAR ... 1982). The peak of radioactive atmospheric fallout occurred in 1963. As for the northern seas, during the period 1945 - 1992 the Barents and the Kara Seas received 2.1 x 10^{15} and 1.4 x 10^{15} Bq respectively of ^{137}Cs and 1.5 x 10^{15} and 1.0 x 10^{15} Bq of ^{90}Sr (Table 2.1) (A survey ... 1992).

During the early time period after a nuclear explosion or nuclear power station accident, the concentration of radionuclides in the atmosphere, especially short-lived radionuclides, exceed significantly the average monthly or average annual radiation levels. After the largest nuclear explosions (1 Mt or more), particles of radioactive dust were thrown to the stratosphere (to a height of 10-15 km), where

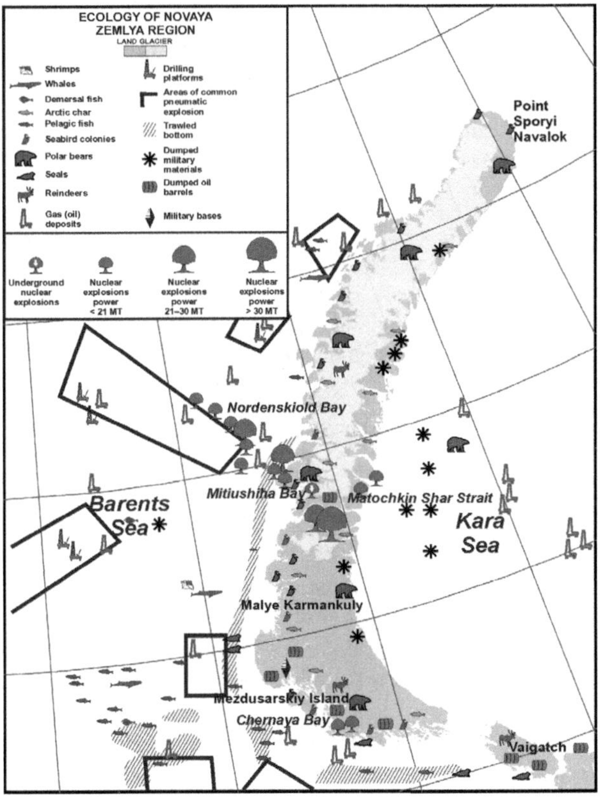

Fig. 2.2. Nuclear weapons tests in the atmosphere and hydrosphere on Novaya Zemlya (1957-1962)

they remained for many months and years slowly descending and scattering over the surface of the earth. Typically about 12 % of radioactive precipitation deposits near the nuclear test site and an additional 10 % within the latitude belt of the test site (Fig. 2.3). The rest (approximately 78 %) represents the global fallout (Izrael 1998; Strand 1998). Together with ^{137}Cs, a large number of particles containing $^{239,240}Pu$ entered the atmosphere as a result of nuclear explosions.

Table 2.1. Sources of anthropogenic radionuclides to the seas of the Arctic Basin (Aarkrog 1994, 1995)

Sources	Comment	Total activity
Global fallout from atmospheric nuclear tests	Between 1957 and 1963 ^{90}Sr input between 0.25 and 0.75 kBq / m^2	^{90}Sr - 4.2 PBq ^{137}Cs - 6.7 PBq
Local fallout on Novaya Zemlya	87 tests in the atmosphere	^{90}Sr - 20 PBq ^{137}Cs - 30 PBq
Sellafield discharges	Total discharge ^{137}Cs to the Irish sea - 41 PBq (decay by 1994 - 28 PBq)	^{137}Cs - 6.7 PBq (the Arctic Ocean)
Atomic submarine «Komsomolets» (07.04.89, depth 1680 m)	Reactor Two torpedoes with nuclear nose-caps (6.9 kg ^{239}Pu)	2.9 PBq ^{90}Sr - 16 PBq ^{137}Cs - 3.1 PBq
Discharges and leakage of liquid and solid radioactive wastes in the former USSR (The Barents and Kara Seas)	17 reactors in the Kara Sea at the time of dumping: at present (actinides) activation products decay products solid waste (1964 - 1990) liquid waste (1960 - 1990)	 90 PBq 0.2-0.9 PBq 4.4-4.8 PBq 4.7 PBq 0.6 PBq 0.9 PBq
Aircraft accident at Thule, Greenland (January 1968)	Unknown quantity of ^{239}Pu (between 400 and 500 g)	1 TBq

In the Arctic, nuclear weapons tests were carried out at three sites on Novaya Zemlya: Chernaya and Sulmeneva Bays and the Matochkin Shar Strait. In 1958, 1961 and 1962 there were at least 30 atmospheric explosions annually, sometimes 7-8 tests monthly, including the most powerful in the world of 54 and 30 Mt (Sykes and Davis 1987). As it is known, the levels of radioactive fallout in the Arctic region exceeded background values from 1 to 11 thousand times. For ^{137}Cs alone, the Novaya Zemlya tests resulted in the discharge of 150 to 560 PBq (MacDonald and Bevers 1996).

The density was rather variable for radioactive fallout or the summary β-activity in the different northern territories during 1961-1989. The greatest daily fallout (577 mKi / km^2) was registered in Amderma in 1962 (Gromov and Spitsyn 1975; Chelyukanov and Saveljev 1992; Izrael 1998). During the period 1962-1992 in Murmansk the highest level of artificial radioactivity in precipitation (to 400 Bq / m^3) was noted in September 1962. High levels of the summary β-activity of precipitation were characteristic for 1963-1964. The variability from month to month was from 120 to 5 Bq / m^3. After the Chernobyl disaster in April 1986 the concentration of β-radiation in precipitation was about 2 Bq / m^3. During the 1990s monthly rain and snow samples in the Murmansk area had low levels of β-activity (from 0.05 to 0.35 Bq / m^3).

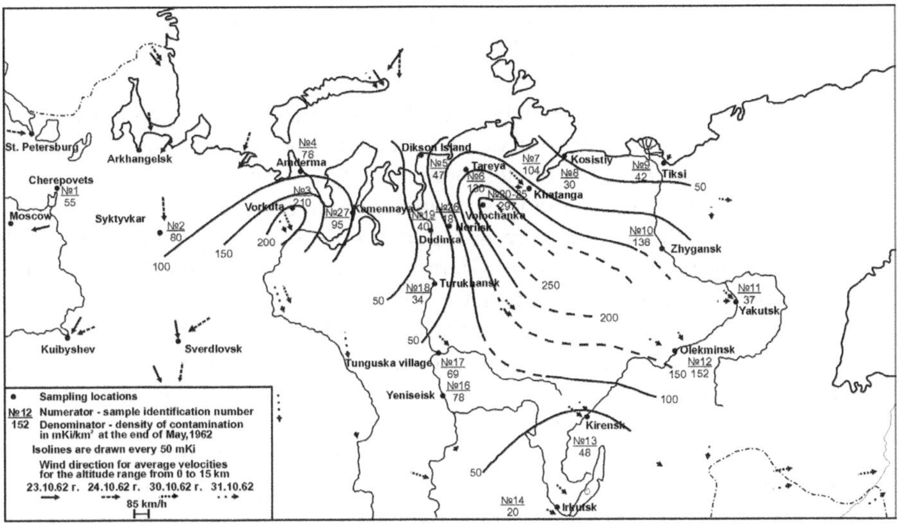

Fig. 2.3. Zirconium + niobium-95 fallout from nuclear explosions carried out in autumn 1962 on Novaya Zemlya (Izrael, 1998)

Surface and underwater tests were carried out on the shelf near the coast of Novaya Zemlya. Several nuclear bombs (of 20 and 25 Mt) were exploded in the coastal zone to the west and the east of Matochkin Shar Strait. About ten nuclear explosions were also carried out offshore in a southwestward direction from the North Island of Novaya Zemlya.

As a consequence, pelagic ecosystems in the Barents and Kara Seas were regularly exposed to radioactive contamination during the period of atmospheric and underwater nuclear tests. Bird bazaars of Novaya Zemlya (the most populated in the Arctic at the time) and seal rookeries were contained in the impact zone. The environment of the Chernaya Bay was most heavily impacted by radiation.

The immediate vicinity of the Novaya Zemlya nuclear test sites represents the epicenter of local contamination. For example, an anomalous quantity of radionuclides is contained in the bottom sediments of Chernaya Bay, the epicenter of contamination from underwater tests of nuclear bombs in 1955 and 1957. These sediments contain 9 000 Bq / kg (about 300 000 Bq / m^2) ^{239,240}Pu, and also ^{240}Am, ^{137}Cs and ^{60}Co from 100 to 500 Bq / m^2 (Matishov D. 1993 b). The areas of impact were quickly extended as a result of the diffusive and advective ocean transport of radionuclides (Matishov D. 1993 a, 1994; Matishov et al. 1994 a, c; Smith et al. 1995 a, b).

The quantity of plutonium in Chernaya Bay is similar to such places as the most polluted parts of the Bailot Strait (the site of the crash of the American B-52 aircraft) and the Irish Sea near Sellafield reprocessing plant. The total quantity of $^{239,240}Pu$ in aleurite and clay sediments of the bay is on the order of 3 $x10^{12}$ Bq (Smith et al. 1994, 1995 a). These quantities are similar to the amount of plutonium found at the American nuclear test sites on Bikini and Enivetock islands in the South Pacific.

At the present time, contamination of Chernaya Bay bottom sediments is a potential source of radioactive admixtures far beyond the limits of the Pechora Sea. It is quite probable that the grounds, anomalously saturated by transuranium elements (e.g. $^{239,240}Pu$), are continuously supplying the bottom waters with radionuclides. Polluted bottom waters are transported out of Chernaya Bay and into the Pechora Sea by tidal and stationary currents where they are eventually incorporated into the general circulation pattern of the Barents Sea.

The radioactive releases associated with the Chernobyl nuclear power station accident resulted in radioactive contamination on a global scale (Fig. 2.1). The Chernobyl disaster once again increased the ^{137}Cs content of the atmosphere by several orders of magnitude (Izrael 1998; Strand 1998). As a result of the accident, about 100 PBq of ^{137}Cs entered the atmosphere. The fallout was mainly concentrated in Europe. About 1 PBq reached the Arctic region directly (70-90 °N). Another 5 PBq was transported by the Norwegian current to the polar seas from the North and Baltic Seas (MacDonald and Bewers 1996).

It is possible to envision the spread of radioactive particles after the explosion of a nuclear bomb on Novaya Zemlya using the scenario of events following the Chernobyl disaster (Fig. 2.4, 2.5). Fallout of ^{90}Sr, ^{137}Cs with rain was detected in many regions of Europe and in the northern seas soon after the explosion (Information ... 1986; Radioecological consequences ... 1991; Impact ... 1996; Abaturov et al. 1996). The general contaminated area within the 20 μR isoline was about 200 000 km^2 in the first days. The cloud of radioactive fallout reached the Arctic as well. Wind patterns distributed radioactive products first to the Baltic areas, then to Sweden, Norway and Finland, and within Russia - to the Arkhangelsk District, to the southern part of the Kola Peninsula and to the locality of Salekhard in the Ob mouth. The levels of radioactive pollution in the northern part of Scandinavia was about 2000 Bq / km^{2}, close to the level of global fallout.

There are two main pathways for the influx of anthropogenic radionuclides into the Black Sea waters: atmospheric fallout and river discharges. During the period of nuclear weapons testing the most intensive fallout was observed in 1963-1964 reaching in total more than 29 TBq of ^{99}Sr. The summary activity of fission products falling on the surface of the Black Sea was in total 2.5 mKi during 1960-1977, including 0.0175 mKi of ^{90}Sr (Batrakov et al. 1997). The total discharge of ^{90}Sr by Dnieper from May 1986 to August 1992 was 50 TBq, or about 2200 Ki, 0.5 % of the total amount of strontium released by the reactor.

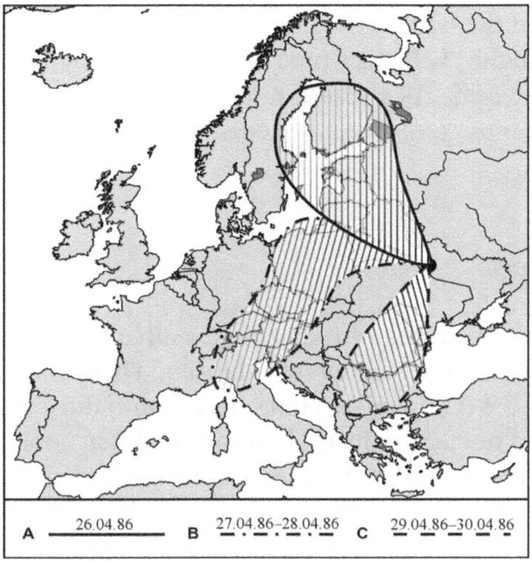

Fig. 2.4. Main directions of transfer of the radioactive Chernobyl dust in the first days of the accident (STUK-Radiation and Nuclear Safety Authority, Finland)

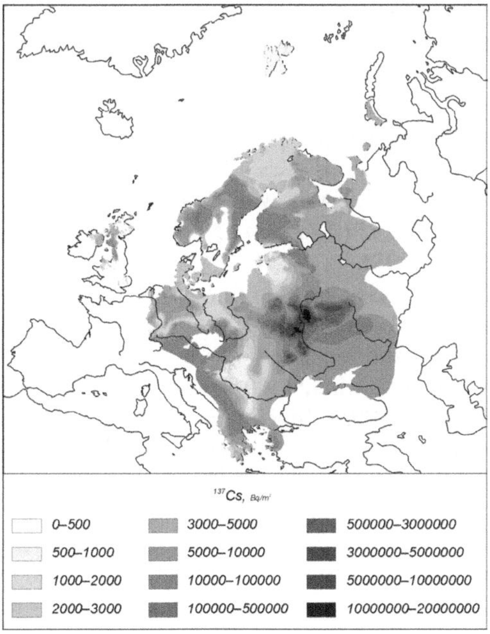

Fig. 2.5. Contamination of European territories by ^{137}Cs (Crane, Galasso, 1999)

In 1990s the Chernobyl radioactivity from the Baltic and other internal European seas began to penetrate into the Barents Sea (Nies et al. 1998). The role of sea transportation in the transfer of radionuclides from the Baltic and other basins of primary accumulation will over time become more significant in the balance of sources in the Arctic seas. As a result, radioactive Chernobyl fallout will also contribute to the contamination of ecosystems of the Arctic coast and the northern seas.

2.1.2 Sea Dumping

The second and the third types of radioactive sources are connected with the practice of dumping radioactive wastes into the sea. These include releases of liquid radioactive wastes, the dumping of containers with radioactive wastes into the sea and discharges from nuclear industrial facilities. When considering these types of discharges on a regional and local scale, it is necessary to consider the transport of radionuclides by major rivers of the Arctic region. This includes releases from nuclear plants into Siberian rivers but also global fallout in river catchment areas. Accidents involving submarines and airplanes with nuclear materials onboard also belong to this category of source type.

Radionuclides are discharged from three Western European nuclear plants and are eventually transferred by ocean currents to the Arctic: Sellafield (earlier Windscale) in the Cumbria on the north-western coast of England, La Hague near Sherbourg (France) and Dunree on the north-east coast of Scotland. The total input of these sources is surpassed only by fallout from nuclear weapons testing. Sellafield first became the most important source for the Irish Sea and eventually for the whole Arctic Ocean basin. The total activity of released artificial radionuclides is 160 x 10^{15} Bq, including ^{137}Cs - 25.6%, ^{3}H - 17.5%, ^{106}Ru - 17.5%, ^{95}Zr – 15%, ^{241}Pu - 13.1%, ^{90}Sr - 3.9% as well a number of radionuclides in lesser release quantities (Fig. 2.6, 2.7) (A Group ... 1990; Izrael et al. 1994; Sivintsev and Kiknadze 1995; Strand 1998).

The average daily releases of liquid radionuclides from Sellafield was for ^{239,240}Pu - 0.1, ^{90}Sr - 2, ^{89}I and ^{144}Ce - 10, ^{106}Ru and ^{137}Cs - 100 Ki. The peak of these releases occurred from the middle to the end of the 1970s. The maximum occurred in 1975 when approximately 5 PBq of ^{137}Cs were released into the Irish Sea (Kershaw and Baxter 1995). Since then the influx of ^{137}Cs has gradually decreased and is now on the order of 0.01 PBq yearly (MAFT / SEPA 1997, 1998).

During the 1970s, the concentration of ^{137}Cs in surface waters in the zone of release reached 200-250 Bq / m^3. The sediment layer most impacted by ^{137}Cs and ^{90}Sr was the 10-20 cm depth zone, corresponding to the time of the maximum discharges. The volume of released wastes from the initial moment of dumping in 1957 to 1988 was 37 x 10^{15} Bq ^{137}Cs and 5.5 x 10^{15} Bq ^{90}Sr (Gueguениat et al., 1997). It may be noted for comparison that the Barents Sea has received during this period two times more ^{137}Cs than received by the Baltic (70 kKi) and Black (45 kKi) Seas after the Chernobyl disaster (Izrael et al., 1994).

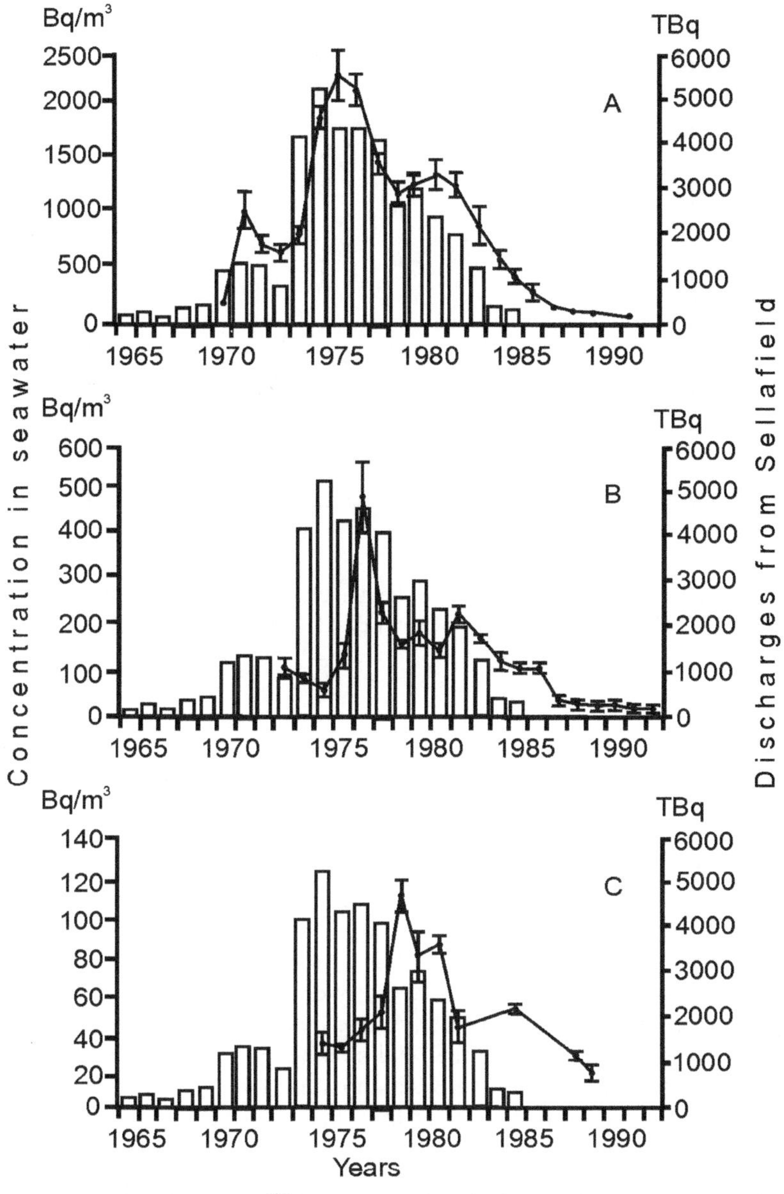

Fig. 2.6. Temporal variations in ^{137}Cs concentrations in the North-West European Seas surface water (Kershaw and Baxter 1995). **A** - Northern Strait (Irish Sea), **B** - Eastern Scotland (57-58 N, 1.5-2 W), **C** - Western Norway (59-61 N, 3.5-5 W)

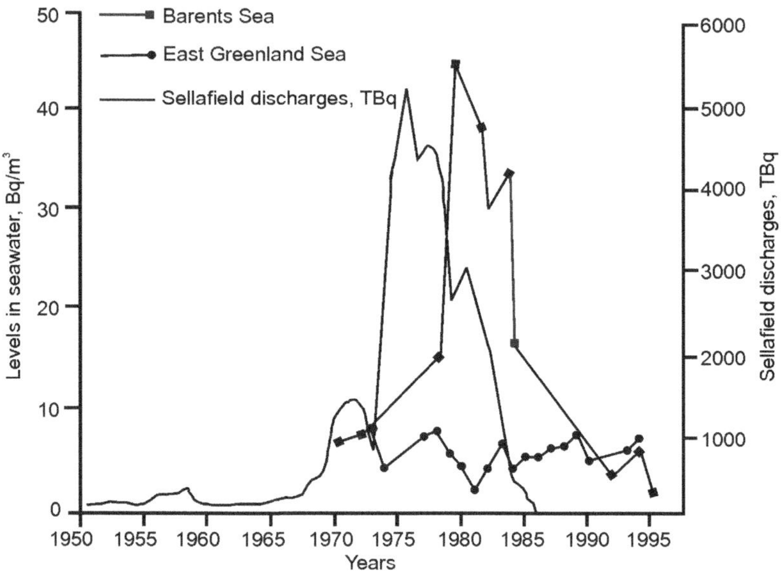

Fig. 2.7. Discharges of ^{137}Cs from the Sellafield nuclear reprocessing facility (10^{12} Bq / m^3) and radioactivity levels in the Barents Sea and East Greenland Current (Bq / m^3)

2.1.3 Large river supplies

The discharge of radionuclides by large rivers of the Arctic region is also a significant source of contamination. A number of the world's largest rivers, with total runoff reaching 3 300 km^3 / year, are connected to the Arctic Ocean basin. These include the Rivers Yenisey (603 km^3 / year), Ob (530 km^3 / year), Lena (520 km^3 / year) and Mackenzie (340 km^3 / year) (Fig. 2.8). The most important from a radioecological perspective are the rivers draining into the Kara and Laptev Seas. These rivers receive annually more than 2000 km^3 of runoff (Aagard et al., 1985) representing 50 % of the total continental runoff to the Arctic Ocean. Associated with these seas are the Siberian rivers (Ob, Yenisey and Lena), which collect water and radioactive particles from the vast catchment area to the east of the Ural Mountains.

The stock of dissolved and suspended organic matter of Lena River is estimated as 3.6 x 10^6 and 1.3 x 10^6 tons per year correspondingly (Strand 1998). The Ob and Yenisey together transport to the seas about 20-30 million tons of sediments yearly. In this connection the input of radionuclides to the river water can take place by two ways: by means of wash-out from the soil of global fallout within the watershed area and from objects containing radioactive materials situated on the river itself.

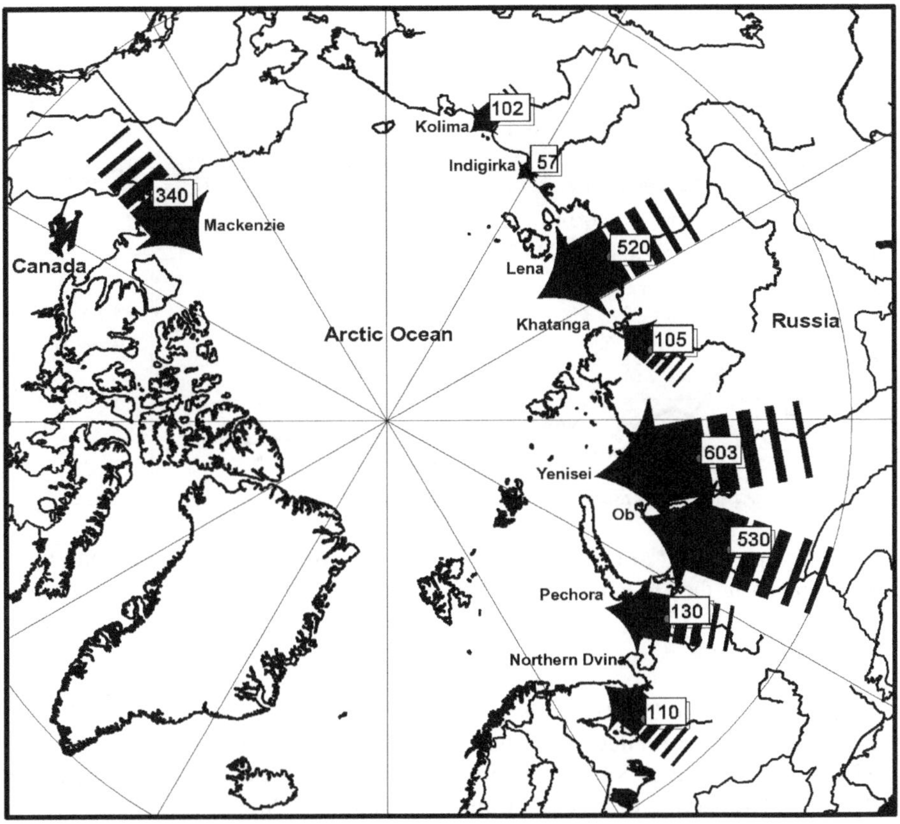

Fig. 2.8. Fresh water (km^3/yr) received by the Arctic Ocean from the largest rivers, which potentially transport pollution to the Arctic from industrial sites located on the vast watersheds (Aagaard et al. 1985)

The northern Rivers Lena, Indigirka, Pechora, and Northern Dvina, have no nuclear reprocessing facilities in their basins so that the input of radionuclides to these rivers is from global fallout. The river transport of ^{90}Sr to the basins of the Barents-White, Laptev and East-Siberian Seas was in 1961-1990, 381 x 10^{12}, 462 x 10^{12} and 170 x 10^{12} Bq respectively (Fig. 2.9) (Chumichev 1995). For the same period the transport of ^{137}Cs to the Barents Sea is estimated as 20 x 10^{12} Bq (Vakulovsky et al. 1993). The concentration of ^{90}Sr, ^{137}Cs, and ^{239,240}Pu in the bottom sediments of estuaries increases normally for rivers with large-scale watersheds (Kuznetsov et al. 1994; Vakulovsky et al. 1994 a; Matishov et al. 1997 b, c, 1998 b, c, 1999 a, c, i, 2000 b, c; Stepanets et al. 1999; Ivanov 1999). The radioecological rehabilitation of these marine ecosystems proceeds through processes of radionuclide transport, migration, and redistribution, and the further loss of radionuclides through radioactive decay.

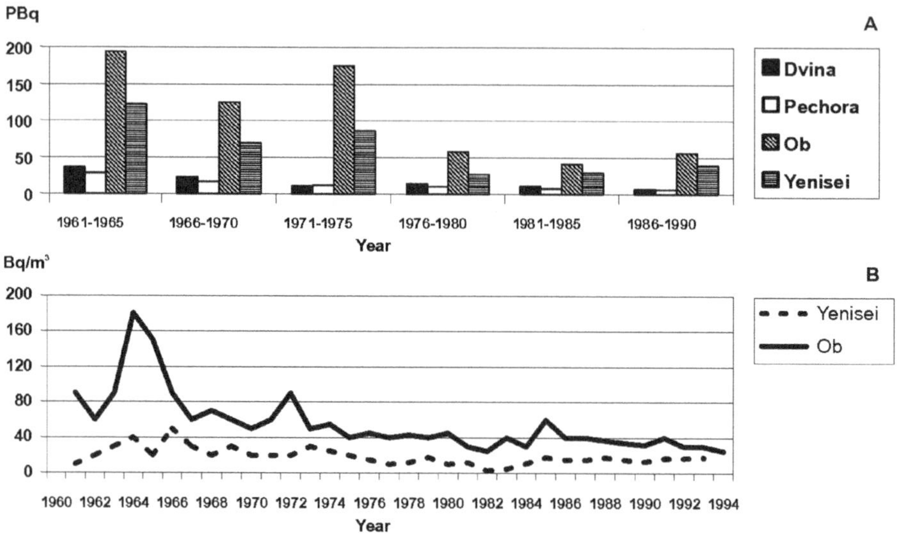

Fig. 2.9. (**A**) Input of ^{90}Sr to the Arctic Basin with river runoff from 1961 to 1990. (**B**) ^{90}Sr concentration in the Ob River (1961-1964) and Yenisey River (1961-1992) (Chumichev 1995)

The Yenisey and Ob Rivers during 1950-1990s were systematically polluted by direct discharges from nuclear industrial facilities. The total amount of radioactive substances released to Siberian rivers is estimated as 63 000 PBq (Brungot et al. 1999 b). Radioactive discharges from installations at Mayak and Tomsk are received by the Ob basin; wastes from the Krasnoyarsk reprocessing plant are released directly to the Yenisey. These sources are transported downstream by the rivers and into the Kara Sea (Kuznetsov et al. 1994; Vakulovsky et al. 1994 a; Novitsky et al. 1999).

Perhaps, the highest levels of ^{90}Sr in the Ob River were up until 1960, as a result of discharges from the Mayak plant from 1949 to 1956. In 1960s the ^{90}Sr concentration in the Ob varied from 50 to 180 Bq / m^3. Modeling the discharge of the highly mobile ^{90}Sr from Mayak during high river flow, it has been shown that ^{90}Sr can reach the sea in approximately half-year. Based on the average annual dissolved concentrations of ^{90}Sr during the period from 1961 to 1990, about 1.4 x 10^{15} Bq of ^{90}Sr entered the Kara Sea through the Ob River system (Fig. 2.10, 2.11) (Chumichev 1995).

In total the Siberian rivers release to the seas about 10 000 Ki annually owing to their vast catchment areas; the Amazon River releases 8 000 Ki, and the Huanghe River releases 6 000 Ki, annually. Present-day releases from Sellafield do not exceed 2 000 Ki / year into the Irish Sea.

The spectrum and quantity of radionuclides in bottom sediments of the Ob and Yenisey estuaries reflect the obvious influence of radioactive releases from nuclear facilities (Matishov et al. 1994 a, c; 1997 c; Lisitsyn 1994). Within the turbidity zone of an estuary, where river and sea water first mix, almost all river suspended matter is deposited due to the influence of sorption and biofiltration processes. Recent investigations have shown enhanced accumulation of ^{90}Sr and $^{239,240}Pu$ in estuaries, as well as for the short-lived radionuclide, ^{60}Co.

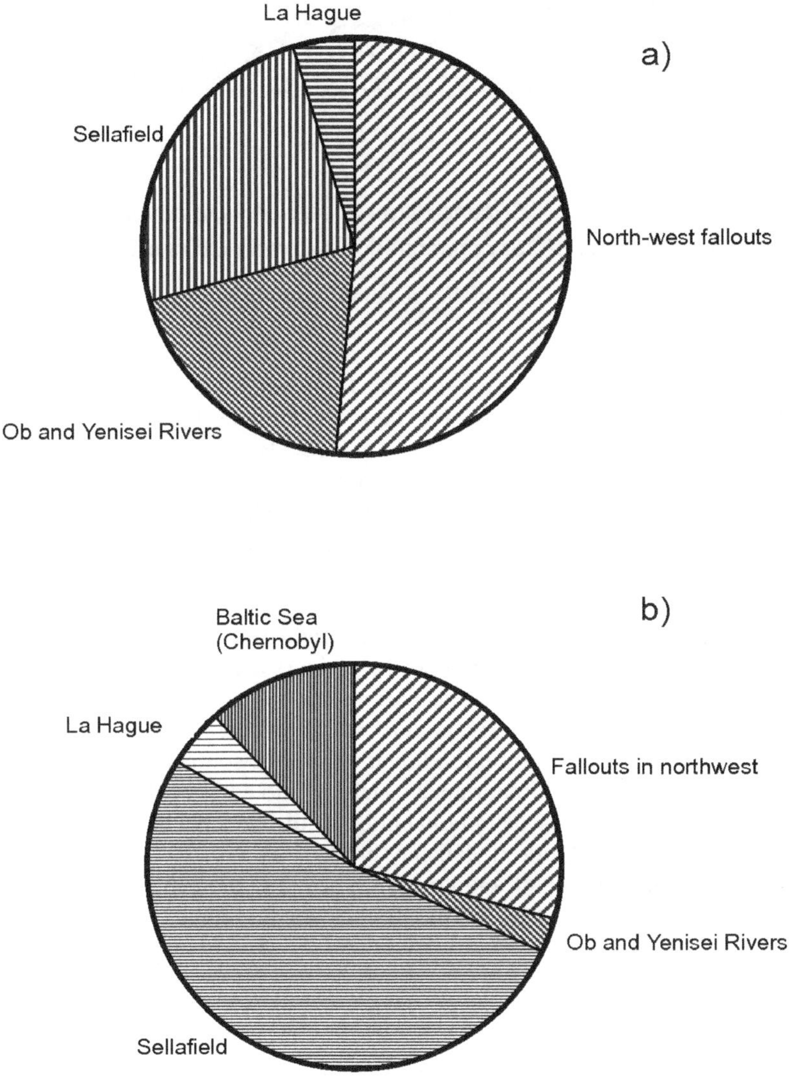

Fig. 2.10. Components of the Kara Sea contamination by ^{90}Sr (**a**) and ^{137}Cs (**b**) (IIASA, 1996)

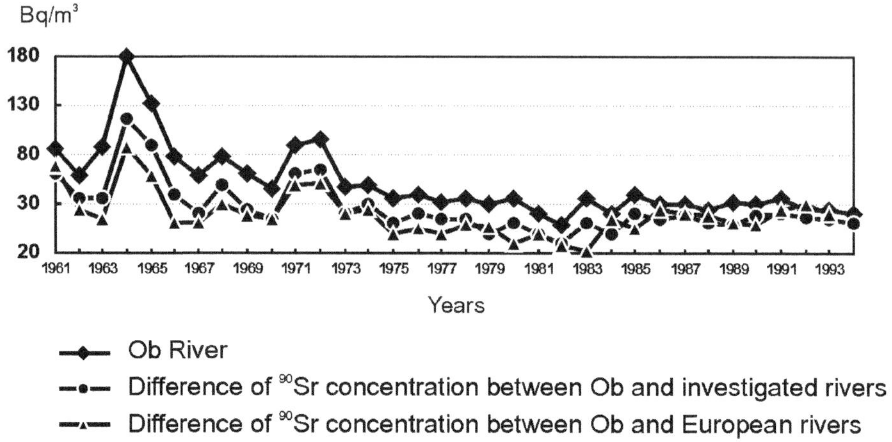

Fig. 2.11. Concentrations of ^{90}Sr in the Ob compared to other Russian rivers ((Chumichev, 1995)

We can note for comparison that the transport of radionuclides by rivers is characteristic for the southern seas as well. The Dnieper River was responsible for transporting into the Black Sea about 18.5 x 10^{11} Bq ^{137}Cs and about 21.83 x 10^{12} Bq ^{90}Sr from June to October 1986, or about 0.1% and 7% respectively of the volume of releases of these radionuclides during the Chernobyl accident ((Yablokov et al. 1993; Pavlyutin and Babitskaya 1996; Gulin et al. 1997). During 1986-1988 the ^{137}Cs content in the Black Sea water increased by nearly four times.

Despite the fact that global fallout, Sellafield releases and releases of global fallout-derived radionuclides from northern rivers are the major sources of radioactive contamination in the Arctic Ocean, the role of industrial sources of radionuclides in West Siberian rivers must not be underestimated. Investigations of artificial radionuclides in bottom sediments from the southern part of the Kara Sea demonstrate the principal importance of taking into account the West Siberian nuclear facilities as sources of radionuclides in the general balance of radioactive contamination of the Arctic ((Matishov et al. 1994 a, c, 1995 a). The anomalous values of plutonium, cobalt and strontium accumulation in bottom sediments of the Ob and Yenisey Bays are confirmation of this principal.

2.1.4 Sea burial

The burial of nuclear waste on the sea bottom occurred in the North Atlantic and Arctic Seas in 1950-1960s (Fig. 2.12, 2.13, 2.14). British nuclear facilities were authorized to annually bury in the Channel 5 000 tons of solid waste with summary activity of 4 kKi of β- and γ-emitters (Dovgusha and Tikhonov 1995; Vartanov and Hollister 1997). Highly radioactive waste was transported to an ocean area located approximately 480 km from Portugal's Viland Cape.

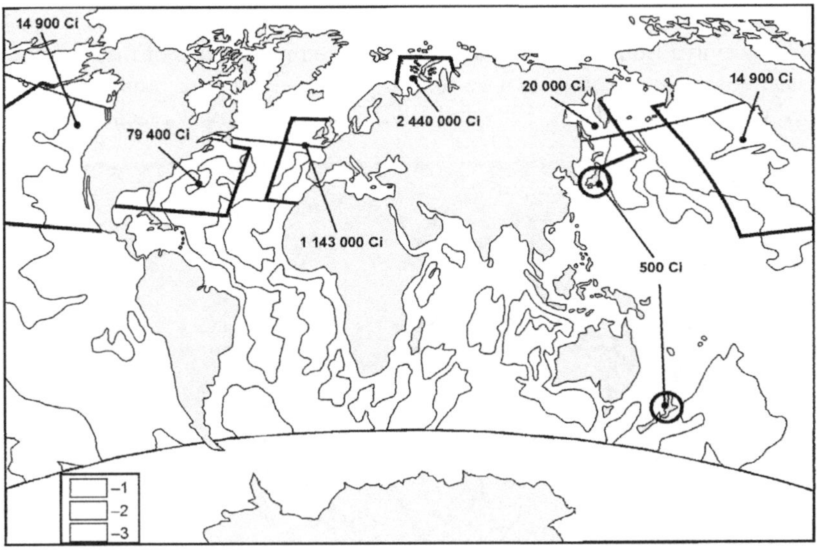

Fig. 2.12. Ocean areas, suitable and unsuitable for the burial of nuclear wastes, adopted by IAEA (Vartanov and Hollister 1997)

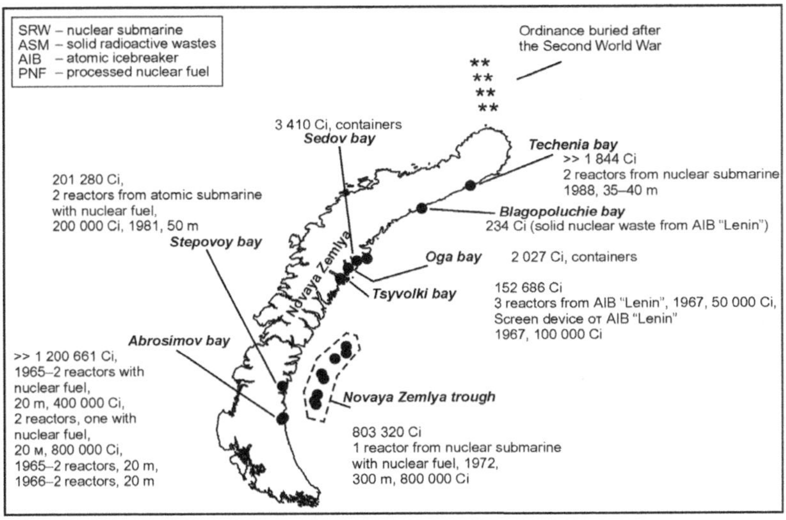

Fig. 2.13. Sites for burial of solid radioactive wastes in the Kara Sea (by: Yablokov et al. 1993)

Five areas for the release of liquid radioactive wastes (RAW) were designated in the Barents Sea for the Russian Northern Fleet and other national organizations (three offshore with depths to 300 m and two in the coastal zone). In addition eight burial areas were designated for solid RAW - in the Kara Sea (one in the Novaya Zemlya trough of 380 m depth and seven in the bays of the Novaya Zemlya east coast with depths from 12 to 140 m (Fig. 26, 27). Liquid RAW were released in 1965, 1975, 1988, 1989 and 1992. The total activity reached 10 000 Ki.

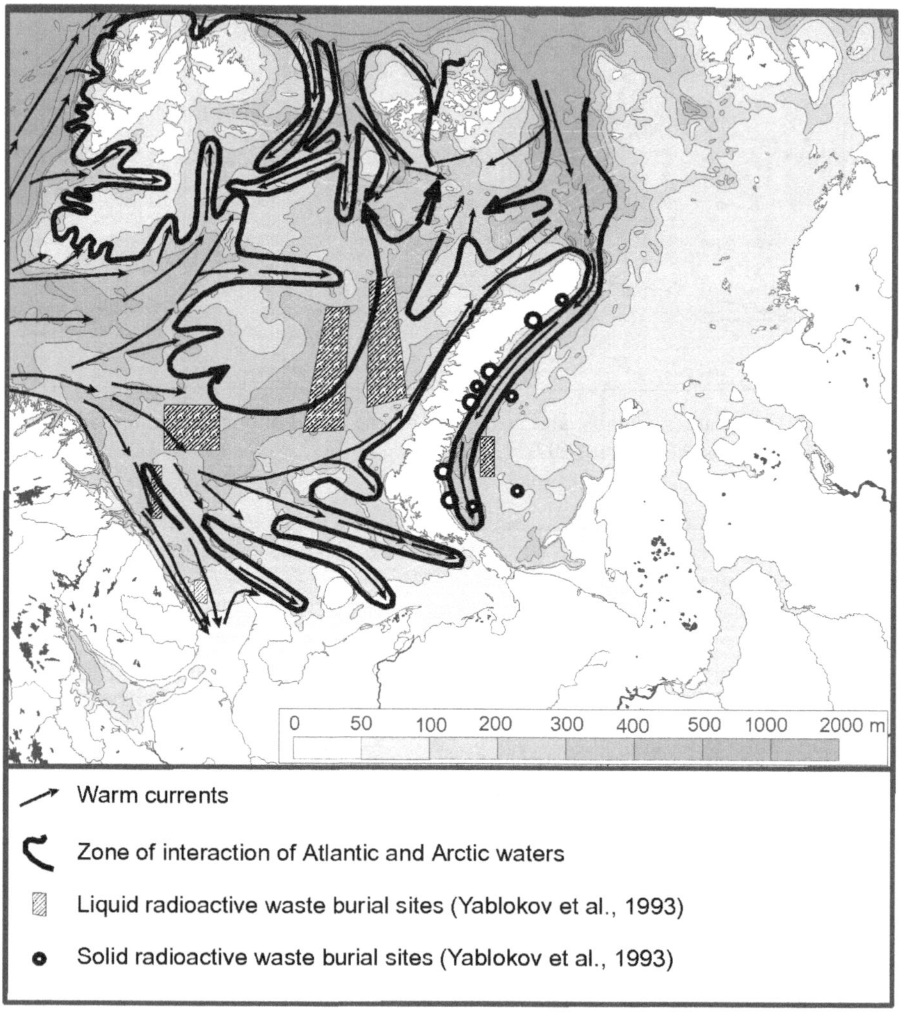

Fig. 2.14. Advection of Atlantic water to the Barents Sea over glacial troughs in comparison to the burial sites of solid and liquid radioactive wastes in the Barents and Kara Seas

The burial of solid RAW in the Novaya Zemlya fjords began in 1964. The volume of waste buried was 6 000-7 000 m^3 / year from the end of the 1970s through the 1980s (Tables 2.2, 2.3). Burial containers held six nuclear reactors from submarines including processed nuclear fuel and the shield from a nuclear icebreaker. The activity of these reactors and fuel has decreased many times as a result of radioactive decay since burial occurred approximately 30 years ago. At present the total activity of these wastes are 115 kKi in the northern seas; about half is connected with the shield and fuel of the nuclear icebreaker. The spreading of artificial radionuclides from these sources is insignificant, no more than several meters from the burial sites (A survey ... 1992; Radioactive ... 1994; Dumping ... 1996).

However, increased levels of radioactivity in the vicinity of the burial sites indicate the presence of local contamination. The released radionuclides are usually sorbed by bottom sediments in the near vicinity of the dumped objects. This fact was confirmed by direct measurements of radiation situation at dumping sites. Any danger posed by these objects will not occur for many decades due to the protective shields that will be breakdown over time through corrosion (Strand et al. 1992, 1993, 1994, 1996, 1997; Foyn and Nikitin 1993; Foyn and Semenov 1993).

Experts estimate that 480 x 10^{12} and 600 x 10^{12} Bq of liquid and solid radioactive wastes, respectively, were dumped in the Barents and Kara Seas during the entire period. In the Kara Sea, 85 x 10^{15} Bq of solid radioactive waste and nuclear fuel was buried representing 85% of the total amount of nuclear waste buried in these seas (Yablokov et al. 1993).

2.1.5 Land discharges

Local sources of radionuclide contamination in the Far North are connected with nuclear objects and test sites in coastal areas (Matishov et al. 1994 a, b, c, 1996 b, f, 1997 b, f, 1998 b; Namyatov 1995, 1998). The environment of the White Sea is influenced by the presence of a nuclear submarine facility in Severodvinsk and a submarine test site in Kandalaksha Bay (Matishov et al. 1996 b). However, the inlets of Kola and Motovsky Bays, in particular, in Murmansk, Severomorsk, Polayarny, Bays Sajda, Gadzhievo, Olenjya, Pala, Zapadnaya Litsa, Ura and Ara are the main locations where radioactive materials are concentrated. of radiation danger are concentrated in the inlets of the Kola and Motovsky Bays, in particular, in Murmansk, Severomorsk, Polayarny, Bays Sajda, Gadzhievo, Olenjya, Pala, Zapadnaya Litsa, Ura and Ara. Different forms of radioactive waste are found here: spent nuclear fuel, gaseous, solid and liquid radioactive waste (Kolomiets et al. 1993; Nielsen et al. 1996).

The naval bases and ship-repairing yards of the Russian Northern fleet, the storage areas for spent nuclear fuel and decommissioned nuclear ships have existed for nearly forty years in many places of the Kola Peninsula (Fig. 2.15). These enterprises of the nuclear industry result in small planned releases of radiation as part of the industrial process. However, in some years large accidents with discharges of radioactive substances to seawater and to the atmosphere have also

Table 2.2. Burial of solid RAW in the Kara Sea (Dovgusha and Tikhonov 1998)

Location and time of burial	Number	Depth (m)	Number of buried objects			Number of burials	Total activity at the time of burial (in ^{90}Sr equivalent), Ki
			containers	large objects	floating barge with solid RAW		
Novaya Zemlya Depression , 1967-1991	1	380	3174	11	10	22	3520
Sedov Bay, 1982-1984	2	13–33	1108	104	–	8	3410
Oga Bay, 1968-1983	3	up to 24	472	4	1	8	2027
Tsivolka Bay, 1964-1978	4	56–135	1600	6	2	8	2684
Stepovoy Bay, 1968-1975	5	25–27	–	5	–	5	1280
Abrosimov Bay, 1966-1981	6	12–20	8	7	4	7	661

Table 2.3. Burial of solid and liquid RAW by the USSR (1960-1991) (Dovgusha and Tikhonov 1995)

RAW Type	Volume, (m^3)	Number of containers	Number of large objects	Number of the sunken barges	Activity, Ki	Activity (in ^{90}Sr equivalent), Ki
Liquid	190434	–	–	–	23753	–
Solid	31534	6508	155	17	–	15502

taken place (Nielsen et al. 1996). In particular, the reservoir of spent nuclear fuel in Andreev Bay (Zapadnaya Litsa, 1982), resulted in an estimated release of $37 \cdot 10^{12}$ Bq to the marine environment. Another accident occurred in Ara Bay (1989) resulting in the release of $74 \cdot 10^{12}$ Bq from a nuclear submarine to the marine ecosystem.

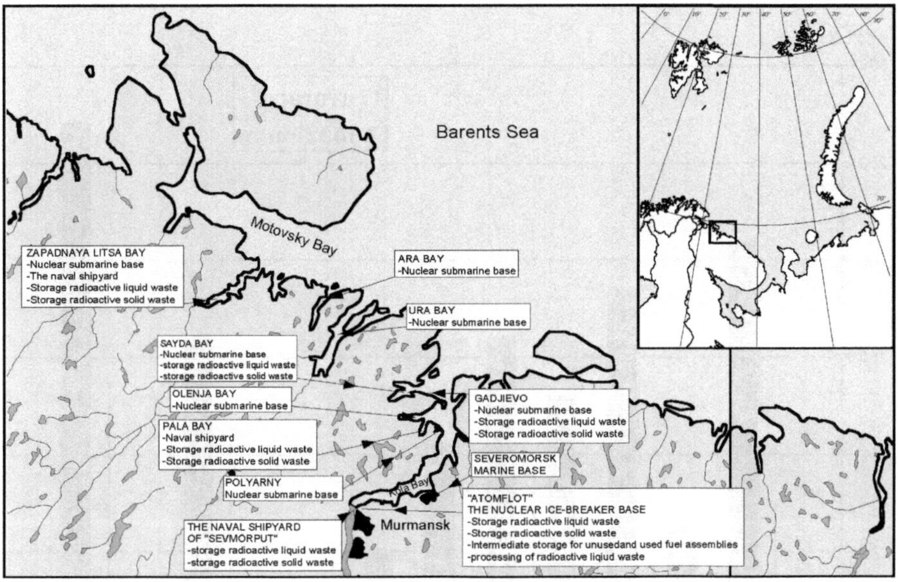

Fig. 2.15. Disposition of the sources of radioactive contamination on the coasts of the Kola and Motovsky Bays

A liquid radioactive waste purification plant, situated on the territory of the «Atomflot» facility (Murmansk), began operating in 1989. This facility is the base of the nuclear icebreaker fleet. The radioactivity released directly into the southern section of the Kola Bay after purification is $(1\text{-}3) \cdot 10^{-10}$ Ki / l. The average volume of ^{137}Cs and ^{90}Sr released from «Atomflot» is estimated at $1.6 \cdot 10^7$ and $7.6 \cdot 10^7$ Bq / year correspondingly (Fig. 2.16; Table 2.4) (Namyatov 1998). The greatest amount of ^{137}Cs and ^{90}Sr was released to the Kola Bay in 1992. The intentional release of liquid radioactive substances with specific activity less than 10-8 Ki / l is permitted into bays where naval bases are situated. The release of liquid RAW with specific activity from 10-8 to 10-5 Ki / l was also permitted in the coastal zone (25 nautical miles from shore).

The emissions from nuclear objects are indicated by the composition and amount of artificial isotopes detected in the environment. Relatively high concentrations of ^{60}Co and even its prevalence over ^{137}Cs is characteristic of the sediments in Kola and Motovsky Bays in the vicinity of sources of radioactivity. Contamination by ^{239,240}Pu isotopes is also observed (Matishov et al. 1996 f, 1997 b, f, 1998 b, c, 1999 e). Local sources are responsible for ^{60}Co, ^{134}Cs, ^{152}Eu, ^{154}Eu and other similar elements in bottom sediments. These radionuclides are also easily detected in algae samples taken in the vicinity of the primary radioactive sources in Kola and Motovsky Bays. Maximum values of ^{137}Cs accumulation in these areas can be 20-40 times more than in offshore areas. Obviously it is necessary to monitor and control the potential risks of radioactive contamination from sources in the region.

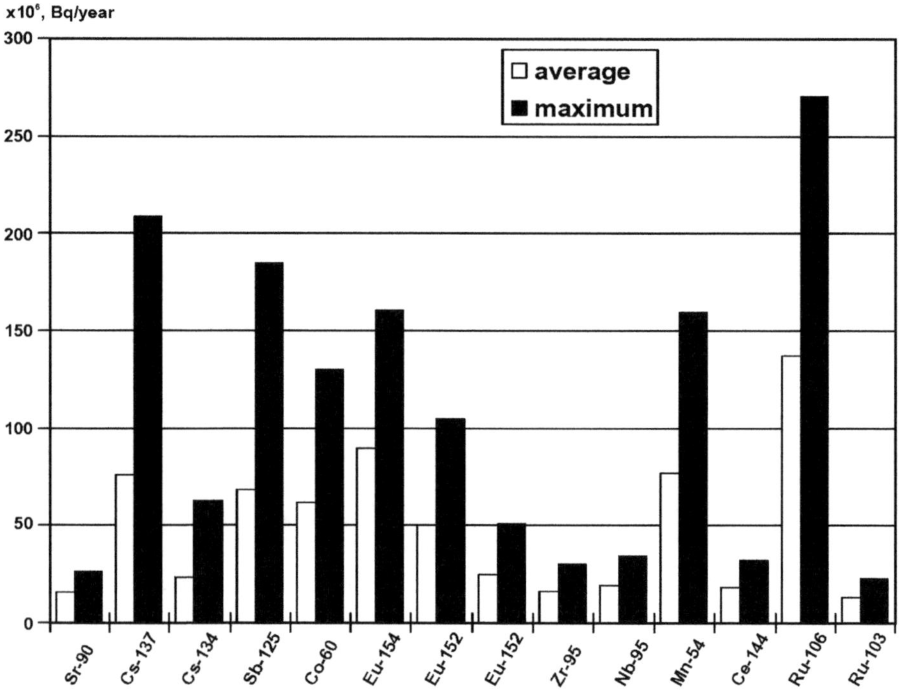

Fig. 2.16. Average annual and maximum releases of artificial radionuclides to the Kola Bay from the «Atomflot» nuclear reprocessing facility, Kola Bay, Northwest Russia (Matishov et al. 2000)

2.1.6 Nuclear accidents at sea

Accidents with nuclear submarines and airplanes carrying nuclear materials are rather widespread in the ocean and particularly in the Arctic Seas. The levels of radioactive contamination resulting from nuclear submarine accidents are well known. For example, environmental contamination at the site of the nuclear submarine «Komsomolets» accident is described in numerous works (JRNEG 1993; Saetre 1994; Hoibraiten and Thoresen 1995; Kuznetsov et al. 1996, 1997 a). These investigations indicate only small environmental contamination has

Table 2.4. Total discharge of artificial radionuclides (106 Bq/year) to Kola Bay of purified liquid radioactive waste from the «Atomflot» facility (data of the «Atomflot» laboratory)

Nuclide	1989	1990	1991	1992	1993	1994	Total	Average	Ratio of nuclide: ^{137}Cs
^{90}Sr	7.4	5.6	2.5	26.3	15.5	37.0	94.0	15.7	0.21
^{137}Cs	20.7	31.1	5.56	208.7	94.0	97.0	457.0	76.2	–
^{134}Cs	6.3	9.2	3.3	62.8	31.5	27.0	139.1	23.2	0.30
^{125}Sb	28.1	39.2	26.6	185.0	92.1	40.7	411.8	68.6	0.90
^{60}Co	31.8	44.4	10.0	130.0	72.2	81.0	369.6	61.6	0.81
^{154}Eu	47.0	65.1	40.0	160.0	139.5	84.4	536.9	89.5	1.18
^{152}Eu	20.0	27.8	18.1	105.0	79.2	50.3	299.3	50.0	0.65
^{95}Zr	10.7	12.6	5.2	35.5	51.4	30.0	147.6	24.6	0.32
^{95}Nb	9.2	12.5	5.2	30.3	22.6	17.0	96.9	16.2	0.21
^{54}Mn	10.4	14.1	10.4	34.4	25.5	22.9	117.7	19.2	0.26
^{144}Ce	25.9	29.6	35.9	159.8	130.9	81.0	463.2	77.2	1.01
^{58}Co	9.6	13.3	8.1	32.6	27.8	16.3	107.7	18.0	0.24
^{106}Ru	81.4	111.0	46.6	270.8	199.4	115.8	825.1	137.5	1.82
^{103}Ru	6.3	8.5	5.2	20.4	22.9	12.6	77.7	13.0	0.17

resulted from submarine accident. For the present, the threat caused by radionuclides from the perished submarine «Komsomolets» does not pose any significant risk to human health or the environment. Even when the protective housing for the submarine nuclear reactor is destroyed through corrosion, the majority of the radioactive products will have already decayed resulting in minimal releases into the environment. The long-lived radionuclide $^{239,240}Pu$, is expected to remain tightly bound to mineral and organic particles in the bottom sediments in the vicinity of the submarine. The release of radionuclides to the marine environment will cause much less radiation than a similar release on land (Strand 1998).

2.2 Secondary sources

The fourth type of contamination source is known as secondary sources. Secondary contamination is the accumulation of technological releases of radionuclides into ice, river water, or marine biota with subsequent transport to distant areas over the course of the ensuing days, months and even years after incorporation into the various transport media. Glaciers and icebergs from the polar archipelagoes and pack ice are release radionuclides in areas of intensive sea ice melting (e.g. Fram Strait). In areas of bird rookeries and fish shoals, macrophytes and other biota are enriched in artificial radionuclides as a result of bioaccumulation and biosedimentation processes.

Sea ice and icebergs are repositories of artificial radionuclides received from the atmosphere with rain and snow. The primary redistribution of radioactive particles on the surface of sea ice and continental glaciers is caused by surface winds. The active migration of artificial radionuclides to the lower ice surface is possible during the period of the spring phytoplankton bloom at marginal ice zones. The behavior, distribution and effects from radionuclides in pack ice and icebergs are not well investigated.

The glaciers of Greenland, Franz-Josef Land and other polar archipelagoes have accumulated long-lived radionuclides from the atmosphere with rain and snow beginning with the first nuclear explosions (Matishov et al. 1994 a, c; Strand 1998). The total area of continental glaciation in the Arctic is about 2.3 mln. km^2. Numerous icebergs break off these glaciers and drift in the open sea for 5-10 years (Fig. 2.17). From 50 to 100 icebergs move each year into the Central Polar Basin (Matishov 1987) and replenish their stocks of radionuclides by aerosol deposition.

Radioactive substances from nuclear fallout are deposited constantly on the pack ice of the Arctic Ocean. The Central Polar Basin is ice bound all year round with ice ranging in age from 1-2 to 3-5 years (Fig. 2.12) and from 1-2 to 4-6 m thick. The contemporary pack ice (covering the area about 0.3 mln. km^2) contains ^{137}Cs, ^{90}Sr,and $^{239,240}Pu$ (Matishov et al. 1994 a, c; Nies et al. 1998). These radionuclides were accumulated in the 1990s from aerosols.

In accordance with the Roshydromet data in 1993 the atmospheric precipitation in the Kara Sea contained about 2.1 Bq / m^3 of ^{137}Cs and 1.4 Bq / m^3 of ^{90}Sr. The concentration of ^{137}Cs in the water and ice of the Chukchi Sea oscillated around 1 Bq / m^3 (Pavlov et al. 1999).

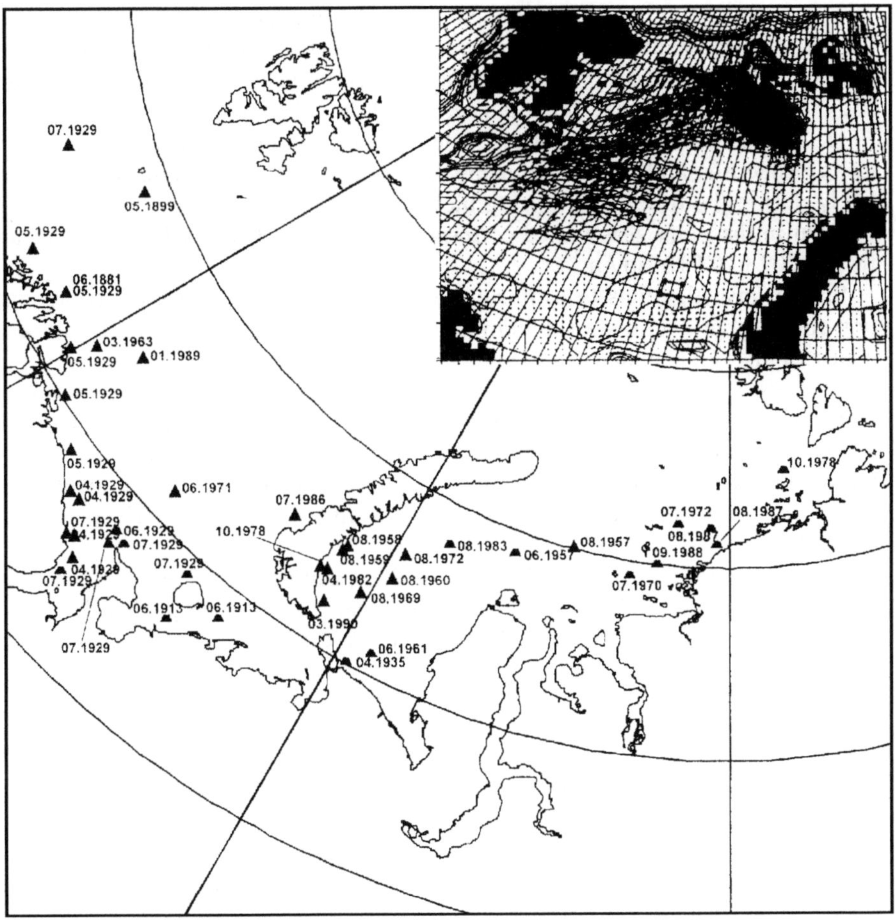

Fig. 2.17. Sea ice drift (1995-1996) and anomalous cases of the appearance of icebergs in the southern part of the Western Arctic (Abramov 1985)

Seasonal ice (age=5-9 months) of the Laptev, Kara, Pechora and other seas concentrates radioactive particles both from aerosols and from river water. Probably, radionuclides are taken up by ice in the mixing zone where turbid river and seawater interact. The salinity of surface waters along these hydrological fronts is from 5 to 25 ‰. The content of radionuclides in such ice has not been directly measured. The magnitude of this secondary source can be estimated from the average content of sediment in the Laptev Sea (3-5 mg / l) and the ice thickness during the freezing period about 2 m (Lisitsyn 1978).

During sea ice formation sediments are taken up by ice, especially in estuaries. In particular, the clay sediments from the lower part of the Chukchi Sea ice contains ^{137}Cs concentrations of 4-6 Bq / kg (Meese et al. 1995).

The seasonal and pack ice is thought to accumulate radioactive substances from the water during ice formation (Pfirman et al. 1997). The clay particles, contained in sea ice are able to absorb radionuclides. In some years part of the Lena, Yana and Indigirka Rivers enter the initial section of the Siberian Drift.

Ice associated zooplankton play a role in concentrating radionuclides in drifting ice. Crustaceans and other organisms of sub-glacial fauna filtrate the water actively during the time of the spring bloom. In 1990s the level of ^{137}Cs, ^{90}Sr, and ^{239,240}Pu in the polar seas varied correspondingly from 3 to 30, from 1 to 12 and from 0.002 to 0.018 Bq / m^3.

Special investigations to estimate the total quantity of radioactive contamination currently in Arctic sea ice and icebergs, has not been carried out. By the estimates of many specialists (Klenova 1948, 1960; Lisitsyn 1978; Pfirman et al. 1997; Nies et al. 1998; Harms et al. 1999) the concentration of terrigenous sediments in the Arctic sea ice varies from 800 to 3 000-10 000 tons / km^2. The quantity of clay sediments in ice that is transported annually into the Arctic Ocean through the Fram Strait, is approximately 7 to 150 mln tons. This amount of radioactivity in drifting ice is similar to radioactive releases from rivers into the Arctic Ocean Basin.

There is one direct piece of data about the quantity of ^{239,240}Pu scattered over the ice surface. In 1968 the crash of an American aircraft with nuclear bombs on board occurred in the vicinity of Thule (North-Western Greenland). This disaster resulted in 0.1 TBq of plutonium being deposited on the ice surface (Aarkrog 1995; Strand 1998). Immediately after the disaster the content of plutonium in seawater of the Bylot Strait increased (about 0.2 Bq / m^3) relative to other places along the Greenland coast. The levels returned to background values in 1970.

Radionuclides transported in Arctic waters usually are attached to mineral or organic particles (Fig. 2.18). Part of the solid ice containing radionuclides attached to particles melts locally on the shelf and does not contribute to the long distance transfer of contaminants. A fraction of the radionuclides penetrates under the ice along channels of salt-water percolation and with the runoff of melted waters (Pfirman et al. 1997). However, the main part of ice-transported pollutants, evidently, exits the Arctic Ocean. The ice melts gradually in the polar hydrological front zone of the Norwegian-Greenland Basin and the Barents Sea and releasing particles containing radionuclides and other contaminants to the pelagic zone. More work is needed to obtain better estimates of the volume of artificial radionuclides incorporated and transported by ice and icebergs.

Thus, sea ice and icebergs act both as a secondary source of radioactive contamination and as one of the transport mechanisms. Contaminated particles from Yenisey, Ob, Lena and other rivers and radioactive dust from continental glaciers are redistributed over the Arctic Ocean with drifting ice (Matishov et al. 1994 a, c;

Pfirman et al. 1997; Strand 1998). The Kara Sea is the largest supplier of ice (to 200-400 km^3 annually) (Zakharov 1981).

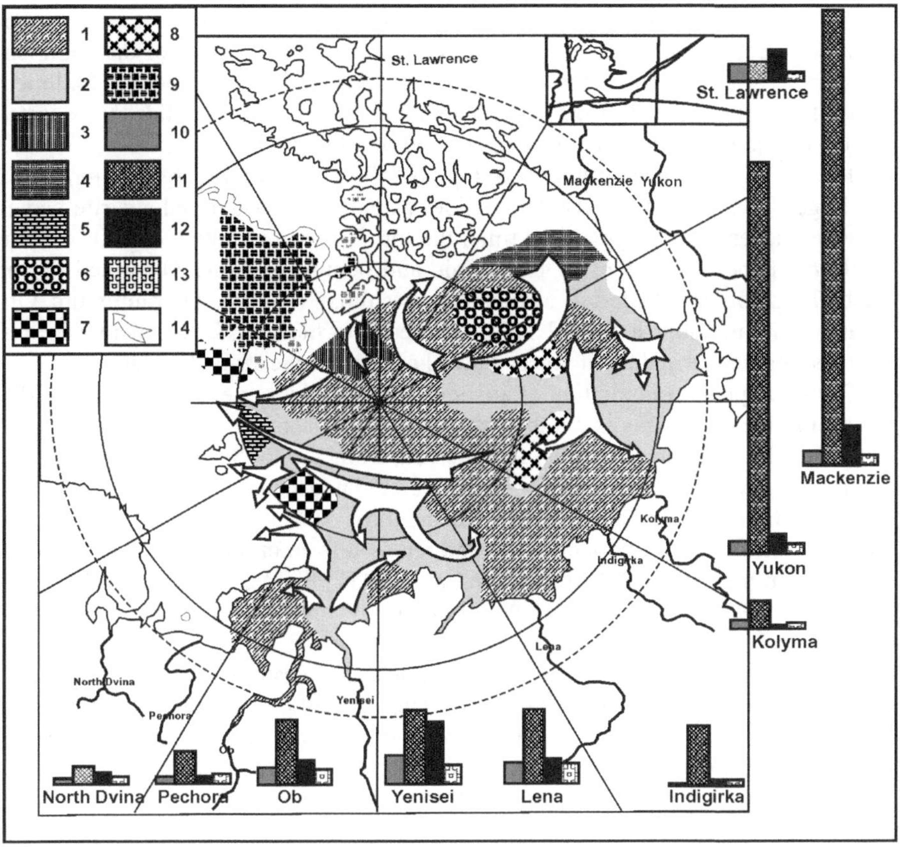

Fig. 2.18. Mineralogic provinces of the Arctic, sea ice drift patterns and Arctic sedimentation (Belov and Lapina 1961 et al.) *Provinces: 1 - epidote-amphibolic; 2 - ore-epidote-piroxene; 3 - chloritoide; 4 - dolomite; 5 - granate; 6 -ore; 7 - piroxene; 8 - piroxene-amphibolic; 9 - areas of contemporary land glaciation; 10 - Freshwater runoff from major rivers; 11 - particulate runoff; 12 - ok runoff; 13 - Organic carbon (C_{org}); 14 - drift pathways*

Underwater landslides are another significant source of pollutants for the continental base and abyssal plains of the Arctic Ocean, Norwegian-Greenland Basin and North Atlantic. In modern times unconsolidated sediments are formed along the shelf border and displaced downward through the canyons under the influence of gravity (Lisitsyn 1978; Leontjev 1984; Matishov 1987). The landslide and turbid streams contain saturated particle suspensions (about 2 g / cm^3) flow in the direction of the slope of the seabed. As a result, large volumes of sediments accumulate at the base of the continental slope.

Let's presume that the concentration of ^{137}Cs and $^{239,240}Pu$ in the clay sediments of the continental base is on the order of 2-10 and 0.2-2.0 Bq / kg correspondingly. The contemporary layer (1-5 cm) of liquid silts at the base of the continental margin stretches for many hundreds and thousands of kilometers. Sediment deposits ranging in depth from 2-4 km contains artificial radionuclides transported off the continental margins during the period of 1950-1990s.

The turbid streams carried down-slope off the continental margin originates in the large rivers and transport radionuclides as part of river-derived aleurite-clayey particle suspensions. At the same time, the turbid streams capture organic materials (fecal pellets, skeletons of planktonic and benthic organisms) formed on the shelf. These organic materials also contain cesium, strontium, and plutonium. Obviously, bioturbation in seabed sediments leads to the transfer of chemical compounds into deeper sediment layers as well as back to the seabed surface.

The biological accumulation of artificial radionuclides over the lifetimes of different marine organisms is another part of the radioactivity cycle of the sea. Sea animals selectively incorporate radionuclides from 1-5 to 20-50 years, depending of the lifespan of the organism. ^{90}Sr is accumulated mainly in skeleton and soft tissues rich in calcium, while ^{137}Cs - for the most part is found in soft tissues and organs. The input of radionuclides to and removal from organisms is controlled by metabolism processes (Policarpov 1964 a, b; 1971; Iljenko 1974; Aleksakhin and Naryshkin 1977; Krivolutsky 1983; Polikarpov and Aarkrog 1993; Lebedeva 1999).

Food ingestion is the primary uptake pathway for cesium, strontium and plutonium for marine organisms (Fig. 2.19). The absorption of these elements from seawater is also an important mechanism for the accumulation of radionuclides by organisms. The decrease of radionuclide content in seawater and producers is accompanied by a decrease in accumulation by consumers. The mortality of older organisms with high concentrations of radionuclides and the development of young organisms with lower concentrations predetermines the environmental purification rate for radioactive fallout occurring in the coming years. The factors influencing radionuclide accumulation in marine organisms are highly complex and only a few of them are taken into account in radioecological investigations.

Whereas marine organisms remove radioactivity from seawater, the nesting of colonial birds every year in limited areas can lead to an increase in the content of radioactive substances per unit of water area because of contamination by droppings. As a result, ^{137}Cs and ^{90}Sr levels in marine fauna living in areas inhabited by large bird populations tend to be higher than in other similar areas. For example, sea birds and fish are the main consumers of zooplankton in the waters of Franz-Josef Land, little auks *Alle alle* and polar cod *Boreogadus saida* being the specialized planktonic species (Timofeev and Shirokolobova 1995). One little auk catches during a 100 day nesting season about 10 kg of zooplankton, filtering incidentally more than 10 000 m^3 of water (Golovkin et al. 1972). By S.M. Uspensky (1956), the whole population of 250 000 pairs of little auks filters not less than 50 km^3 of water. Consequently, the role of these birds in the biogeochemical cycle of matter is rather great: they consume and transport to land 5 000 tons of zooplankton annually. Obviously, the droppings enrich the coastal water with radioactive substances.

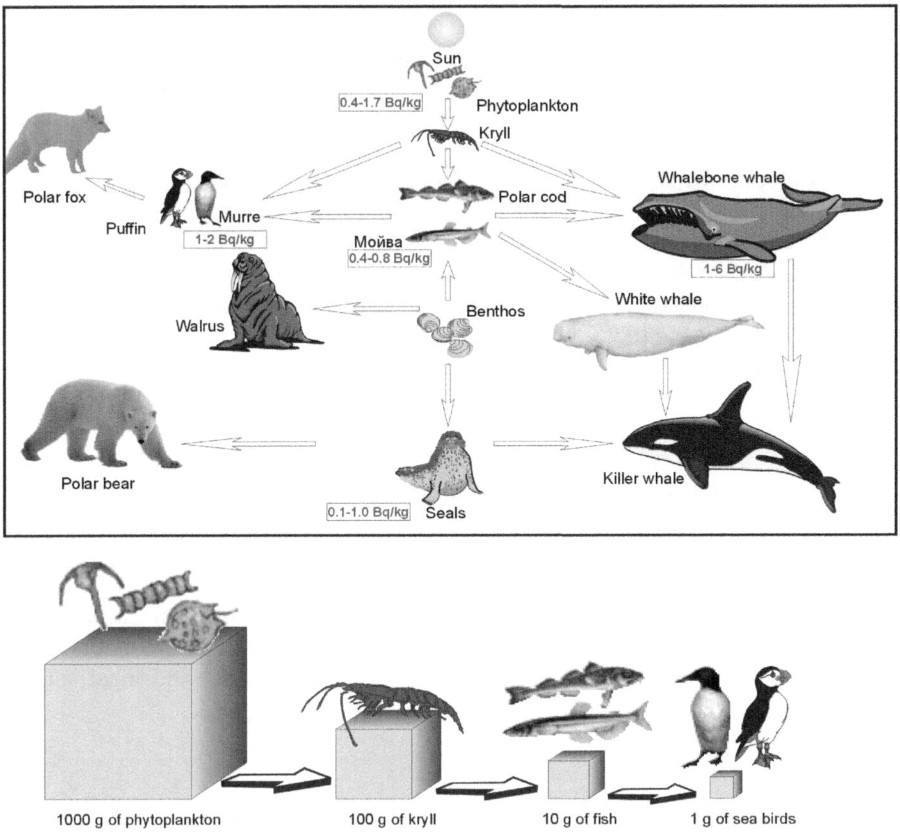

Fig. 2.19. Idealized food web of the polar seas and ^{137}Cs transfer through the food web

The quantity of radionuclides associated with the total mass of marine organisms probably is on the order of one to a few percent of their total content in the ecosystems of the northern seas. Because the incorporation of radionuclides into biota is a function of productivity, it follows that the Barents Sea biota incorporates from the water larger quantities of radionuclide than the biota of the Kara or Laptev Seas. For example, the annual production of zooplankton in the Barents Sea varies from 120 to 300 mln tons (Matishov and Drobysheva 1994) and the average annual biomass of pelagic fish (capelin, polar cod, herring, cod fry) - from 12 to 30 mln tons (Seliverstov et al. 1977; Anon ... 1985). The level of

^{137}Cs, ^{90}Sr and $^{239,240}Pu$ in these marine organisms is correspondingly within range 0.4-2.0, 0.3-1.0 and 0.001-0.005 Bq / kg.

The global scale of the consequences of the Chernobyl disaster have increased the interest in the investigation of the transport of radionuclides by migrating birds and of the influence of radiation on bird populations. The vast marshy areas and bird bazaars, which are the habitat and reproductive areas of many bird species, were contaminated as a result of the Chernobyl accident. As a consequence, ^{90}Sr levels have increased noticeably in the eggshells, food and bodies of birds.

An important task of radiobiology is the continued development of models for estimating radionuclide uptake and incorporation into the biomass of plankton, benthos, fishes, birds and sea mammals.

Considering the sources and pathways of penetration of radioactive contamination to the organisms of the Arctic Seas, one must take into account the behaviour of different organisms. For example, the migration patterns of birds in the northern seas, which cover many hundreds of thousands of kilometers. Vertical migrations are characteristic also for zooplankton, fish, and sea mammals. The range of daily and seasonal vertical migrations varies from a few meters up to many hundreds of meters. As a result, contaminants migrate repeatedly through the food chain and return to the seawater after the death of marine organisms or are removed from the photic layer of the Ocean through biogenic sedimentation processes.

Fish, birds and sea mammals also eat a variety of prey. The rate of accumulation of artificial radionuclides in these organisms increases when the bottom-feeding sea worms or bivalve mollusks are food sources. It is obviously necessary to take into account the locations of bird bazaars, spawning places of fish, and seal rookeries in relation to local and regional sources of artificial radionuclides. Further investigations are needed to determine the accumulated amounts of ^{137}Cs and ^{90}Sr in biota of the northern seas in relation to regional and local sources, Chernobyl fallout and Sellafield releases.

2.3 Transboundary transfer of radioactive substances

2.3.1 Water circulation within the Arctic Ocean

Cross-border migrations of radioactive substances received by marine ecosystems from major sources has attracted the attention of researchers (Kuznetsov 1971; Vakulovsky et al. 1985, 1988, 1994 a; Matishov et al. 1992 c, 1994 a, b, c, 1997 b, c, f; Strand et al. 1992, 1994; Strand and Nielsen, 1995; Schlosser et al. 1995; Rissanen et al. 1995, 1998, 1999 a; Kershaw 1997; Strand 1998; Nies et al. 1998).

The difficulty posed by investigating radionuclide transformation and transfer mechanisms in the pelagic zone arises from the dependencies of chemical sedimentation, chemical diffusion, formation of complexes, ionic exchange and extraction on numerous biogeochemical reactions. The specific features of the mode of entry and distribution in the Ocean of the radionuclides themselves are also important factors, including the main characteristics of the source function, e.g. the source capacity, frequency of releases, spectrum of admixtures etc. The influence of advective transport of pollutants and oceanic currents can be readily traced using some of the radionuclides associated with global radioactive contamination (^{137}Cs, ^{90}Sr, $^{239,240}Pu$). These radionuclides exist in seawater primarily in the dissolved state, are found in sufficient quantities and have relatively long half-lives. ^{134}Cs, for example, has been used as a tracer of the cross-border transfer to the Arctic Basin of water originating from the Sellafield radioactive waste reprocessing facility (Nikitin et al. 1991).

Large rivers and permanent sea currents are the main arteries by which terrigenous radioactive materials are transported to the sea. The geographical features of the river basins and characteristics of the river watersheds determine the mode of washout of terrigenous substances to the Ocean (Fig. 2.20). The majority of radionuclide contamination transported to the sea by river suspended matter is unloaded in the coastal and estuarine zones. A considerable portion of the remain

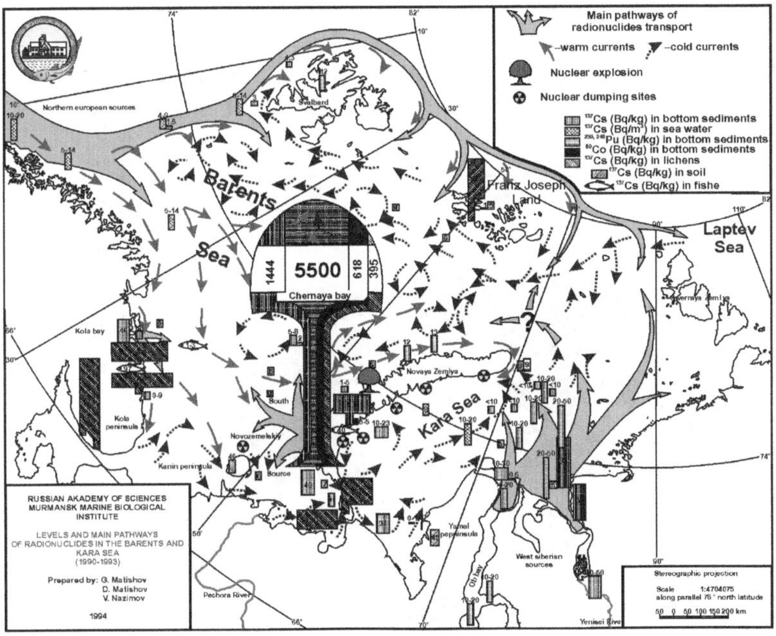

Fig. 2.20. Levels and main pathways of radionuclide transfer in the Barents and Kara Seas (1990-1993)

ing contaminant load is transported to the continental shelf with ice, currents, and turbid streams where the dissipation and differentiation of the sediment suspension depends on both the hydrological regime and sea bottom relief (Fig. 2.21).

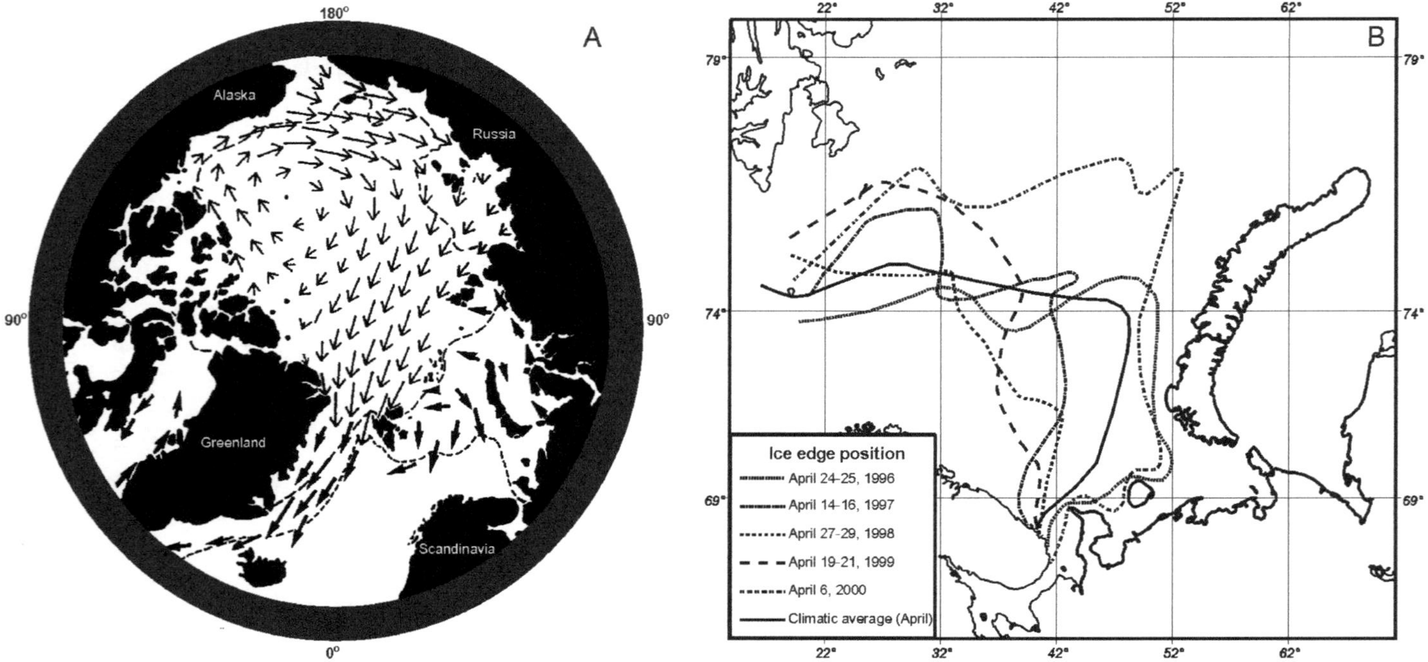

Fig. 2.21. Ice distribution in the Arctic Ocean: **A** - maximum (short lines) and minimum (long lines) distributions and general directions of ice drift (Aagaard and Carmack 1990); **B** - ice edge position in the Barents Sea (April 1996-2000)

Radioactive substances also deposit directly from the atmospheric onto the sea surface (Izrael and Tsyban 1989; Izrael 1998).

The seawater surface, in contrast to land, does not maintain high concentrations of artificial radionuclides as a result of vertical mixing which are related mainly to the thermal convection. As a result, background levels of artificial radionuclides in the sea are several orders of magnitude less than on land (Matishov et al. 1994 a, b, c; Strand 1998).

Current knowledge of the redistribution, transfer and accumulation of radioactive substances in the sea is based on the general circulation patterns of oceanic water masses. In the western Arctic, the pathways of radionuclide transfer are seen on the map, 'Levels and main pathways of radionuclides in the Barents and Kara Seas' (Matishov et al. 1994 a).

The hydrological and hydrochemical characteristics of the Barents and Kara Seas are known mainly in the southern and southeastern parts of the basins. In contrast, the areas adjoining Franz-Josef Land, Novaya Zemlya, and the eastern part of Spitsbergen are still poorly investigated. Because of this fact, oceanic modeling of these areas is based not on hydrological data but on theoretical parameterizations of oceanographic structure.

The results of MMBI oceanographic surveys conducted during the 1990s and other published data allow the characterization of the hydrodynamic and hydrochemical regime and also the thermohaline structure of the Arctic Seas (Matishov et al. 1992 a, 1996 c, 1998 d; Matishov D. 1994). Shelf topography is especially important. The waters of Atlantic and Arctic origin, follow the topographic troughs, transferring and redistributing biogenous, radioactive and other substances.

The oceanographic regime determines the conditions and pace of the transfer of radioactive substances in the Arctic Seas. Atlantic waters enter the Barents Sea from west and north, and enter the Arctic Basin from Fram Strait. These relatively warm waters along with the large freshwater runoff of the Siberian rivers are important to the thermohaline structure and ice conditions of the Arctic Ocean (Fig. 2.22). The influence of the Atlantic waters is greatest in the Barents, Kara and Laptev Seas (Chaplygin 1790; Nikiforov and Shpajkher 1980).

About 20 000 km^3 of warm, salty water is transferred annually through the Faroe-Shetland Strait. These waters bring about $7 \cdot 10^{18}$ kJ of heat to the Norwegian and Greenland Seas. About 120 000 km^3 and $9 \cdot 10^{17}$ kJ enter annually the Arctic basin through the Fram Strait. Approximately $17 \cdot 10^3$ km^3 / year of water of Atlantic genesis passes annually through the Kara Sea and through the strait between Novaya Zemlya and Franz-Josef Land (Nikiforov and Shpajkher 1980). About 30 000 km^3 of warm Pacific water passes annually through the Bering Strait (Schlosser et al. 1995; Kershaw 1997). Consequently, the exchange of water between the Eurasian Basin and Atlantic is greater than between the Canadian Basin and Pacific.

The Arctic Ocean is the source of freshened waters and artificial radionuclides for the North Atlantic owing to the permanent ice drift with the East Greenland current, approximately 2 800 km^3 / year (Aagaard et al. 1985) and to the abundant river runoff. About 5 000 km^3 of river water enters the Arctic Basin annually (Ivanov 1976). The freshened and very cooled waters of the Arctic Basin enter the Atlantic Ocean as the cold surface East Greenland Current.

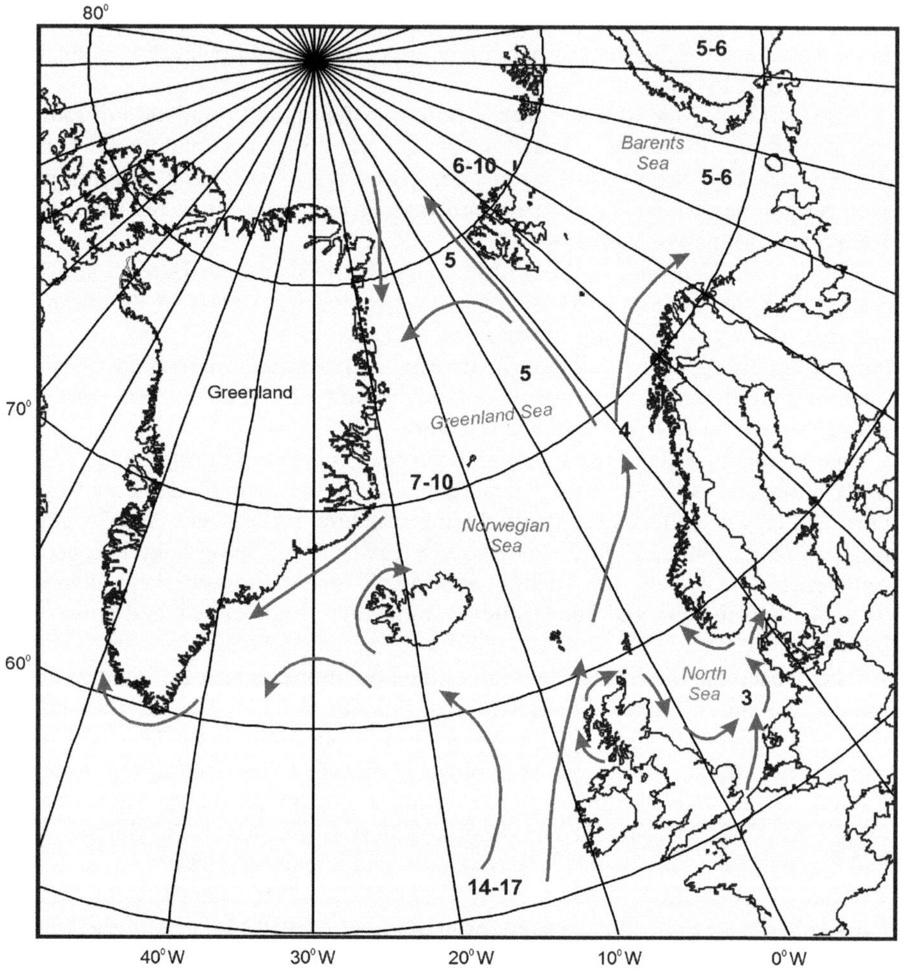

Fig. 2.22. Radionuclide transfer times and pathways from Sellafield (Dahlgaard 1993)

The water stratification of the Arctic Ocean consists of three main layers. The upper layer (to 100-200 m) is characterized by cold and freshened waters covered by sea ice. The active layer of the Arctic Ocean is divided from the lower layer by a transitional zone known as the halocline showing an increase in salinity and density (Volkov and Lukin, 1977; Lebedev 1992; Kershaw and Baxter 1995). Below the halocline layer (200-600 m) is followed by a layer of warm and salty waters of Atlantic origin.

The surface water circulation is determined by the Eastern anticyclonic vortex and Transarctic Current (Fig. 2.22). These main large-scale phenomena form the most effective pathways of dissolved contaminants in the surface and mixed oceanic layers. The circulation in the intermediate and deep layers of the Arctic Ocean is slower and more complicated than the surface layer. The water masses, carried by the West Spitsbergen Current, turn eastward and subduct under the surface waters. They form the core of the Atlantic layer and are involved in the recirculation as several cyclic loops around the main ridges and slopes of the Eurasian and Canadian Basins.

Global fallout and radioactive releases from large northern rivers were the main sources of direct radionuclide influx to the marine ecosystems of the polar basins (Fig. 2.23). However, the European reprocessing facilities, especially Sellafield on the Irish Sea coast, are the main source of artificial radionuclides (Vakulovsky et al. 1985; Nies et al. 1998; Strand 1998). The system of currents transfers radioactive substances from European reprocessing facilities along the Norwegian coast to the Arctic Ocean. Approximately 6-8 years later, radionuclides are carried out of the Arctic by the East Greenland Current, completing the transport cycle (Table 2.5).

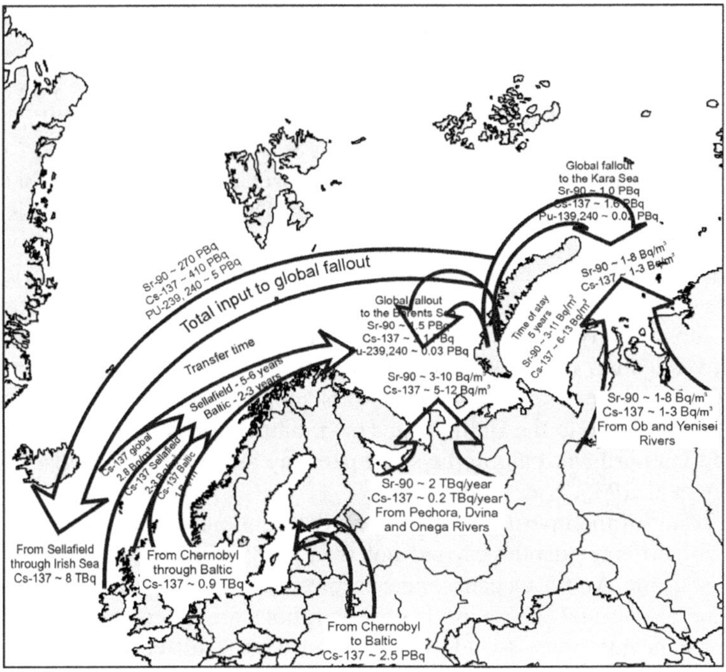

Fig. 2.23. Main sources of radioactive contamination in the Barents and Kara Seas (Baklanov et al. 1996)

Table 2.5. Approximate transport time of radionuclides to the North Atlantic and Arctic Seas (Kershaw and Baxter 1995)

Region	Transport time (years)	Tracer used	Data source
North Strait	1	^{137}Cs	Kershaw and Baxter 1995
East Scotland	2-3	^{137}Cs	"
West Norway	3-4	^{137}Cs	"
North-eastern Barents Sea	4-5	^{137}Cs	"
Nordcape	5-6	^{134}Cs:^{137}Cs	Kautsky 1988
Barents Sea	5	^{137}Cs	Livingston et al. 1984
Spitsbergen	5	^{137}Cs	Dahlgaard and Aarkrog
	7	^{134}Cs:^{137}Cs	1986; Kautsky 1988
East Greenland	7	^{137}Cs, ^{99}Tc	Dahlgaard and Aarkrog
	7-9	^{134}Cs:^{137}Cs	1986; Kautsky 1988
Baffin Bay	8	^{137}Cs, ^{99}Tc	Aarkrog et al. 1987
Arctic Basin, depth 1500 m, LOREX	8	^{137}Cs	Livingston et. al. 1984

The maximum contamination of the Irish, North and Norwegian Seas was observed in the mid-1970s (the greatest release was in 1975) when ^{137}Cs discharges from European reprocessing facilities reached a level of 3-5 PBq / year (Kershaw and Woodhead 1991; Kershaw 1992, 1997; Kershaw and Baxter 1995; Nies et al. 1998). The ^{137}Cs concentration in the North Sea reached 150-200 Bq / m^3 in this period. The signal of these West European discharges crossed the Barents Sea, reaching the Central Polar Basin 6 years later. The ^{137}Cs content in this basin increased to 10-50 Bq / m^3 (Nies et al. 1998). Probably, about 40 % of ^{137}Cs in the water mass of the 0-250 m layer of the Arctic Ocean are of industrial origin.

Despite the fact that the volume of radioactive wastes discharged from Sellafield is substantially lower today in comparison with the maximum during 1974-1977, investigation of the radioactive contamination of the Arctic Ocean is still important. At present radioactive cesium is supplied by the accident at the Chernobyl power station to the Baltic Sea. This radionuclide is transferred to the North Sea and then northward along the same pathway as the Sellafield discharges (Vakulovsky et al. 1985; Nies et al. 1998).

As a result of the investigation of ^{137}Cs distributions in the suspension-solution system for different depths carried out from 1985-1987, it became obvious that the ^{137}Cs in the Arctic Ocean waters is present primarily in the dissolved state (Nikitin et al. 1991). ^{137}Cs particle concentrations are several tenths of percent of seawater concentrations. In addition, no substantial differences in radionuclide concentrations are observed for suspended particles collected from different water depths. Thus, the approach to the estimation of possible pathways and accumulation of radionuclides in the Kara, Barents, Norwegian and other polar seas cannot be uniform. The migration of radionuclides, depending on their individual properties, is regulated to a great extent by the processes of sedimentation transfer, bioaccumulation, or mixing of radionuclides in water masses during their movement for hundreds and thousands of kilometers (Matishov et al. 1994 a, c, 1997 a).

2.3.2 Deep-water troughs as pathways for shelf-ocean exchange

Knowledge of the geomorphological characteristics of individual areas is necessary to understand cross-boundary transport of radionuclides. The regional setting is especially important in the northern seas as a result of the system of deep border and transverse troughs. The troughs divide the shelf in regular intervals and are open to the continental slope. Such peculiarities of the bottom topography provide conditions for the free influx and active circulation of the North Atlantic Current waters on the glacial shelves from the North to the Kara Seas (Fig. 2.24).

The system of shelf troughs controls the main Atlantic water flows that transport to the Arctic the suspended and dissolved radionuclides of different origin (Matishov et al. 1992 a, c, 1994 a, c, 1996 f, g, 1998 b, c, d). The closed depressions on the bottom of troughs (250-700 m) are filled with clayey sediments. These are areas where radioactive contamination accumulates on the continental shelf.

The influence of bottom relief on the regional and general oceanography is considered the Barents Sea as an example. The typical feature of the Barents Sea shelf is the existence of large raisings, such as Murmanskaya, Central, Nordcape, Bear-Hope Heights. Actually these raisings are watersheds of all large currents and circulations of water masses. The height of these sloping (15-60′) bottom raisings is considerably large and sometimes reaches 200-400 m. They are of 200-500 km length and of 100-200 km width. The usually sloping hillsides can reach the steepness of 2-4° on the ledges.

The warm waters of the Nordcape and Spitsbergen Currents flow from west and north along the troughs into the inner part of the Barents Sea and finally join each other in the area between Franz-Josef Land and Novaya Zemlya (Fig. 2.24). In particular, warm (to +3 °C) and saline (near 35 ‰) Atlantic waters flow along deep-water Franz-Victoria and Saint Anna Troughs and also in large straits of the Franz-Josef Land (Matishov et al. 1992 a, 1998 d). Atlantic waters in the 100-400 m layer surround the east and west coast of the archipelago and also enter from the north to the North-West Strait. Here warm waters (to +3 °C) can be traced in the 150-300 m layer. The upper boundary of West Spitsbergen deep waters in the Franz-Victoria Trough is situated at a depth of about 50 m.

It became clear in the last years that the transformed waters (34.8-35 ‰) of the West Spitsbergen Current penetrate the deep straits (400-600 m) inside the Franz-Josef Land Archipelago from the north (Matishov et al. 1992 a, 1994 c). Thus, Atlantic waters, carrying heat as well as contaminants from the North Atlantic occupy the deep and bottom layers of the Barents Sea to the north of 77 °N. The advection of the West European wastes in these water flows toward the Central Polar Basin is one of the main features of radioecological conditions in these seas.

In the southwestern part of the Barents Sea not one branch (Tantsyura 1959) but four separate branches of the Nordcape Current are formed. They are named by corresponding troughs as the Nord-Dupet, Kanin, Gusiny and Moller Currents ((Fig. 2.25, 2.26). These streams are from 20-30 to 50 km wide with a vertical extent from 100-140 to 200 m. The waters of Atlantic origin in the troughs are characterized by a salinity of up to 35 ‰ and a temperature range of 1-4 °C. As a result of these troughs the warm Atlantic waters, transporting radionuclides, penetrate the Barents Sea.

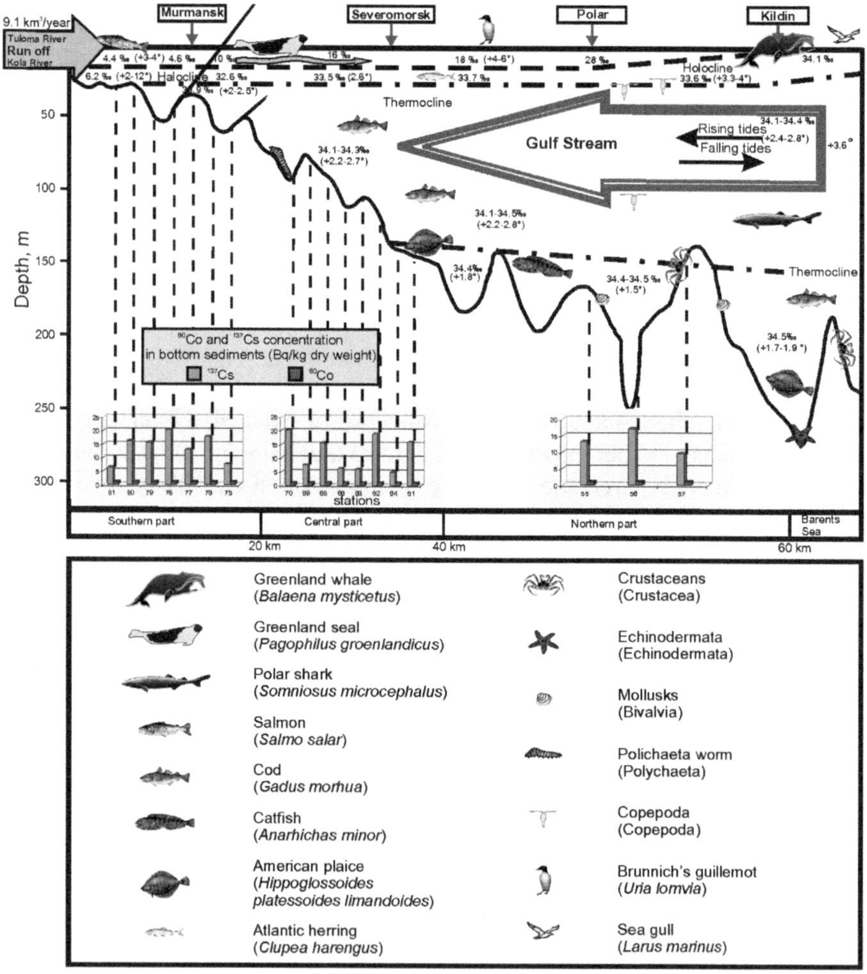

Fig. 2.24. Ecology of the Kola Bay and distribution patterns of radionuclides on the seabed (Spring 1996)

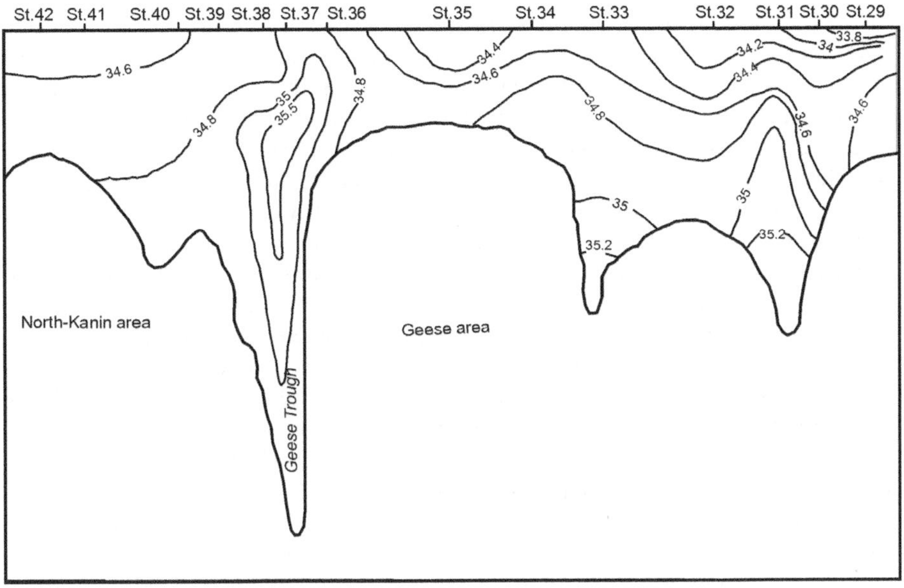

Fig. 2.25. Salinity pattern in the southeastern Barents Sea (MMBI, 1992)

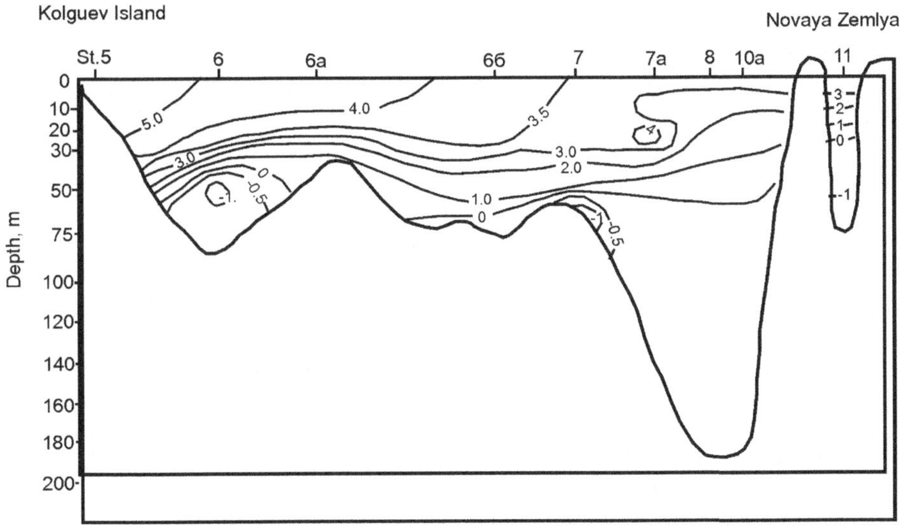

Fig. 2.26. Sea temperature from Kolguev Island - Novaya Zemlya (MMBI, 1992)

The salinity of the Pechora Sea is comparatively low because of the runoff of the Pechora River and other rivers. To the east of 53 °E salinity decreases from 31 to 20-25 ‰, and in the estuarine areas to 10 ‰ or less. The upper water layer of

20-25 m thickness in the southern part of the sea and to 100 m - in the northern part tends to move generally in an eastward direction. The field of surface currents is formed by two powerful streams of warm and salty waters of Atlantic origin (Kanin and Kolguev-Pechora Currents), the cold Litke Current, and the White Sea and Pechora Currents of runoff origin (Potanin et al. 1989). Three water masses are clearly distinguished in the area adjoining the Pechora mouth: freshened (of salinity less than 25 ‰), seawater of decreased salinity (25-30 ‰) and higher salinity waters (more than 30 ‰). The salinity of the rest of the Pechora Sea increases steadily from south to north (to 34.5 ‰ in the upper layer), as the influence of the west Barents Sea waters, deriving from the Atlantic water mass, becomes more intensive. The highest salinity is characteristic for the deep and intermediate waters of the South Novaya Zemlya Trough - more than 34.96 ‰. The vertical thermal structure of this trough is more complicated. In the deep layers the temperature regime is maintained by heat advection with the transit of Barents Sea waters from west to east. In the northern, relatively deep part of the sea near-bottom water temperatures reach negative values (to -1 °C).

Atlantic waters inflow through the troughs entering almost all fjords and bays of the Barents Sea coast, including the Novaya Zemlya, Spitsbergen, Kola Peninsula. The transportation and redistribution of radionuclides in particular, in the Kola and Motovsky Bays, is governed, apart from the Atlantic advection, by tidal currents. The tides control almost all water masses of the fjords. At the same time, freshwater runoff currents are also important in the Kola Bay and to some extent in the Motovsky Bay, during the spring.

In spring in the pelagic zone of bays (to 15-20 m) the distinctly pronounced halocline is formed. The velocity of the freshened waters (melted and river), gliding on the surface of salty ocean waters, can reach 1-4 km / hour. These conditions are typical for not more than two or three months. Probably, anthropogenic radionuclides, which are brought by rivers or enter the bay basins during the deliberate discharges of pollutants, are partly carried out to the Barents Sea coastal zone descending from the surface into the halocline.

The existence of the surface and bottom waters advection from bays, for example, from the Chernaya bay, is clearly confirmed by the high concentration of ^{137}Cs and $^{239,240}Pu$ in the grounds of the South Novaya Zemlya Trough (Matishov et al. 1992 c, 1994 a, c, 1996 f). D. Smith (Smith et al. 1995 a), who had analyzed the ratio coefficients of $^{239,240}Pu$ and ^{238}Pu for bottom sediments, came to the conclusion about existence of a certain emission of $^{239,240}Pu$ from the Chernaya Bay. Then the radioactive waters are seized by the coastal Litke Current, and the plutonium is spread throughout the Barents Sea and other polar seas.

The concentrations of radionuclides are low in the coastal shelf to the west of Novaya Zemlya. At the same time, this area was more then once subject to radiation fallout from nuclear explosions and also to the transfer of radioactive elements by the western branch of the Novaya Zemlya current. The explanation is that rapid accumulation of the fine fraction of abiotic and biotic material along the west boundary of Novaya Zemlya is not possible because of the hydrodynamic regime. Strong currents dilute the surface runoff from Novaya Zemlya removing radionuclides to the Kara Sea. The analogous dilution of the surface runoff is also possible along the southern boundary (the Kara Gate Strait) where the deposition of ^{137}Cs in bottom sediments is not observed (Matishov et al. 1994 b, c; Yudakhin et al. 1998).

Atlantic waters, moving through the large troughs and depressions of the sea bottom, especially in the winter season, are subject to major heat loss and an increase in density because of cooling and brine formation during freezing. The intensive formation of cold bottom waters takes place in the Central Depression and in the area between Novaya Zemlya and Franz-Josef Land (Tantsyura 1959; Novitsky 1961; Hydrometeorology ... 1992; Lebedev 1992; Matishov et al. 1998 d).

High density waters accumulate near the sea bottom and flow along the deep troughs toward the outer shelf boundaries. Finally they descend along the continental slope and penetrate below the Arctic halocline. The Santa Anna Trough, where the Barents water masses merge with the branch of Atlantic waters entering from the Fram Strait, is the main exit for the dense bottom waters. Through this conduit radionuclides are transported from the Barents Sea into the Central Polar Basin (Matishov et al. 1994 a, c; Nies et al. 1998).

Thus, on the vast Barents Sea shelf with depths to 300-500 m, the Atlantic waters (temperature to 10 °C, salinity to 35.1 ‰) interact with the Arctic, Sub-arctic and Barents water masses with salinities varying from 31 to 34‰. This wide (20-50) km) zone of the polar hydrological front is divided into a large number of meandering fronts of second order in which the thermal gradients reach approximately one degree per kilometer. Besides the peculiar meandering of the polar front under the influence of troughs and banks, the Barents Sea waters are distinguished by the seasonal thermocline at water depths of 10-40 m. The appearance of a summer thermocline is controlled to a great extent by the freshened layer of surface waters near the ice edge and in the zone of melt water runoff from the coasts of Franz-Josef land, Novaya Zemlya and Spitsbergen.

The convective intermixing of water masses, the volume ratios between volumes between different water masses, and the residence time of radionuclides in the basin all influence the total amount of radioactive contamination observed in pelagic ecosystems and the mean concentrations of radionuclides in the marine environment.

During the 1990s, ^{137}Cs levels in the Barents Sea varied from 4 to 12 Bq / m3. The Barents Sea receives about 10-20 % of ^{137}Cs and about 30 % of ^{90}Sr discharged from Sellafield (Nikitin et al. 1991; Nies et al. 1998). Ultimately the wastes reach the Arctic Basin. The main input of radionuclides to the Arctic Basin at present is atmospheric fallout. ^{137}Cs discharges from Sellafield reach the Barents Sea in 4-5 years. During the 1980's, the cesium concentration of Arctic Ocean waters increased as a result of the Chernobyl disaster. Cesium continues to be delivered by water exiting from the Baltic Sea.

The ratio between ^{137}Cs and ^{90}Sr in the Barents Sea can be used for the determination of contamination sources. This ratio is approximately 1.5 in offshore Ocean waters (UNSCEAR 1982). In the Norwegian Coastal Current it is about 2.0-2.1 (Nies et al. 1998). However, the ratio is 1.9 and less in the central part of the Norwegian-Greenland Basin owing to the fact that global fallout remains the predominant source in this region.

The ratio $^{137}Cs : ^{90}Sr$ for the Barents Sea is 2.2. The main sources of contamination are both Sellafield discharges and global fallout. In comparison with earlier data, the Sellafield contribution to the sea has been significantly diminished (Kershaw and Woodhead 1991; Kershaw 1992, 1997; Kershaw and Baxter 1995). However the introduction of radioactivity to the environment as a result of the Chernobyl accident discharges is partly responsible for the relative decrease in the Sellafield contribution.

The ratio $^{137}Cs : ^{90}Sr$ in the Greenland Sea is above 2.3. This indicates the influence of Sellafield discharges from earlier years when the ratio was higher than 5.9 until 1985 and reached 11.3 in 1976 (Nies et al. 1998). At present the input of Chernobyl fallout to the general balance of radioactive contamination of the Barents and Kara seas is on the order of 10-20 % level.

So the current directions and boundaries of zoogeographical provinces in the polar seas follow bottom topography. The complicated system of border and transverse troughs, divides the surface of the Barents Sea shelf (to 500-600 m) providing open access of substances and heat energy from the North Atlantic to the high Arctic. In particular, the Nordcape Current as it moves eastward, divides into tens of large and small streams of warm waters. Knowledge of these and other new oceanographic features leads to a more detailed understanding of the transport and accumulation of radioactive elements.

2.3.3 Transfer by ice drift - the Kara and Laptev Seas

Anthropogenic radionuclides, such as $^{239,240}Pu$, ^{137}Cs, ^{90}Sr, which are present in the pelagic zone are absorbed actively by organic and inorganic particles. During the period of ice formation in the coastal zone and in the open sea, radionuclides are incorporated into the cryopelagic ecosystem (Matishov et al. 1994 a, c; Pfirman et al. 1997; Nies et al. 1998; Pavlov et al. 1999). Thus, sea ice provides a potential pathway for the transfer of radionuclides from the seas of the Siberian shelf to adjoining parts of the Arctic Ocean and further to the Norwegian-Greenland Basin (Fig. 2.27). The drift of pack ice takes place mainly through Fram Strait and in the West Greenland Current.

In the Kara and Laptev Seas a sharp boundary of turbid river water exists as a result of runoff from land. The boundary extends for hundreds of kilometers into the open sea. All major Siberian rivers discharge 30-65 % of the total annual runoff during one summer month (Treshnikov 1984). In autumn, suspended terrigenous material is incorporated into ice in the process of sea water freezing. In winter, water circulation is less intensive when wind velocities and horizontal density gradients are small (Harms et al. 1999). The main direction of water mass flux and radionuclide advection from the Kara Sea is eastward to the Laptev Sea and northward to the center of the Arctic Ocean. The resulting water exchange between the Kara and Laptev Seas through the Vilkitsky Strait is directed eastward and is estimated to be from 4.9 to 11.0 km^3 / year (Pavlov et al. 1999).

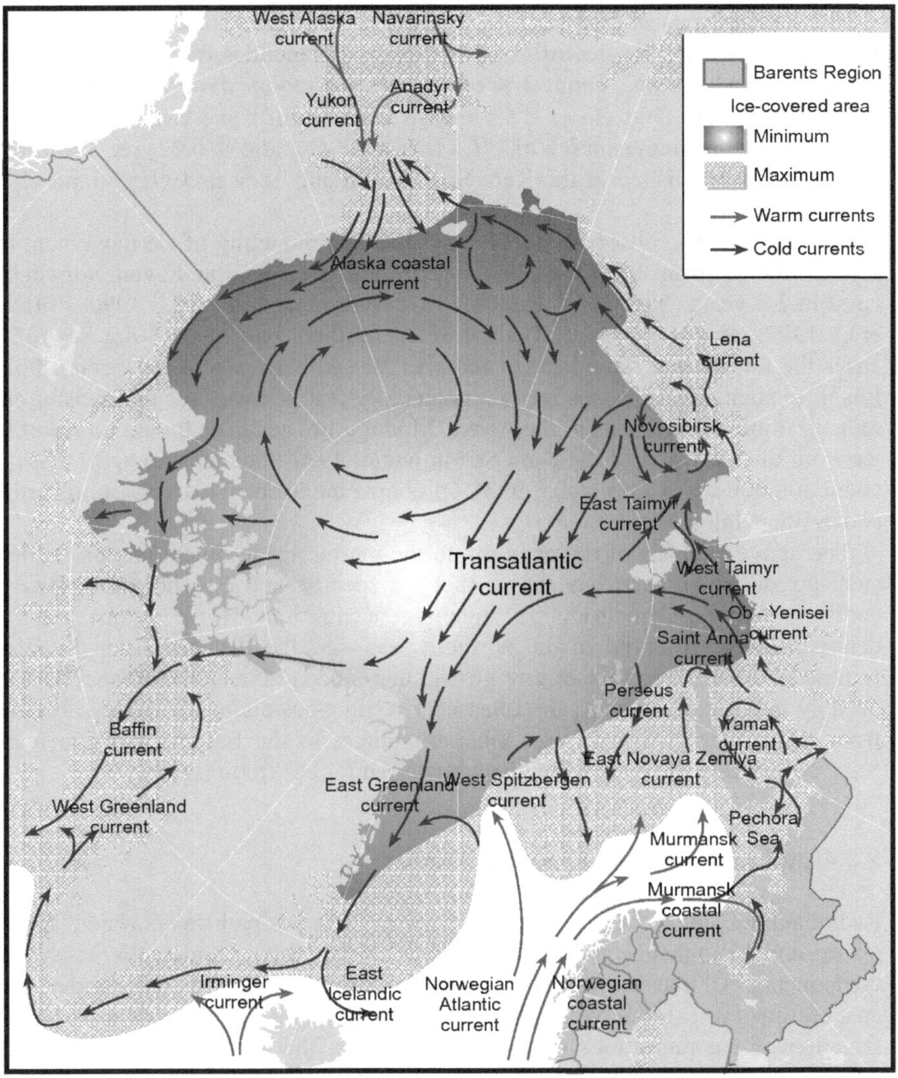

Fig. 2.27. Sea surface currents and the area of sea ice extent (by AMAP/GRID-Arendal)

During the 1990s, relatively low concentrations of ^{137}Cs (3-8 Bq / kg) and ^{90}Sr (3-12 Bq / kg) were formed in the Kara Sea owing to the influx of transformed Atlantic and freshened Pechora waters through the Karskie Vorota Strait and also to Siberian river runoff. ^{137}Cs and ^{90}Sr concentration in the surface waters of the

Laptev Sea in summer 1993 were of the same order as in the Kara Sea in summer 1992 (Pavlov et al. 1999). In the beginning of the 1980s the ^{137}Cs content in the surface waters reached maximum values (to 30-40 Bq / kg).

The transfer time of radionuclides from West European sources to the Barents Sea is 5-7 years. As for the influx of artificial radionuclides from West European sources to the Kara Sea, about 2 % of Barents Sea radioactivity reaches the Kara Sea, or about $150 \cdot 10^{12}$ Bq of ^{137}Cs and $33 \cdot 10^{12}$ Bq of ^{90}Sr (A survey ... 1992). Chernobyl is the single source of ^{134}Cs (^{134}Cs : ^{137}Cs ratio is 0.02, see Table 2.6) able to be distinguished in the Kara Sea (Strand and Nielsen 1995; Strand et al. 1997).

Satellite surveys, observations of buoy drift and modeling of ice movement indicates that ice from the Kara Shelf reaches the Greenland and even Norwegian coast in 2-3 years (Vise 1937; Rudels 1987; Hydrometeorology ... 1990; Pfirman et al. 1997; Nies et al. 1998). The ice of the northern part of the Kara Sea influences the Laptev and Barents Seas and also the southern part of the Central Polar Basin. In such situations sea ice is a potentially fast pathway for the transfer and redistribution of pollutants in the Arctic. Model estimates of sediment transport by ice from the west part of the Kara Sea in winter 1997, indicate an export mass of about 300 000 tons (Nies et al. 1998). Of course the ice drift also transports artificial radionuclides.

The above analysis of radioactivity in the marine environment (water, suspension, bottom sediments) has shown that the specific activity and the density of contamination vary across a wide range of concentrations. However, one common denominator is observed in the northern seas - as the distance from the West European sources of radioactivity grows, the concentration of radionuclides noticeably decreases. Besides, the release of terrigenous and radioactive substance from Novaya Zemlya and West Siberian sources to the Barents and Greenland Seas, radioactive substances are also transferred to the Arctic Basin.

2.3.4 Biological transfer mechanisms

Radionuclides enter the sea basins by many different pathways; water, atmospheric and biological - and then distribute among abiotic and biotic ecosystem components. All marine organisms from microorganisms to fish, birds and cetaceans, are links in extremely complicated trophic food chains. Therefore the reconstruction of the migrations of marine organisms, the accumulation of radionuclides by different species of marine flora and fauna, and obtaining quantitative estimates of bioaccumulation, is a complex task. The form of life, habitat characteristics, migration patterns, trophic relations, and the feeding behavior of different marine organisms must be taken into account in such an analysis.

Species number, population size, dominant species, and species interactions especially trophic interactions characterize the structure of marine communities and ecosystems. Owing to the trophic chains and passive and active migrations of organisms for many hundreds and thousands of kilometers, large fluxes of matter

Table 2.6. Radionuclide ratios in sea water in the Kara Sea (Strand et al. 1993)

Station	Depth, m	^{137}Cs:^{134}Cs	^{90}Sr:^{137}Cs	^{238}Pu:$^{239,240}Pu$	^{90}Sr:^{129}I	^{99}Tc:^{129}I	$^{239,240}Pu$:^{90}Sr	$^{239,240}Pu$:^{137}Cs	^{238}Pu:$^{239,240}Pu$	^{99}Tc:^{137}Cs	^{239}Pu:^{241}Am
1	2	3	4	5	6	7	8	9	10	11	12
1	0	0.027	0.41	0.075	1471	75.8	0.00105	0.00043	0.075	0.024	6.7
	28	0.033	0.53	–	–	–	–	–	–	–	–
	70	0.025	0.74	–	–	–	0.00177	0.00131	–	–	–
	170	0.023	0.81	0.014	–	–	0.00132	0.00107	0.014	–	8.2
2	0	0.024	0.53	0.057	–	–	0.00095	0.00050	0.057	0.0168	–
	80	0.030	0.63	0.111	–	–	0.00146	0.00092	0.111	–	4.5
	190	0.022	0.37	–	–	–	0.00226	0.00083	–	–	–
3	0	0.030	0.40	–	1755	–	0.00100	0.00040	–	0.019	–
	80	0.034	0.44	0.068	–	–	0.00155	0.00068	0.068	–	8.1
	323	0.022	0.33	0.088	–	–	0.00250	0.00082	0.088	–	8.9
4	0	0.029	1.85	–	–	–	0.00038	0.00071	–	0.022	5.5
	70	0.033	0.47	0.110	–	–	0.00138	0.00065	0.110	–	2.8
	250	0.031	0.27	0.067	–	–	0.00175	0.00048	0.067	–	–
5	0	0.029	2.38	0.039	6931	53.3	0.00093	0.00220	0.039	0.017	5.1
	7	0.034	3.83	–	–	–	–	–	–	–	–
	85	0.028	0.49	–	–	–	–	–	–	–	–
	120	0.031	0.48	–	–	–	–	–	–	–	–
6	0	0.032	2.49	–	5611	52.5	0.00064	0.00159	–	0.016	8.6
	16	0.026	0.50	–	–	–	0.00179	0.00088	–	–	10.2
7	0	0.026	–	–	–	–	–	–	–	–	–
	75	0.030	–	–	–	–	–	–	–	–	–
	170	0.033	–	–	–	–	–	–	–	–	–

Table 2.6. (cont.)

1	2	3	4	5	6	7	8	9	10	11	12
8	0	0.026	0.56	0.056	2939	90.5	0.00032	0.00018	0.056	0.021	–
	70	0.030	0.31	–	–	–	0.00133	0.00041	–	–	–
	124	0.033	0.41	–	–	–	0.00158	0.00065	–	–	–
9	0	0.028	0.48	–	–	–	0.00065	0.00031	–	–	6.0
	10	0.026	0.44	–	–	–	0.00148	0.00065	–	–	–
	20	0.021	0.42	–	–	–	0.00172	0.00072	–	–	12.7
10	0	0.030	0.76	–	3530	75.0	0.00047	0.00036	–	0.021	–
	50	0.034	0.58	–	–	–	0.00054	0.00032	–	–	–
	115	0.027	0.27	–	–	–	0.00231	0.00063	–	–	–
11	0	0.022	0.36	0.065	2007	86.1	0.00086	0.065	0.20	–	–
	40	0.031	0.22	–	–	–	–	–	–	–	–
	87	0.030	0.46	0.074	–	–	0.00084	0.074	–	–	–
12	0	0.027	0.42	–	–	–	0.00055	0.00023	–	–	–
	35	0.022	0.46	0.019	–	–	0.00165	0.00076	0.019	–	–
	85	0.019	0.53	–	–	–	0.00221	0.00117	–	–	37.0
13	0	0.025	0.66	0.027	–	–	0.00176	0.00117	0.029	–	10.7
	100	0.020	0.59	0.087	–	–	0.00315	0.00185	0.87	–	57.5

and energy occur - from plankton and benthos through fish fauna to birds, seals, whales and white bears.

The eastern migrations of plankton and, in particular, ichthyo-plankton, as well as feeding migrations of many fish species, provide an example of the large-scale transfer of organic and radioactive substances from the Norwegian to the Barents Sea ecosystem. There is a natural migration of fish roe and larvae from the Lofoten Islands and Northern Norway towards West Spitsbergen and then to Franz-Josef Land and Novaya Zemlya (Fig. 2.8, 2.9). Many benthic organisms have larvael plankton, for example, polycheates, crustaceans, mollusks, and echinoderms, also can drift far northward. The rotation of Atlantic waters in the Gulf Stream system, described above, also contributes to such interrelations. This rotation influences the migration patterns of biota and determines the structure and dynamics of ecosystems in the northern parts of the Norwegian, Kara and Barents Seas. As the influence of the Atlantic waters diminishes further to east and north, the shelf flora and fauna become quantitatively limited (Zekevich 1963; Matishov and Pavlova 1990).

It is important to note that the bioproductivity of all ecosystem components is subject to significant changes under the influence of climate (Galkin 1988; Matishov and Pavlova 1990; Matishov and Denisov 2000). The most general feature of the Barents Sea, for example, is the widening to the east and northeast of the spawning and feeding areas and also of migrations of marketable fish species (cod, halibut, capelin, polar cod) in warm period and the reciprocal process - in cold periods.

The vertical migrations of marine fauna, in particular, of nekton, are no less important than horizontal migrations. Actively swimming animals in seawater are able to accomplish daily and seasonal migrations to depths of 100-700 m, and some species - to 1-2 km. It is characteristic, first of all, for fish and mammals (cetaceans, seals etc.), large Cephalopod mollusks (squids and others), and some bird species to spend most of their life in floating (Zenkevich 1963; Birds of the USSR 1990; Sokolov and Arsenjev 1994; Krasnov et al. 1995; Timofeev 1996).

The spreading and accumulation of radionuclides in marine ecosystems, evidently, occurs more intensively than in freshwater environments. This results from the dynamic properties of sea water masses, the long distance migrations of marine organisms, more intensive biological production and loss, and in general, faster rates of assimilation into the components of the marine ecosystem. There exists a wide range of sensitivity and resistance to these controlling factors due to the diversity of marine plant and animal species in northern Europe's marine ecosystems.

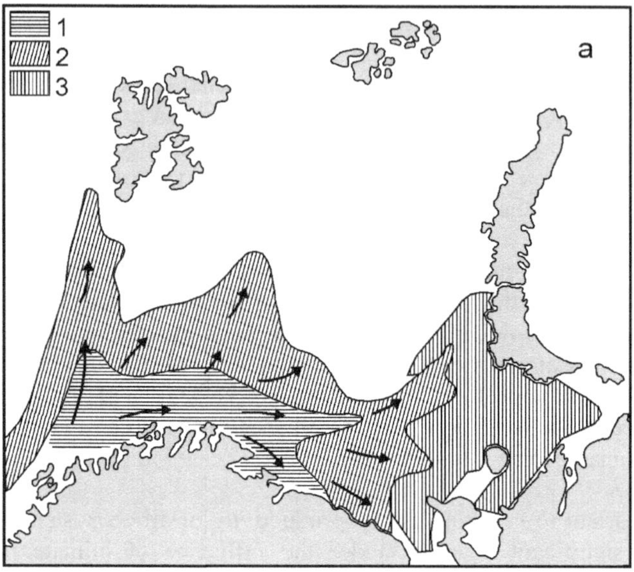

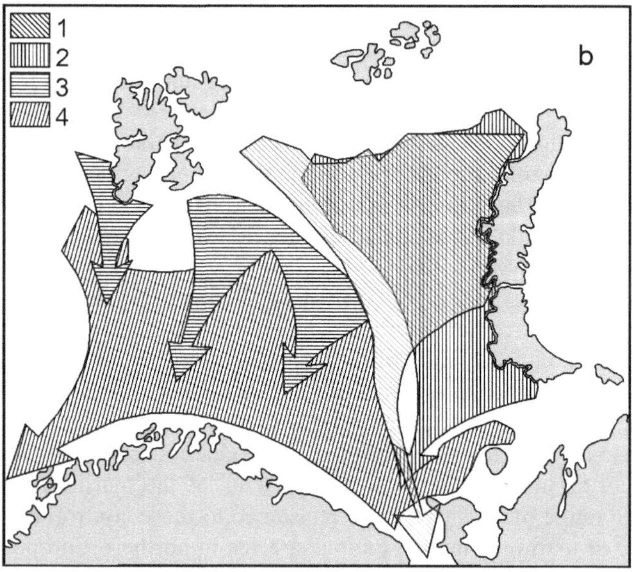

Fig. 2.28. Migration of biological and radioactive substances in the Barents Sea: **a** - *transfer and spreading of micro- and macroplankton (1-2), spreading of benthos (3)*; **b** - *migration of: 1 - seal; 2 - polar cod; 3 - cappelin; 4 - cod*

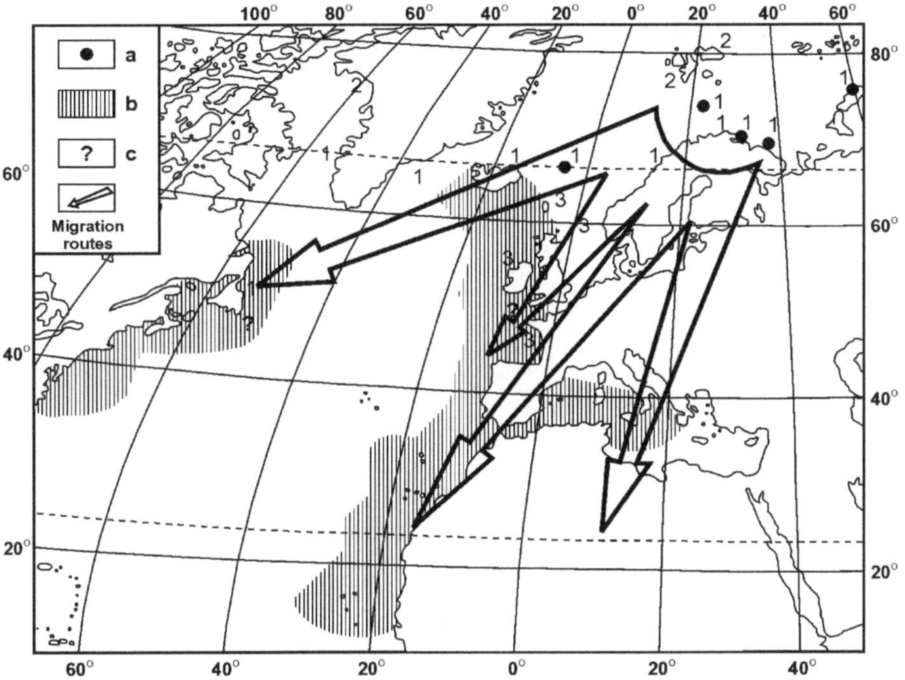

Fig. 2.29. Puffin migration routes: **a** - nesting areas; **b** - wintering and migration areas; **c** - possible nesting sites. Subspecies: *1 - Fratercula arctica arctica; 2 - F. a. naumanni; 3 - F. a. grabae* (Belopolsky 1957)

3 The pelagic ecosystem

3.1 Patterns of radionuclide distribution

Seawater is a solution of compounds consisting mainly of oxygen and hydrogen (96.5 %). Eleven elements with concentrations of more than 1 mg / liter (Cl, Na, Mg, S, Ca, K, Br, C, Sr, B, F) dominate in the salt composition of seawater (99.98 %) (Vernadsky 1926; Vinogradov 1967). The natural radioactivity of seawater is dominated by the radionuclide ^{40}K. The contribution of other radionuclides (e.g. ^{232}U, ^{226}Ra and others) is small, totaling about 1 % of the ^{40}K content.

Particles transported with seawater descend rapidly to the seabed while others drift slowly, traveling long distances from their origin. The vertical migration of particles depends on the size of suspension, the rate of water exchange, the composition of sediments and plankton, and the bioproductivity of the basin. The pelagic ecosystem is the primary filter of radioactive particles. The spectrum of artificial radionuclides, considered in our work, includes ^{134}Cs, ^{137}Cs, ^{90}Sr, ^{60}Co, $^{239,240}Pu$.

3.1.1 Baltic, North and Irish Seas

During the previous decades, the level of radioactive contamination in mid—latitude seas (Baltic, the North, the Irish) was controlled by releases from coastal sources and also by the transfer and sedimentation of Chernobyl fallout. As a result of monitoring throughout the 1970s-1990s, the dynamics of γ- and α-activity of water, bottom sediments and some fish species has been widely investigated (Nies et al. 1998). Detailed data on the releases during the Chernobyl accident were collected by the Finnish Centre of Radiation and Nuclear Safety (Ilus et al. 1988a, b, 1993; Ilus and Ikaheimonen 1989, 1991).

Before Chernobyl, the ^{137}Cs content in the Baltic Sea (1974-1985) ranged from 20 to 45 Bq / m^3 (Ilus et al. 1993). In 1986-1987, as a result of the Chernobyl accident, the quantity of radiocesium in the offshore waters increased almost by an order of magnitude to 100-200 Bq / m^3. The Finnish and Bothnia Bays were saturated by ^{137}Cs to 500-970 Bq / m^3 (Table 3.1). These are the highest radiocesium levels ever detected in a coastal sea where the waters are measurably diluted with freshwater coming from the surrounding land areas. In the beginning of the 1990s the contamination of Baltic Sea surface waters decreased to 90-100 Bq / m^3 (Fig. 3.1).

Table 3.1. Concentration of ^{89}Sr, ^{90}Sr and γ-emitting radionuclides in the Baltic Sea water from 1986-1987 (Ilus and Ikaheimonen 1991), Bq / m^3)

Station, date	H, m	S, ‰	^{89}Sr	^{90}Sr	^{40}K	^{95}Zr	^{95}Nb	^{103}Ru	^{106}Ru	^{125}Sb	^{110}Ag	^{131}I	^{134}Cs	^{137}Cs	^{140}Ba	^{134}Cs:^{137}Cs
1	2	3	4	5	6	7	8	9	10	11	12	13	14	15	16	17
								Baltic Sea, 1986								
CZS 126	0	7.18	–	–	2500	0	0	0	–	0	–	16000	53.0	100	0	0.53
30.04																
CZS 137	0	7.43	–	–	2900	0	0	0	–	6.00	–	5800	40.0	86	0	0.47
02.05																
BW 189	0	7.59	–	–	2800	0	0	0	–	0	–	0	23.0	53	0	0.43
07.05	60	7.69	–	–	3100	0	0	0	–	0	–	0	13.0	41	0	0.32
	100	10.84	–	–	3700	0	0	0	–	0	–	0	6.9	26	0	0.27
BY 15	0	7.45	–	–	3300	0	0	0	–	0	–	0	33.0	74	0	0.45
09.05	45	7.34	–	–	2900	0	0	0	–	0	–	0	8.6	28	0	0.31
	100	9.56	–	–	3600	0	0	0	–	0	–	0	17.0	48	0	0.35
Teili-1	0	6.83	460	22	2500	0	0	45	–	0	–	0	24.0	55	0	0.44
17.06	60	8.57	500	19	3400	0	0	12	–	0.41	–	0	3.4	25	0	0.14
	100	10.13	–	–	3500	29	150	42	–	21.00	–	0	5.5	27	0	0.20
	153	10.24	340	22	3800	0	0	0	–	0	–	0	0.0	21	0	–
B-4	0	8.00	–	18	–	–	–	–	–	–	–	–	30.0	65	–	0.45
19.07	100	11.00	–	19	–	–	–	–	–	–	–	–	<10.0	28	–	<0.45
B-6	0	8.30	–	17	–	–	–	–	–	–	–	–	<25.0	40	–	<0.62
21.07	90	9.30	–	17	–	–	–	–	–	–	–	–	<9.0	23	–	<0.37
B-37	0	8.30	–	18	–	–	–	–	–	–	–	–	<13.0	41	–	<0.31
23.07	200	12.70	–	21	–	–	–	–	–	–	–	–	<6.0	22	–	<0.26
B-46	0	8.30	–	19	–	–	–	–	–	–	–	–	<22.0	44	–	<0.50
24.07	100	10.70	–	20	–	–	–	–	–	–	–	–	<6.0	36	–	<0.17

Table 3.1. (cont.)

1	2	3	4	5	6	7	8	9	10	11	12	13	14	15	16	17
B-50	0	8.60	–	18	–	–	–	–	–	–	–	–	35.0	87	–	<0.39
	80	8.90	–	19	–	–	–	–	–	–	–	–	16.0	34	–	<0.47
Bothnia Bay, 1986																
C VI 02.09	0	3.20	120	16	1200	–	–	11	0	0	0	–	17.0	42	–	0.40
GA51 24.10	0	3.67	–	–	1500	–	–	0	0	0	0	–	62.0	140	–	0.44
GA56 25.10	0	4.86	–	–	2000	–	–	0	20	6.80	0	–	120.0	260	–	0.46
Bothnia Sea, 1986																
EB 1 27.08	0	5.00	140	28	200	–	–	44	71	16.00	0	–	290.0	580	–	0.50
GA 63	0	5.96	–	–	2200	–	–	0	0	18.00	3.2	–	270.0	570	–	0.47
26.10	130	6.05	–	–	2200	–	–	14	36	8.60	3.1	–	240.0	490	–	0.49
Riga Bay, 1986																
R-1 17.06	0	7.10	–	24	–	–	–	–	–	–	–	–	27.0	68	–	0.40
R-2	0	6.80	–	20	–	–	–	–	–	–	–	–	<15.0	48	–	<0.33
R-3	0	5.70	–	25	–	–	–	–	–	–	–	–	49.0	79	–	0.62
R-4 18.06	0	3.50	–	26	–	–	–	–	–	–	–	–	23.0	52	–	0.44
R-5	0	5.80	–	25	–	–	–	–	–	–	–	–	36.0	64	–	0.56
R-6	0	6.30	–	17	–	–	–	–	–	–	–	–	40.0	73	–	0.54

Table 3.1. (cont.)

1	2	3	4	5	6	7	8	9	10	11	12	13	14	15	16	17
R-7	0	6.40	–	19	–	–	–	–	–	–	–	–	61.0	95	–	0.64
17.06																
							Bothnia Bay, 1987									
C VI	0	2.86	–	14	1100	–	–	–	–	–	–	–	37.0	98	–	0.38
18.08																
EB 1	0	6.00	–	25	2300	–	–	–	–	–	–	–	160.0	420	–	0.38
01.08																
							Finnish Bay, 1987									
F-10	0	1.78	–	24	–	–	–	–	–	–	–	–	32.0	81	–	0.39
30.07	25	3.58	–	34	–	–	–	–	–	–	–	–	74.0	220	–	0.34
F-12	0	2.42	–	24	–	–	–	–	–	–	–	–	36.0	111	–	0.33
	20	5.22	–	35	–	–	–	–	–	–	–	–	66.0	194	–	0.34
F-13	0	4.16	–	30	–	–	–	–	–	–	–	–	75.0	225	–	0.33
29.07	25	5.03	–	32	–	–	–	–	–	–	–	–	96.0	262	–	0.37
F-14	0	4.02	–	28	–	–	–	–	–	–	–	–	61.0	174	–	0.28
13.09																
F-15	0	4.43	–	33	–	–	–	–	–	–	–	–	84.0	242	–	0.35
29.07	25	5.48	–	31	–	–	–	–	–	–	–	–	77.0	235	–	0.33
LL 3a	0	4.81	–	23	1700	–	–	–	–	–	–	–	85.0	230	–	0.37
5.08																
F-19	0	4.83	–	30	–	–	–	–	–	–	–	–	114.0	295	–	0.38
8.07	30	6.87	–	26	–	–	–	–	–	–	–	–	97.0	263	–	0.37
F-20	5	5.26	–	33	–	–	–	–	–	–	–	–	86.0	266	–	0.32

Table 3.1. (cont.)

1	2	3	4	5	6	7	8	9	10	11	12	13	14	15	16	17
	25	7.87	–	28	–	–	–	–	–	–	–	–	53.0	150	–	0.35
F-5	0	5.15	–	31	–	–	–	–	–	–	–	–	88.0	276	–	0.32
9.07	50	7.77	–	30	–	–	–	–	–	–	–	–	56.0	166	–	0.37
F-8	0	5.06	–	30	–	–	–	–	–	–	–	–	96.0	265	–	0.36
	75	8.05	–	–	–	–	–	–	–	–	–	–	53.0	172	–	0.31
F-24	0	6.69	–	29	–	–	–	–	–	–	–	–	82.0	225	–	0.36
28.07	75	8.77	–	33	–	–	–	–	–	–	–	–	38.0	112	–	0.34
F-11	0	7.13	–	19	–	–	–	–	–	–	–	–	65.0	207	–	0.30
14.07	90	9.68	–	20	–	–	–	–	–	–	–	–	10.0	70	–	0.13

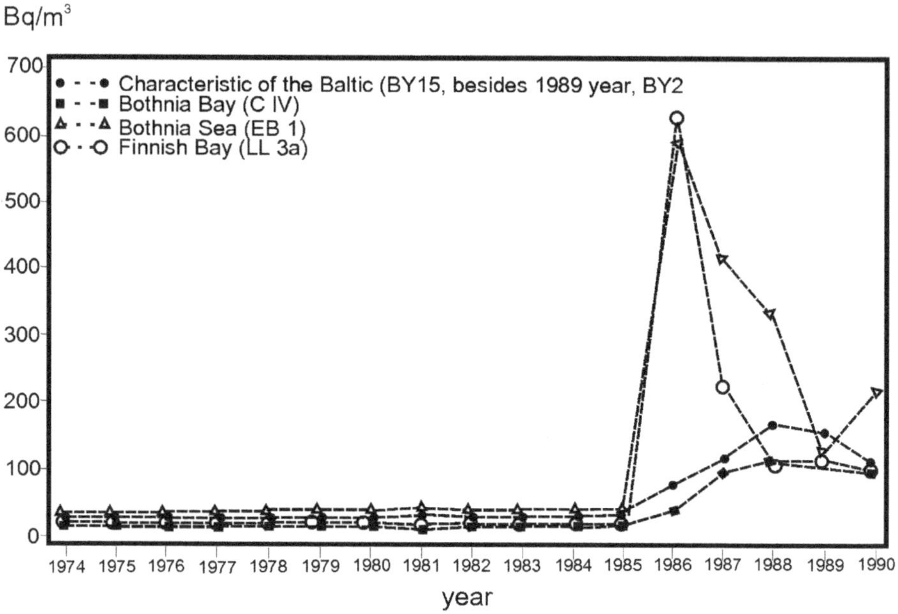

Fig. 3.1. Seawater ^{137}Cs concentrations in the Baltic Sea (1974-1990) (Ilus and Ikaheimonen 1991), Bq / m^3

The dynamics of contamination of the Baltic waters by ^{90}Sr is similar to ^{137}Cs. Before the Chernobyl impact these waters contained about 8-16 Bq / m^3 of ^{90}Sr. After the disaster surface waters in different areas ranged from 15-40 Bq / m^3 (Ilus et al. 1993).

The effects of river wash out and radionuclide migration from catchment basins is quite obvious. The removal, transformation and sedimentation of Chernobyl radionuclides from the restricted, low salinity (3-8 ‰) Baltic Sea occurred at a relatively slow rate. During the 1990s the concentration of ^{137}Cs in the Skagerrak area was unusually high (50-100 Bq / m^3) because of the permanent advection of Baltic waters to the west (Nies et al. 1998). The ratio of ^{134}Cs and ^{137}Cs clearly indicated Chernobyl as the origin of the contamination.

The restricted Azov Sea, like the Baltic Sea, received significant contributions of Chernobyl fallout. The accident resulted in several different periods of radionuclide releases to the environment. These can be distinguished in the history of radioactive contamination of the Azov Sea (Timoshchuk 1970; Ryabinin et al. 1992; Vakulovsky et al. 1994b).

The maximum content of artificial radionuclides in the marine ecosystem corresponded to the period of maximum atmospheric fallout. The ^{90}Sr content in seawater varied from 125-231 Bq / m^3, with a mean value of 168±35 Bq / m^3. The content of this radionuclide in fish was very insignificant - about 0.8 Bq / kg wet weight.

Concentrations of artificial radionuclides in the Azov Sea were decreasing until 1986. After the Chernobyl disaster radionuclide concentrations increased sharply as a result of atmospheric fallout and wash-out from the catchment area of rivers flowing into the Azov Sea. Average ^{137}Cs concentrations in seawater (Fig. 3.2, 3.3) decreased during this period from 70.4 Bq / m^3 in 1986 to 10-11 Bq / m^3 in 1991-1992 (Table 3.2). After the accident, maximum concentrations reached 250 Bq / m^3. Before Chernobyl, the ^{90}Sr content in seawater was 3 times lower than in 1966, varying in the range 33.3-95 Bq / m^3, with a mean of 57.1 Bq / m^3. The Azov Sea is connected to the Black Sea through a the narrow Kerch strait. In June - July 1986 the ^{137}Cs concentration in the Black Sea was 40-480 Bq / m^3. In September - October 1993 these values were 15-40 Bq / m^3 (Medinets et al. 1993, cited by: Cross-Border ... 1995).

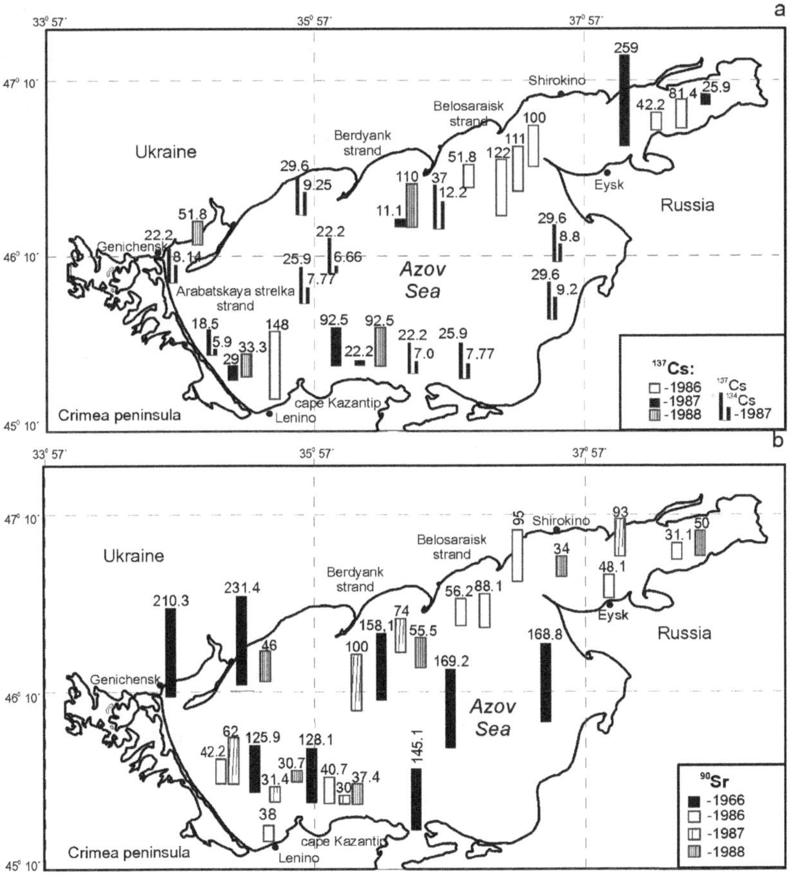

Fig. 3.2. Content of ^{137}Cs (**a**) and ^{90}Sr (**b**) in the Azov Sea, Bq / m^3 (Matishov et al. 2000)

Anthropogenic radionuclides have been measured in the Irish and North Seas since the 1950s. The main source of contamination to these seas are discharges of reprocessed nuclear waste from the Sellafield site into the Irish Sea. Maximum ^{137}Cs concentrations (to 120-400 Bq / m^3) in the North Sea were detected in 1979 (Nies et al. 1998). The contamination of the North Sea s in and around Scotland since the 1970s is as follows. Beginning in the 1970s the levels of ^{137}Cs in seawater varied from 20 to 130 Bq / m^3 depending on the geographical location (Fig. 3.4). In the 1980s the surface waters contained 30-240 Bq / m^3 (Dahlgaard and Aarkrog 1986; Kautsky 1988; Aarkrog 1989; Hunt and Kershaw 1990; Kershaw 1992, 1997).

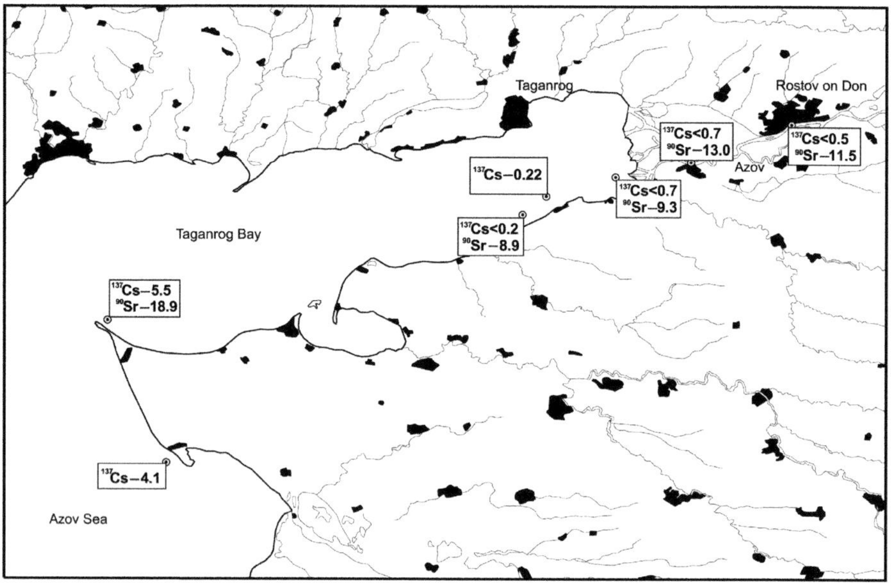

Fig. 3.3. Content of artificial radionuclides in the Azov Sea water (MMBI data, June 1999), Bq / m^3

At the end of the 1980s, ^{137}Cs contamination of the waters surrounding the British Islands decreased sharply to 10-60 Bq / m^3. Only near Sellafield in the Irish Sea did the ^{137}Cs concentration remain about 80-120 Bq / m^3. During the mid-1990s ^{137}Cs concentrations in the surface layer of the North Sea varied from 3 to 25 Bq / m^3 (Fig. 3.5). In coastal waters of the eastern portion of the sea ^{134}Cs was detected in the range 0.5-1.0 Bq / m^3 in 1995 (Kershaw and Baxter 1995; Nies et al. 1998). The distribution of radionuclides in surface waters is controlled by the prevailing current patterns which transport pollutants from their sources.

Table 3.2. ^{137}Cs and ^{90}Sr in the Azov Sea, Bq / m^3

Date of sampling	Area of investigations	^{90}Sr	^{137}Cs	^{90}Sr: ^{137}Cs
June 1986	Kazantip Cape	40.7	–	–
	Arabatsky Bay (Lenino Settlement)	38.5	148	0.26
September 1986	Kazantip Cape	33.3	100	0.33
November 1986	Belosarajskaya Spit: inner side	56.2	51.8	1.08
	outward side	95.0	111	0.86
	Taganrog Bay (Ejsk Town)	48.1	42.2	1.14
	Arabat Arrow Peninsula (outward side)	42.2	22.2	1.90
	Taganrog Bay	31.1	25.9	1.20
December 1986	The lighthouse	88.1	122	0.72
June 1987	Berdyansk Spit area	74.0	29.6	2.50
	Berdyansk Spit area	100	11.1	9.01
	Arabat Arrow Peninsula	62.0	7.4	8.38
July 1987	Arabat Bay	31.4	29.6	1.06
	Cazantip Cape	30.0	22.2	1.35
November 1987	Taganrog Bay (Ejsk Town)	93.0	259	0.36
May 1988	Cazantip Cape	37.4	33.3	1.12
	Arabat Bay	30.7	40.7	0.75
	Berdyansk Spit area	53.3	33.3	1.60
		55.5	110	0.50
	Taganrog Bay	50.0	40.7	1.23
	Taganrog Bay (Shirokoye Settlement)	34.0	77.7	0.44
	Taganrog Bay	46.0	81.4	0.57
	Genichesk Town area (behind the Tonky Strait)	42.9	51.8	0.83
June 1988	Taganrog Bay (Shirokoye Settlement)	34.0	44.8	0.76
July 1988	Cazantip Cape	44.0	92.5	0.48
November 1988	Arabat Bay	–	29.6	–
August 1989	Cazantip Cape	36.6	25.9	1.41
April 1990		32.5	29.2	1.11

As for ^{90}Sr, surface seawater concentrations in the North Sea and Norwegian-Greenland Basin, are low (from 2 to 21 Bq / m^3) (Vakulovsky et al. 1985; A survey ... 1992; Kershaw and Baxter 1995). In 1981 the Norwegian Current contained 2 - 20 Bq / m^3 of ^{90}Sr, and those of the Nordcape Current - 4-6 Bq / m^3.

The transuranic elements in the Irish and North Seas are also derived from Sellafield (Kautsky 1986; Brungot et al. 1999b). In 1980 the concentration of $^{239,240}Pu$ in surface waters of the North Sea oscillated around 20 mBq / m^3. From 1981 to 1989 levels of this radionuclide in the coastal waters around the British Islands increased from 40 to 260 mBq / m^3. At present, $^{239,240}Pu$ levels have decreased to 2-3 mBq / m^3.

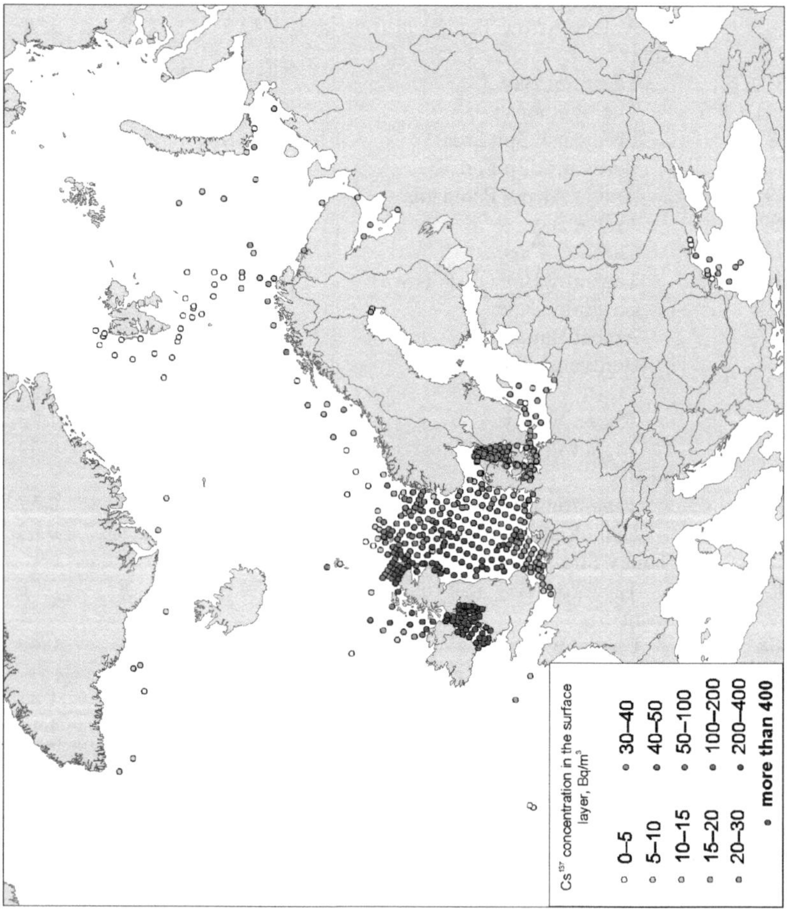

Fig. 3.4. ^{137}Cs in surface waters of European Seas (1970s) (by Crane and Galasso 1999)

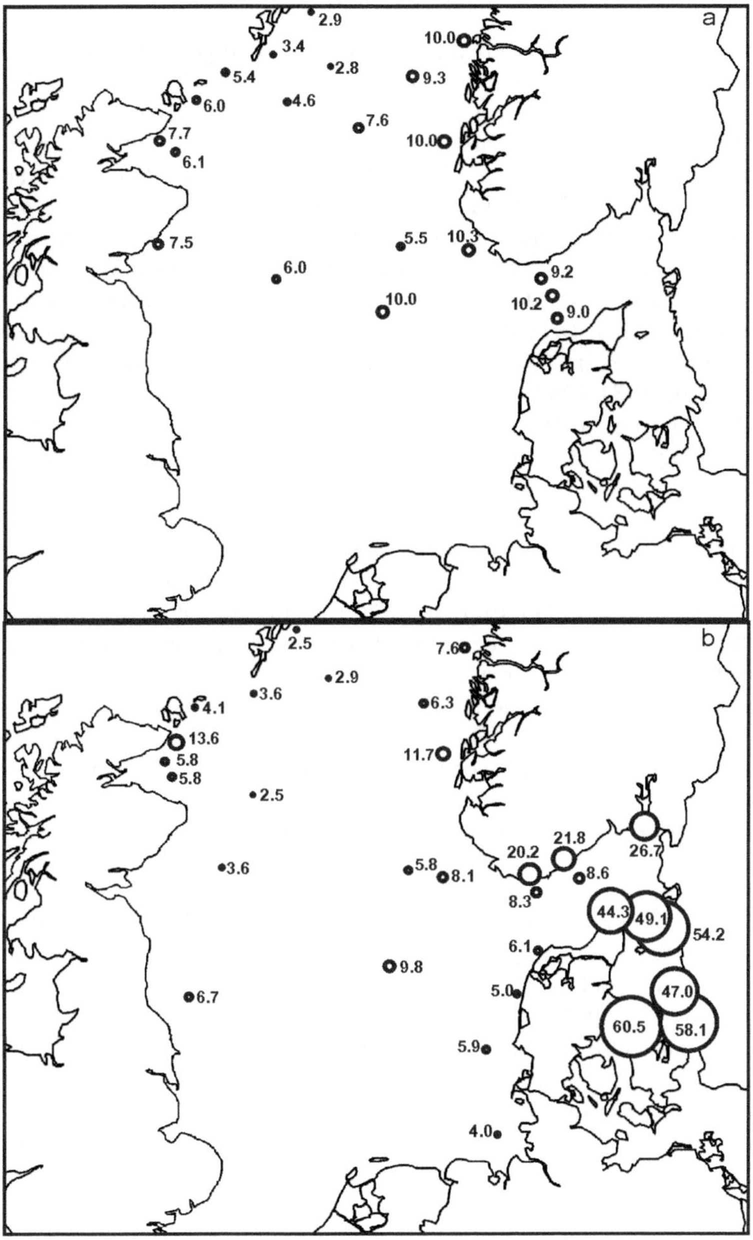

Fig. 3.5. ^{137}Cs concentrations (Bq/m^3) in the North Sea (**a**) 1996 and (**b**) 1997 (Nies et al. 1998)

3.1.2 Norwegian, Barents and White Seas

During 1970-1990s numerous radioecological investigations were carried out in the Norwegian and Barents Seas (Vakulovsky et al. 1985; Kaytsky 1988; Izrael and Tsyban 1989; Strand et al. 1992, 1993, 1994; Kuznetsov et al. 1993; Matishov D. 1993a; Dahlgaard 1993, 1995; Izrael et al. 1994; Matishov et al. 1994a, c; Kershaw and Baxter 1995; Johannessen et al. 1996; Gueguegeniat et al. 1997; Nies et al. 1997; Strand 1998). As a result, changes over time in radionuclide concentrations in these Arctic marginal seas is well-known. The changes reflect the fluctuations in the main contamination sources operating during this period. There are very few data on ^{137}Cs in Arctic waters before the 1970s.

In the *Norwegian-Greenland Basin*, where depths reach 4.0-4.8 km, the salinity of Atlantic waters is 35.2-35.4 ‰ and of cold (-1°) bottom waters - 34.9-34.94 ‰. In 1981-1989 ^{137}Cs concentrations varied from 2 to 29 Bq / m^3 (Fig. 2.6). During the 1990s radiocesium levels in surface waters were on the order of 2-7 Bq / m^3. Only the Arctic waters of the West Greenland Basin (salinity 30-34 ‰) were slightly more contaminated (5-12 Bq / m^3) (Schlosser et al. 1995; Nies et al. 1997).

During 1970-1990s the waters to the southwest of Norway contained higher ^{137}Cs concentrations (Fig. 2.6). In 1981, 137Cs concentrations varied from 70 to 110 Bq / m^3 and in 1988 - from 30 to 70 Bq / m^3 . In 1991 concentrations varied from 8 to 28 Bq / m^3 (Gueguenіat et al. 1997). These measurements show a pronounced concentration decrease to 8-15 Bq / m^3 as the waters are transported by coastal currents northward to the Barents Sea. In recent years the activity of ^{137}Cs in the Norwegian-Greenland Basin has beem low - from 2 to 5 Bq / m^3 (Fig. 3.6). These values decrease to between 2.5 to 0.8 Bq / m^3 in the deeper areas of the basin (more than 1000 m) (Nies et al. 1998).

Surface waters east of the Norwegian-Greenland Basin were characterized by ^{239,240}Pu concentrations of 4-17 Bq / m^3 in 1981 and 8-10 Bq / m^3 in 1989. The distribution of plutonium in different water masses is seen in vertical profiles of dissolved ^{239,240}Pu. Evidently, the distribution of ^{239,240}Pu reflects the original signal of global fallout augmented through the transfer of plutonium from the Arctic Ocean basin back to the northern seas (Holm et al. 1993; Strand et al. 1996, 1997; Landa et al. 1999).

^{137}Cs is present throughout the Barents Sea. During the 1970s, the Barents Sea contained ^{137}Cs in concentrations of 5 - 30 Bq / m^3 except in the south-west where 1^{137}Cs concentrations were 25-60 Bq / m^3.

From maps of ^{137}Cs and ^{134}Cs distributions in the 0-100 m depth range in September 1979, concentrations varied from 8 to 80 and from 80 to 40 Bq / m^3 correspondingly (Kautsky 1988). The maximum radiocesium concentrations are found in coastal waters supplied by the Norwegian Coastal Current (NCC). At the entrance to the Barents Sea NCC waters contained up to 83 Bq / m^3 of ^{137}Cs and near the Varanger Peninsula coast - about 41 Bq / m^3 (Fig. 3.7). The water on the shallow Bear Bank contained in 1979 8-15 Bq / m^3 of ^{134}Cs and ^{137}Cs. These radionuclides are derived from Sellafield and La Hague nuclear reprocessing plants

(Kautsky 1988). The activity of ^{137}Cs, associated with radioactive fallout from atmospheric nuclear bomb tests, is still about 3 Bq / m^3 in surface waters (Nies et al. 1998).

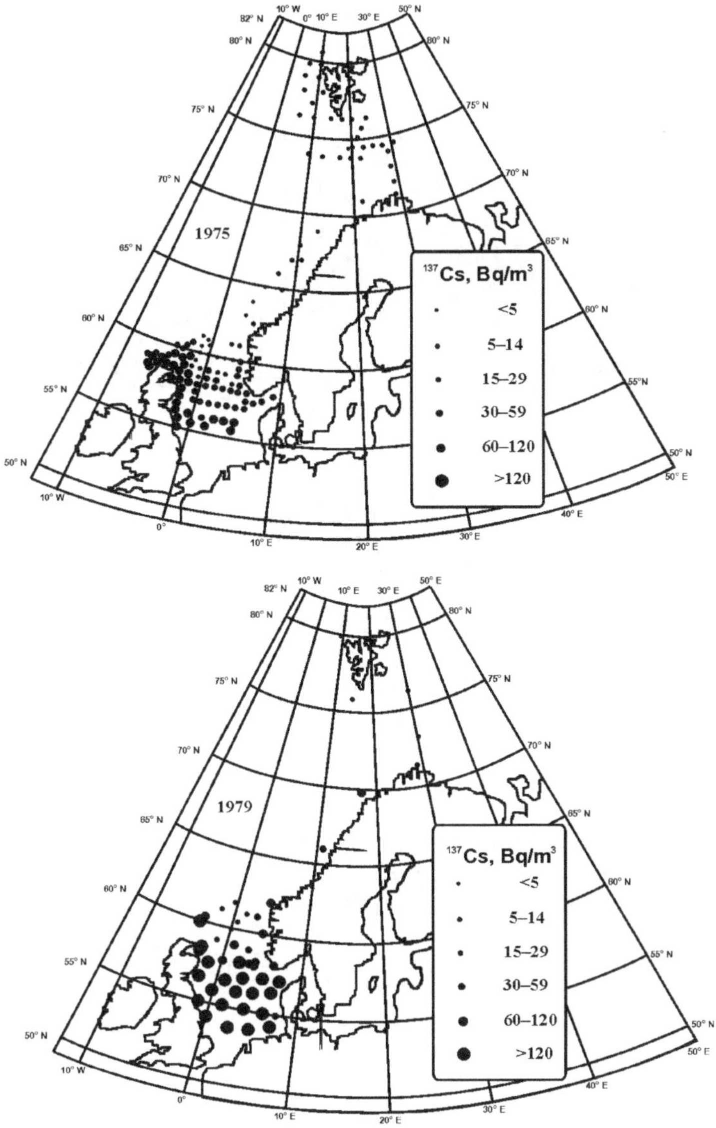

Fig. 3.6.a. ^{137}Cs concentrations in the North, Norwegian, Greenland and Barents Seas (1975-1979) (Kershaw and Baxter 1995)

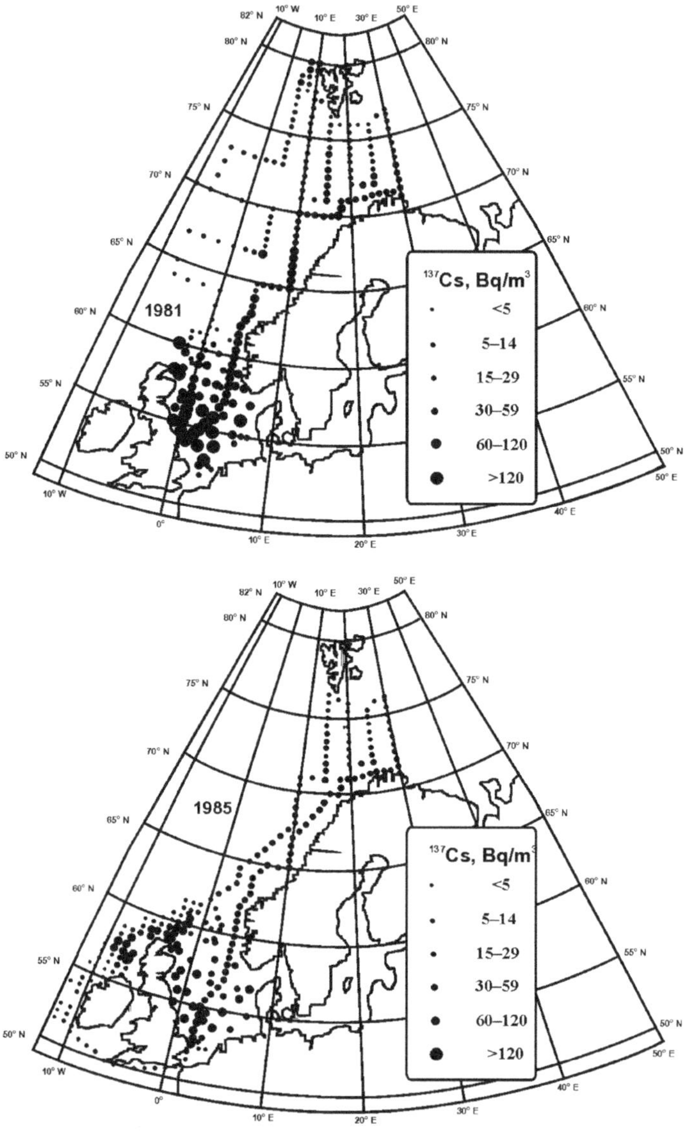

Fig. 3.6.b. ^{137}Cs concentrations in the North, Norwegian, Greenland and Barents Seas (1981-1985) (Kershaw and Baxter 1995)

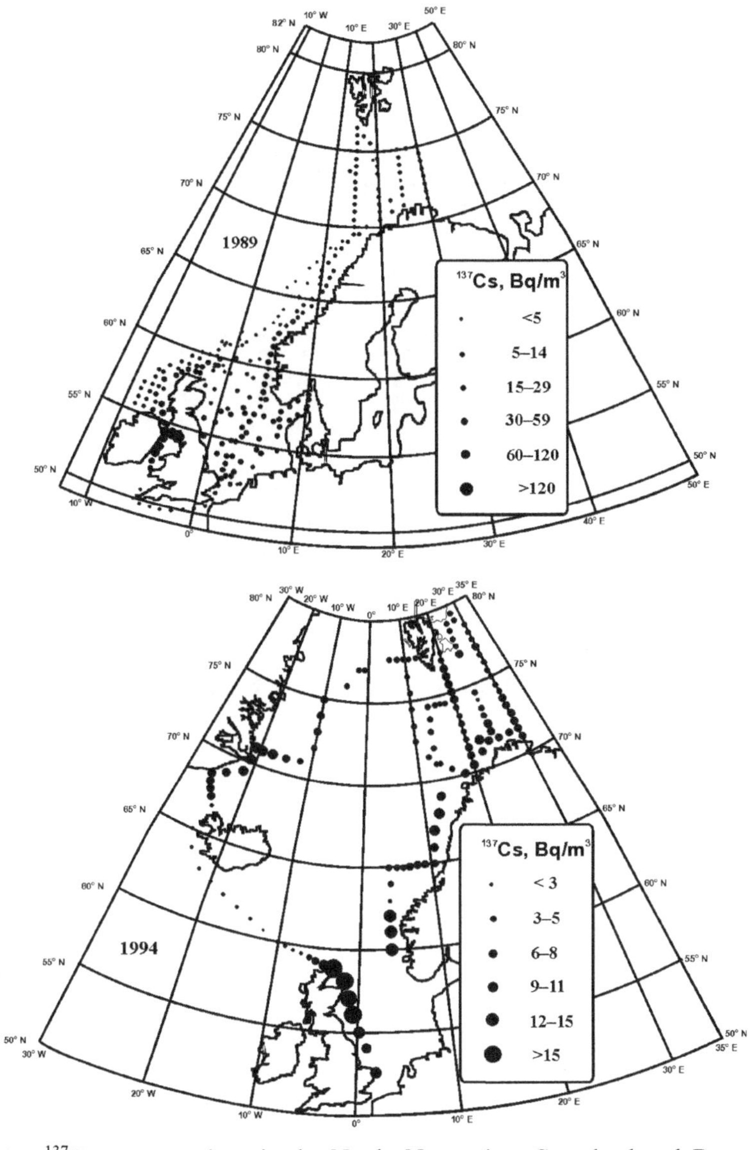

Fig. 3.6.c. ^{137}Cs concentrations in the North, Norwegian, Greenland and Barents Seas (1989-1994) (Kershaw and Baxter 1995)

During the 1980s the Barents Sea accumulated the highest concentrations of radiocesium - from 10 to 40-90 Bq / m^3 in the history of measurements in this sea. The content of radionuclides in the different water masses of the Barents and Kara Seas was studied in detail for the first time in August - October 1982 in the cruise of research icebreaker «Otto Schmidt» (Fig. 3.8) (Vakulovsky et al. 1985). The highest ^{137}Cs levels in the Barents and Kara Seas (20-37 Bq / m^3) were found to the west and east of Novaya Zemlya and to the north of the Kola and Motovsky Bays. Dissolved ^{137}Cs concentrations decreased sharply to 6-8 Bq / m^3 in a north-west direction from Novaya Zemlya to Spitsbergen. The concentration of ^{137}Cs in the Murmansk coastal current decreased to a depth of 300m from 33 to 21 Bq / m^3. Similarly for ^{90}Sr from 10 to 6 Bq / m^3. ^{134}Cs was present in the Barents Sea in small quantities (from 0.9 to 0.09 Bq / m^3).

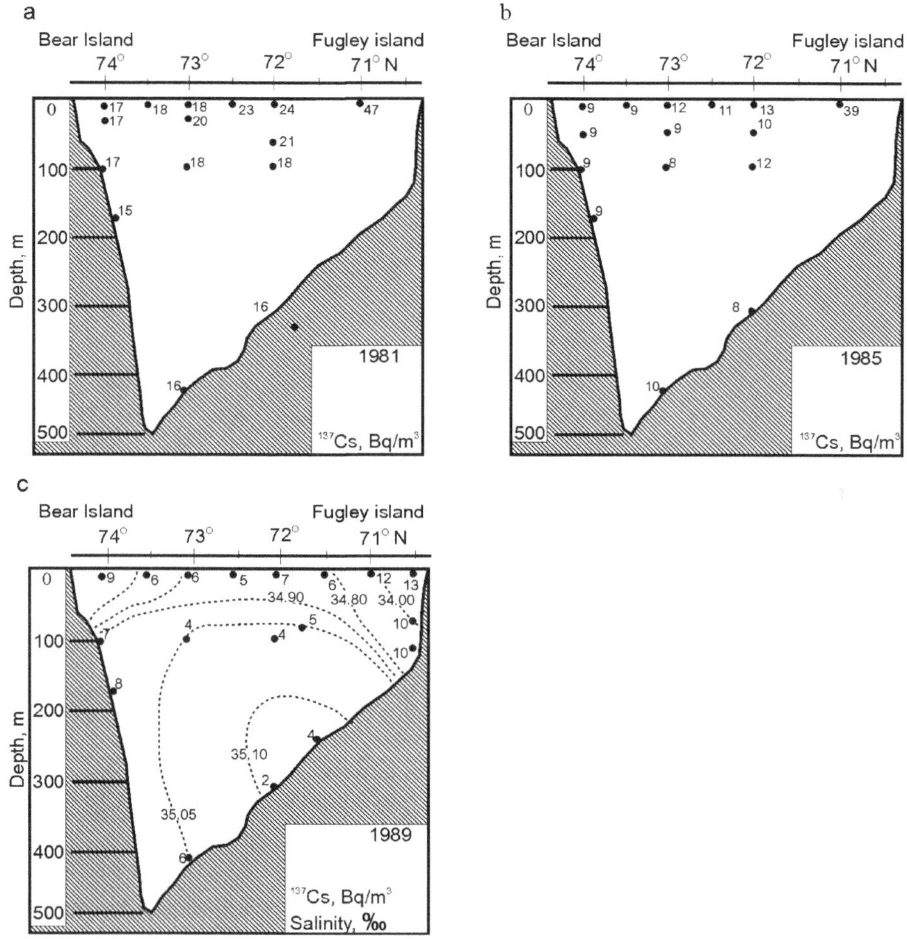

Fig. 3.7. ^{137}Cs distribution with depth between Bear Island and Fugley Islands (20E) (Kershaw and Baxter 1995)

From 1981-1985 the coastal branch of the Nordcape Current was characterized by a ^{137}Cs level from 40 to 90 Bq / m^3. The water masses along the Nordcape section contained in this period from 15 to 50-60 Bq / m^3 of radiocesium in the Sudcape Trough area and 14-30 Bq / m^3 in the area of the Bear Bank. A sharp decrease in concentration in the south of the Barents Sea shelf to 5-15 Bq / m^3 was observed beginning in 1989.

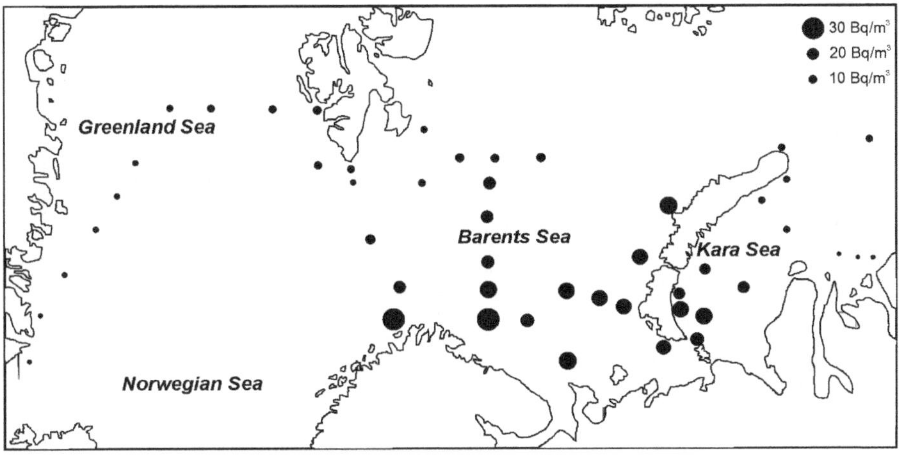

Fig. 3.8. Sampling stations and ^{137}Cs concentrations in the surface waters of the Barents and Kara Seas (Vakulovsky et al. 1985)

During the 1990s radiocesium concentrations were comparatively low. In autumn 1992 the upper layers (to 100 m) to the south-east of the Barents Sea contained from 4.4. to 7.4 Bq / m^3 of ^{137}Cs (Foyn and Semenov 1993).

Overall ^{137}Cs concentrations in surface waters have decreased more than an order of magnitude and currently does not exceed 2-15 Bq / m^3 anywhere in the Barents Sea. In 1995 ^{137}Cs concentrations in the upper water layer of the southwestern part of the Barents Sea and to the west of Cape Hope (Novaya Zemlya) varied from 2 to 6-8 Bq / m^3. Surface and bottom water layers in offshore waters had no significant differences in contamination. Slightly higher concentrations were discovered in the surface waters of the Norwegian Current to the north of Nordcape and near the Kola Peninsula (4.0-15.0 Bq / m^3). Both the Sellafield discharges and the polluted Baltic Sea were the sources of the increased concentrations.

There are very few data on the radioactive content of the bays and inlets bordering the southern Barents Sea. In 1996 water samples were collected from Motovsky and Kola Bays (Fig. 3.9). The ^{137}Cs content in a sample from Zapanaya Litsa Bay was (4.9±0.6) Bq / m^3, in the Ara Bay - (4.7±0.4) Bq / m^3 and at the entrance to the Kola Bay (8.0±0.8) Bq / m^3 (Matishov et al. 1997b,f). The ^{134}Cs level was very low (0.2-0.4 Bq / m^3). In 1998 the surface waters of the Kola Bay near the Atomflot nuclear reprocessing facility contained about 3-9 Bq / m^3 (Nikitin et al. 1999). The levels of radionuclides are slightly higher than those for the open part of the Barents Sea at the end of the 1990s.

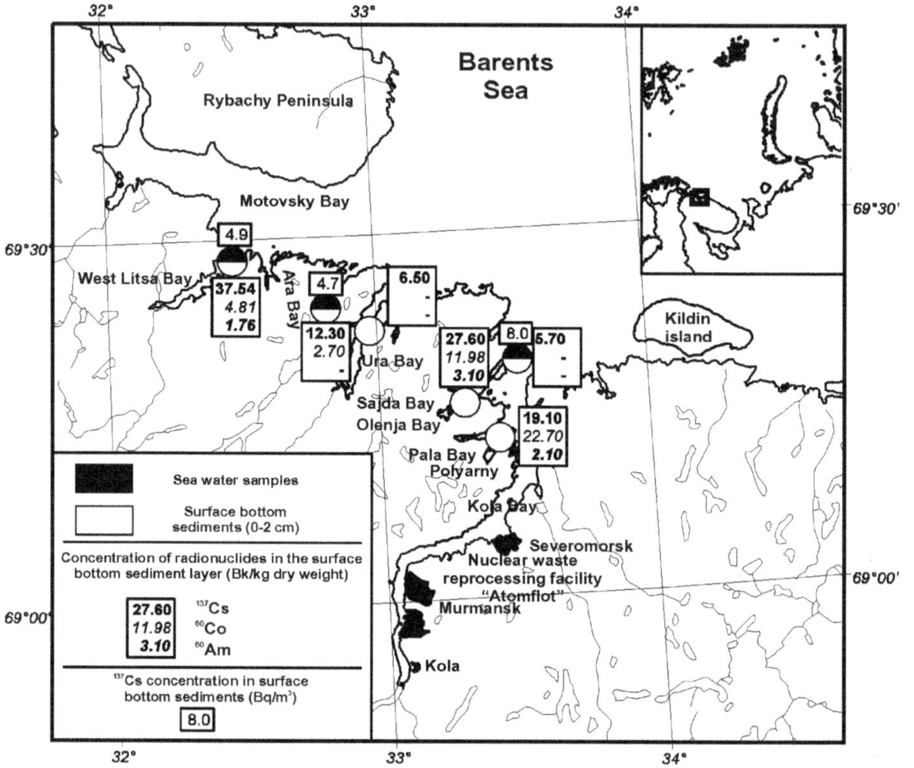

Fig. 3.9. Radionuclides in the Barents Sea coastal zone (MMBI, May - June 1996)

The ecosystem of Chernaya Bay on the southern coast of Novaya Zemlya is an exception. During the 1990s ^{137}Cs concentrations in Chernaya Bay were 14-190 Bq / m^3. This contamination level is a legacy of the underwater atomic explosions previously conducted in the bay.

^{90}**Sr** is also widely observed throughout the Barents Sea (Vakulovsky et al. 1985; A survey ... 1992; Foyn and Semenov 1993; Matishov et al. 1994a,b,c, 1996f,g, 1998b; Kershaw and Baxter 1995; Krosshavn 1995; Izrael 1998; Nikitin et al. 1999). During the 1960s the highest activity of strontium in the sea, to 24 Bq / m^3, was noted. A small decrease in the concentration of this radionuclide, to 7-17 Bq / m^3 occurred between the 1970s and 1980s. Seawater containing ^{90}Sr from 9 to 15 Bq / m^3 was occurred near the west and east coasts of Novaya Zemlya, and also in the Pechora Sea (Vakulovsky et al. 1985).

During the 1990s the level of ^{90}Sr in the Barents Sea decreased, varying from 1 to 7 Bq / m^3. During 1992 surface waters near Nordcape contained 4.3 Bq / m^3 and to the east of the Pechora Sea - 3,7 Bq / m^3. In the coastal zone of Novaya Zemlya (to the west from Hope Cape) in 1995, the entire water column from surface to bottom contained concentrations of 1.4-1.6 Bq / m^3 of ^{90}Sr (Krosshavn 1995). The highest level of this radionuclide (26-140 Bq / m^3) remained in Chernaya Bay. The ^{90}Sr content in Kola Bay water during the period of atmospheric nuclear testing was more than an order of magnitude higher than at present: approximately 54 Bq / m^3 in 1953.

During 1960-1970s, strontium concentrations decreased to 20-30 in the Kola Bay and 9-15 Bq / m^3 in the Barents Sea. As a result, ^{90}Sr concentrations were higher in the Kola Bay than in the open parts of the Barents Sea until the 1980s. Direct atmospheric fallout and the additional transport of fallout radionuclides from river catchment areas into Kola Bay caused the observed concentration difference. The ^{90}Sr content in seawater around the sunken submarine «Komsomolets» in 1991-1993 also was low, 0.5-4.0 Bq / m^3 (Kutsnetsov et al. 1993). The analysis of available strontium concentration data in surface samples taken along the «Kola meridian» section of the Barents Sea in 1963-1998 indicates a rapid decrease in the concentration of this radionuclide in response to the cessation of atmospheric nuclear tests (Fig. 3.10).

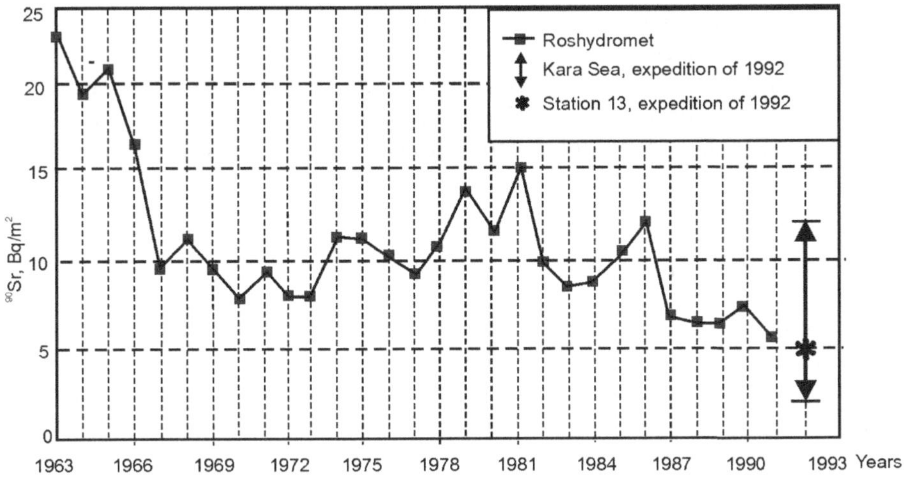

Fig. 3.10. ^{90}Sr concentration in the Barents Sea surface seawaters (Strand et al. 1993)

^{239,240}**Pu** is found in small quantities throughout the entire water column of the Barents Sea (Kershaw and Baxter 1995; Strand 1998; Nikitin et al. 1999). During the 1980s the level of this nuclide in seawater gradually decreased from earlier levels. In particular, the following dynamics occurred in the western part of the Barents Sea, : 1981 - 13-18 Bq / m^3; 1985 - 8-16 Bq / m^3; 1989 - 6-11 Bq / m^3. In the beginning of the 1990s the ^{239,240}Pu contamination in surface waters of the southern Barents Sea (to the northwest of Rybachy Peninsula and to the west of Vajgach Island) was only 4-8 Bq / m^3. Concentration levels were 7-11.5 Bq / m^3 within the 100 m depth zone (Foyn and Nikitin 1993). In 1998, Kola Bay waters near Atomflot contained 18-70 Bq / m^3 of ^{239,240}Pu.

The White Sea reaches depths of 350 m and has surface waters, freshened (25-28 ‰) by river runoff. Below the surface layer density increases due to mixing with the higher salinity Barents Sea water (30-33 ‰). However, the distribution of ^{137}Cs, ^{134}Cs and ^{90}Sr is uniform in both surface and deep waters (Vakulovsky et al. 1988). Contamination levels of ^{137}Cs in surface waters varied from 11 to 15 Bq / m^3 and were somewhat lower than deep water concentrations (20-21 Bq / m^3) because of the dilution of upper layers by river runoff. The mean concentrations of ^{134}Cs in the upper and deep layers of the White Sea were 0.14-0.17 Bq / m^3. The distribution of ^{90}Sr throughout the water column was uniform, 10-11 Bq / m^3. These values are essentially higher than those of the Barents Sea.

Further analysis of the data of Vakulovsky et al. (1985) results in some additional conclusions. First, it is quite normal that the deepest portions of the White Sea with relatively high salinity (28.0-29.7 ‰) have the maximum content of the long-lived radionuclide ^{137}Cs (16-22 Bq / m^3). Secondly, more than half (six) of the research stations are in the Dvina Bay, opposite Severodvinsk and North Dvina River. Releases of artificial radionuclides by the releases just within these geographical areas is quite obvious from the data. Evidently, contamination of the White Sea consists both of radionuclides from Sellafield (Vakulovsky et al. 1988) and of regional sources.

It must be noted that the highest concentrations of ^{90}Sr in the White Sea (about 48 Bq / m^3) were observed in 1963 (Chumichev 1995). Later (in 1996) the levels of this radionuclide in surface waters were significantly reduced (about 3.5 Bq / m^3).

3.1.3 Kara and Laptev Seas

The artificial radionuclide content of the Kara Sea has mainly been studied in the southern region (Vakulovsky et al. 1985; Holm et al. 1993; Foyn and Semenov 1993; Krosshavn 1995; Joseffson et al. 1995; Johannessen et al. 1996; Pavlov et al. 1999). The cyclonic water circulation and an increase of salinity by several units to the north near the mouths of the Ob, Yenisey, and Lena Rivers (33-34 ‰) are characteristic of these sea basins. The bottom waters of Atlantic origin in the troughs of the Kara Sea have salinities on the order of 35 ‰.

^{137}Cs was present in the surface waters of the southwest Kara Sea in the beginning of the 1980s from 3 to 27 Bq / m^3 (Fig. 3.11). At the same time, the maximum ^{137}Cs level reached 40 Bq / m^3 (Nies et al. 1998). In 1992-1997, seawater from the western basin contained relatively low concentrations of anthropogenic radionuclides (Fig. 3.11). The level in near-bottom waters was somewhat higher 6-20.4 Bq / m^3 (Fig. 3.12). ^{137}Cs concentrations in freshened surface waters from the central part of the Kara Sea was low, only 3-6 Bq / m^3 near the Ob and Yenisey estuaries. The analysis of depth profiles from the Novaya Zemlya Trough shows waters with relatively high ^{137}Cs content - 20 Bq / m^3 - are found below the freshened surface layer. The ^{134}Cs content of these waters was relatively low - from 0.1 to 0.7 Bq / m^3. The ratio between 134Cs and 137Cs is within the range

0.019-0.034. Near Sellafield during the period of maximum discharges the ratio was 0.08 suggesting that the origin of ^{134}Cs may be connected to local and not long range sources (Foyn and Nikitin 1993).

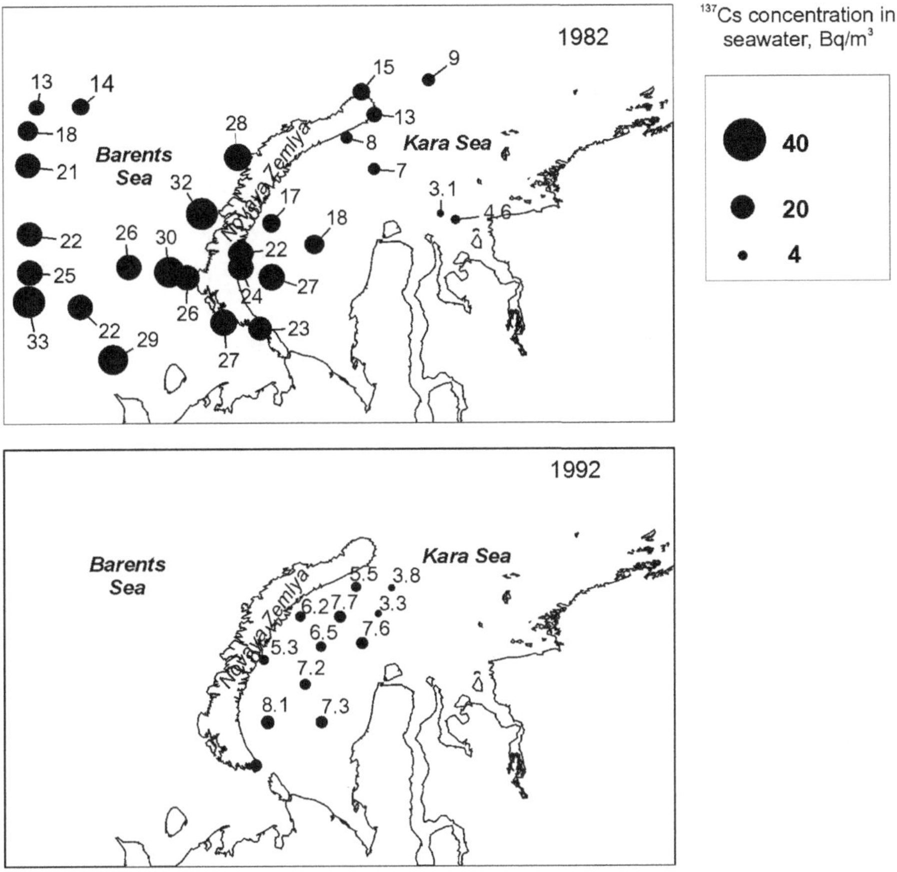

Fig. 3.11. ^{137}Cs distribution in the surface layer of the Barents and Kara Seas in 1982 (Vakulovsky et al. 1985) and 1992 (Strand et al. 1993)

In 1990-1994 the ^{137}Cs level in surface waters of the Kara shelf continued to decrease and varying from 2 to 8-12 Bq / m^3. Very low radiocesium levels (3-5 Bq / m3) were observed in waters to the west of the Yamal peninsula. ^{137}Cs nearly doubles (to 13.4 Bq / m^3) in the near-bottom waters. In 1993 the Chernobyl fallout ratio of ^{134}Cs to ^{137}Cs was 0.074. If Chernobyl is the only source of ^{134}Cs, only 30 % of ^{137}Cs is related to Chernobyl.

Investigations carried out between 1992-1995 in the Abrosimov, Stepovoy, and Tsivolka Bays, revealed radioactive contamination of the environment near the burial sites of solid and liquid nuclear waste containers. The surface waters of the bays contain about 2-6 Bq / m^3 of ^{137}Cs and the bottom layers – up to 9-11 Bq / m^3. In some cases the radiocesium concentration in bottom waters near the containers reaches 12-31 Bq / m^3 (Strand 1998). These concentration levels are likely the result of leakage from containers coupled with long water residence times in the bays.

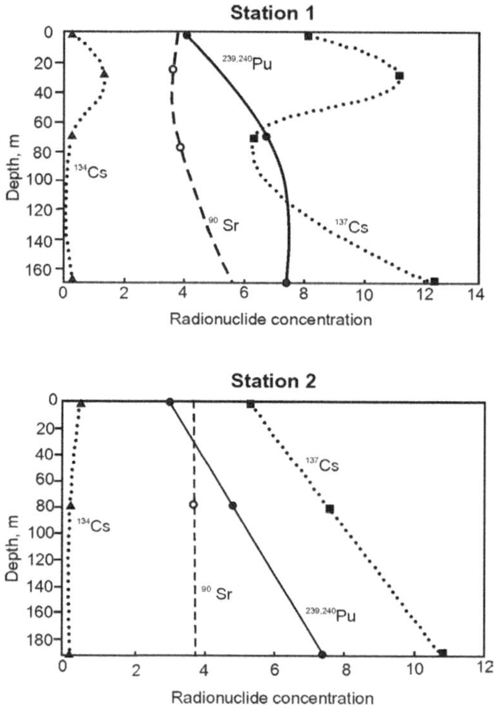

Fig. 3.12. Vertical distributions of dissolved ^{134}Cs, ^{137}Cs, ^{90}Sr (Bq/m^3) and ^{239,240}Pu (mBq/m^3) in Kara Sea water (Strand et al. 1993)

The ^{137}Cs content in the surface waters of the Laptev Sea in 1993 and 1994 varied from 1 to 13 Bq / m^3 (Joseffson et al. 1995; Holm et al. 1993; Pavlov et al. 1999). Higher radiocesium levels were noted to the north of the sea, near the shelf border (8.3-9.2 Bq / m^3) and also in the Vilkitsky Strait (10.6 Bq / m^3). In the southwestern coastal part of the sea the range of ^{137}Cs concentrations was 2.1 to 7.3 Bq / m^3. In the southeastern coastal zone, near the Buor-Khaya Bay in particular, the lowest levels of ^{137}Cs were noted - to 1-2 Bq / m^3 (Fig. 3.13).

Radionuclide concentrations depend to a great extent on the general water circulation and salinity distribution. The Lena and other rivers deliver to the shallow zone of the Laptev Sea (50-150 m) as much as 700 km^3 of fresh water annually. As a result of this runoff, the salinity of seawater along the Lena delta is 1-5 ‰

for many kilometers from the shore. For comparison, the water salinity to the north of the sea is 34 ‰.

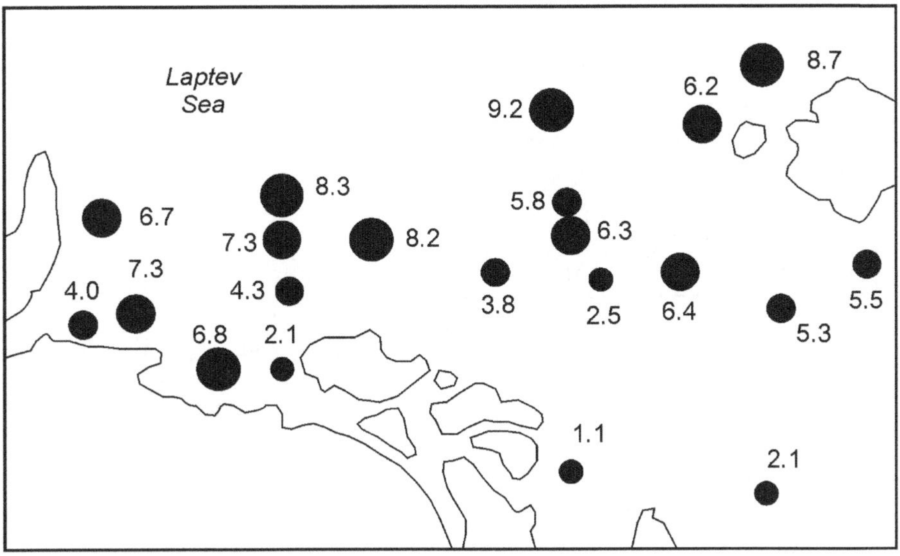

Fig. 3.13. ^{137}Cs concentration (Bq/m^3) in the surface layer of the Laptev Sea (summer 1993) (Pavlov et al. 1999)

The correlation coefficient between the water salinity and ^{137}Cs concentration in the Laptev Sea is 0.95. Thus, the radionuclide transfer by currents and drifting ice from the Kara Sea and Arctic Ocean Basin are the main causes of the increase in ^{137}Cs concentration levels during 1993 and 1994 (Pavlov et al. 1999).

^{90}Sr is the main anthropogenic radionuclide detected in waters of the Kara Sea basin. Its maximum concentration levels reached 56 and 85 Bq / m^3 in 1963 and 1964 respectively (Vakulovsky et al. 1993). During the last 35 years, the ^{90}Sr concentration in the Kara Sea has been decreasing. Such reduction of radionuclide content probably took place as a result of ^{90}Sr radioactive decay (half-life =13.5 years) and the removal of pollutants by currents. In spite of the general downward trend during previous years, a somewhat higher concentration of strontium was observed in the waters of the coastal shelf near the Ob and Yenisey estuaries.

During 1970-1980s the ^{90}Sr content in the Kara surface waters varied from 6 to 28 Bq / m^3 (Fig. 3.14) (Vakulovsky et al. 1985; Strand et al. 1993; Johannessen et al. 1996). In 1982 the highest levels of ^{90}Sr in the Kara Sea (20-21 Bq / m^3) were noted near the Ob mouth and to the east of Matochkin Shar Strait. Even lower concentrations of this radionuclide were determined in the upper water layer in 1992 and 1995 - 3-11 and 1.5-7 Bq / m^3 correspondingly. The highest ratio of ^{90}Sr to ^{137}Cs is observed in water masses on the northeast of the Kara Sea. Probably this situation reflects ^{90}Sr discharges by the Yenisey.

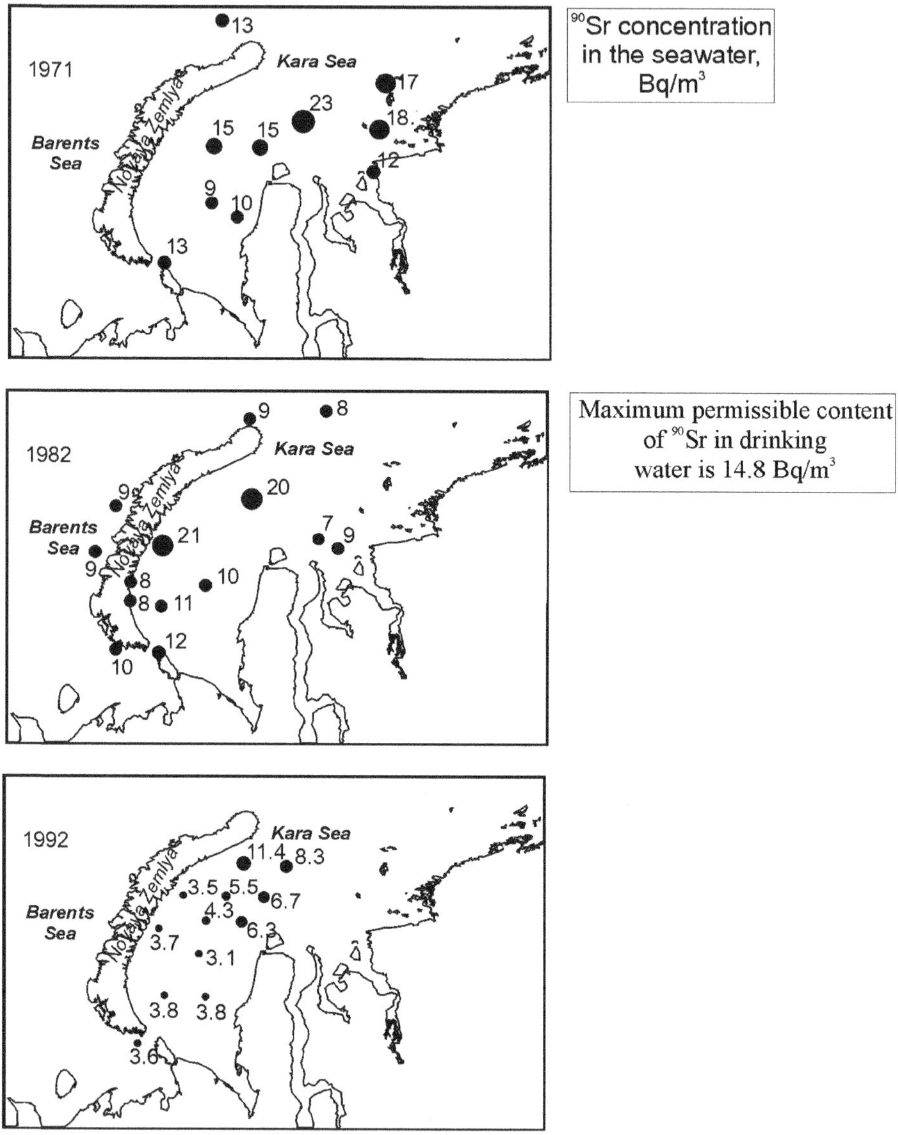

Fig. 3.14. ^{90}Sr distributions in surface water of the Kara and Barents Seas (Vakulovsky et al. 1985; Strand et al. 1993)

The distribution of ^{90}Sr concentrations in surface waters of the Laptev Sea were more uniform, in the range from 3.7 to 6.6 Bq / m^3. The maximum values were registered in the southwest region of the sea (Pavlov et al. 1999). Overall, the ^{90}Sr level in seawater in the middle of the 1990s was lower by an order of magnitude compared to 1964 when the concentration reached 53 Bq / m^3 (Vakulovsky et al. 1993).

^{239,240}Pu in the waters of the Kara Sea is known only in general. In the period 1970-1990s, the concentration of this nuclide both in surface and bottom water varied from 2 to 16 Bq / m^3 (Johannessen et al. 1996; Strand 1998). In 1992, for example, the surface water layer contained 2-8 Bq / m^3 of ^{239,240}Pu and the bottom layer contained 5-16 Bq / m^3.

^{60}Co was observed in the Yenisey estuarine zone during 1985-1991 (Vakulovsky et al. 1994a); the mean concentration was 6 Bq / l.

The somewhat increased levels of ^{137}Cs and ^{90}Sr concentrations in surface waters of the Laptev Sea coastal zone are connected, evidently, with the resuspension of radionuclides from the upper layer of bottom sediments by waves and bottom currents which can attain critical velocity values in this area.

3.1.4 North Atlantic and Central Polar Basins

The radionuclide content in the water masses of the northern part of the Atlantic Ocean is determined mainly by global radioactive fallout (Shvedov and Patin, 1968; Gromov and Spitsyn 1975; Gromov et al. 1985). There is also a small advection of ^{137}Cs, ^{90}Sr and ^{239,240}Pu to the Atlantic from the Norwegian-Greenland Basin. In the 1980s the global contamination of the North Atlantic surface waters by ^{137}Cs was 3-6 Bq / m^3 (~ 0.1 pKi / l) (Vakulovsky et al. 1985; Nikitin et al. 1991). The waters of the Northeast Atlantic to the south of Iceland contained about 2-4 Bq / m^3 of ^{137}Cs in 1994.

In the beginning of the 1980s the surface waters of the Arctic Ocean were characterized by approximately double the contamination level of the North Atlantic because of the input of radioactive waters from Europe's nuclear reprocessing facilities. The ^{137}Cs concentration in the waters of the Arctic Seas was from 10 to 30 Bq / l (Aarkrog 1979; Aarkrog et al. 1985; Vakulovsky et al. 1985, 1988; Holm et al. 1991).

From 1985-1987 the Arctic Ocean waters were contained small amounts of ^{134}Cs (from 0.1 to 0.03 Bq / m^3), ^{137}Cs (2-14 Bq / m^3) and ^{90}Sr (6-8 Bq / m^3). The concentration of radionuclides decreased noticeably from the surface to the deeper (250-550 m) water layers (Nikitin et al. 1991). The radionuclide content associated with suspended material was a few tenths of a percent of the radionuclide content associated with the dissolved water phase of the entire Central Arctic.

The ^{137}Cs and ^{239,240}Pu distributions in the water column of the North Atlantic and Arctic Ocean are similar. In bottom waters the minimum levels of both cesium and plutonium are about 0.5-1.5 Bq / m^3 and 0.5 mBq / m^3 respectively. In the surface waters of the Central Arctic Basin the ^{137}Cs concentration varied in 1979-1994 from 2 to 15 Bq / m^3 (Livingston et al. 1984; Roos et al. 1993; Nies and Nielsen 1996).

In the middle of the 1990s the Arctic Ocean waters had ^{137}Cs levels between 0.8-15 Bq / m^3 (Strand 1998). The distribution of radiocesium in surface waters is in agreement with existing knowledge of the sources and current patterns. A small increase of ^{137}Cs to 14-19 Bq / m^3 is noted in the waters of Atlantic origin above the Lomonosov Ridge. Near Greenland, the highest concentrations are noted along the east coast. The lowest radiocesium concentrations are found in the East Siberian Sea and in the southeastern part of the Makarov Basin. The overall character of the ^{137}Cs distribution is determined by the influence of discharges from nuclear fuel reprocessing plants and by the mixing of water masses originating from different areas.

3.2 Biofiltration of radionuclides

3.2.1 Ecological characteristics of plankton

The bioaccumulation of artificial radionuclides from the marine environment is a central problem of radioecology. Many life forms of the sea possess the selective ability to concentrate from seawater different elements and inorganic compounds.

The levels of radioactive elements in the northern seas vary widely, but in most cases are attributable to a number of global or regional sources representing the current background radiation levels in the environment. However, in a number of concrete cases the contamination of bays and coastal shelves differ from these background levels by an order of magnitude or more (Strand et al. 1993, 1997; Matishov et al. 1994a,b,c, 1997b,c,f, 1998b,c). In order to formulate a clear picture of radionuclide behavior and accumulation in biota, it is necessary understand the biological productivity of basins, the structure of biological communities on the shelves and adjacent continental margins.

The tight coupling between key ecosystem components is important to the formation of high bioproductivity of the Norwegian, Barents, Greenland and some other seas. These key ecosystem components comprise littoral (phytal), pelagic, cryopelagic, and benthic communities (Zenkevich 1963; Matishov and Pavlova 1990; Savchuk and Matishov 1990; Timofeev 1996). The structure of these ecosystems is determined by interactions of abiotic and biotic factors and these interactions are further driven by incoming solar radiation. The highest biological productivity is observed in the zone of Arctic polar hydrological fronts, along the edges of drifting sea ice and near icebergs, in the estuarine zones and also in the waters surrounding bird bazaars where the content of biological substances are 10s-100s of times the normal background concentrations.

Here we shall consider the specific features of the pelagic and cryopelagic ecosystems. The high biological productivity of certain areas in the Norwegian, Barents and other marginal seas depends on the currents, ice conditions and temperature fluctuations. The primary production of organic matter is one of the most important functions of the pelagic ecosystem. Primary production in some marginal sea areas has been estimated to be 100-200 g C / m^2 annually (Koblents-Mishke 1985; Savinov 1992). These levels of productivity drive the high latitude food web and hence support the life cycles of species harvested for subsistence living and commercial purposes.

Plankton is characterized by a short life span, a low species diversity and the seasonally varying trophic relationships. The annual primary production of phytoplankton in the Barents Sea coastal zone is estimated as 30-130 g C / m^2 (Kuznetsov 1980). In the Atlantic and Arctic waters the annual primary production varies between 90-200 and 70-130 g C / m^2 correspondingly (Bobrov 1982; Savinov 1992). The gross value of primary production in C equivalent for the whole Barents Sea is approximately 10 millions tons (Nesvetova and Chepurnov 1989).

Almost all phytoplankton is concentrated in the upper water column (50-60 m) (Fig. 3.15). The most prevalent species are the diatomaceous, Peridinea and Coccolithophoridae (unicellular calcareous algae). Thin shells of these organisms are deposited relatively quickly on the sea bottom through the formation and sinking of zooplankton fecal pellets (Lisitsyn and Vinogradov 1982; Matishov and Pavlova 1990). These waste products which also contain radionuclides are removed from the biochemical cycle and are accumulated in the bottom sediments.

Zooplankton organisms including primitive copepods, worms, Coelenterata, some mollusk species, and also the eggs and larvae of different marine animals in certain periods of their life cycles are very important to the pelagic ecosystem. They are central to the process of organic matter transformation. Zooplankton communities, Calanus, Euphaysiacea, Giperiidae, northern shrimp (Zenkevich 1963; Drobysheva and Timofeev 1990; Timofeev 1996) are characterized as being large and are found across a wide geographical region. In this connection the planktivorous fish facilitate the large-scale transfer of substance from one ecosystem to another.

The life cycle of marine Copepoda is includes the filtration of large volumes of water from the surface layer. For example, the northern part of the Barents Sea is inhabited by a plankton community dominated by *Calanus glacialis*. The Calanus feeds on phytoplankton from February to August - September, consuming 60-80 % of the daily production of algae (Tande 1989). In spring the biomass of the Barents Sea copepods and Euphaysiacea can reach 500-1000 mg / m^3 (Timofeev 1996). Several zooplankton species (*Pseudocalanus elongatus, Calanus Finmarchicus, Oithona similis, Microcalanus pucillus*) dominate among the fjord fauna, in particular, in the Kola Bay. Their abundance and biomass are the highest in the middle part («elbow») of the bay. The maximum quantity of zooplankton reaches 8-10 thousand specimens / m3, which is characteristic for the 0-10 m layer. The species diversity of zooplankton is enriched noticeably by benthic larvae (Bivalvia and others).

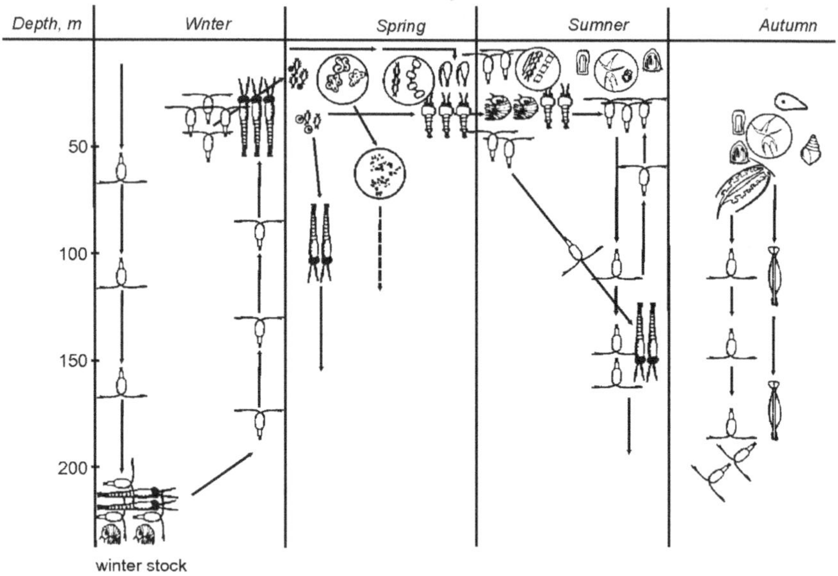

Fig. 3.15. Spreading and migration of plankton organisms with water depth during different seasons: ***Winter:*** *formation of Calanus masses before spawning.* ***Spring:*** *reproduction of Copepoda with eggs and planktonic larvae present, development of diatomaceous algae, growth of the Calanus fry, sand hoppers and others, descent of «sand hoppers» after spawning and decay of spring bloom.* ***Summer:*** *rapid development of diatomaceous algae; Calanus and «sand hoppers» fry; deily vertical migrations of Calanus.* ***Autumn:*** *mass development of jellyfishes and comb-bearers, the formation of the «winter stock»* (Pavshtiks 1956)

The cryopelagic zone and its role in the functioning of marine ecosystems is not known in detail. Drifting and pack sea ice (i.e. cryal zone), are an important feature of the polar seas involved in the cycles of material and energy of the ecosystem. In winter the ice layer, extending many thousands of kilometers, covers most of the pelagic zone. The radioactive atmospheric dust falls mainly on the ice surface, 6-9 months each year.

The structure and function of the cryal zone are controlled in many cases by conditions in the polar hydrological front. The spring and summer melting which occurs at the ice edge influences the entire water column. It marks the beginning of the seasonal productivity and the associated hydrophysical and hydrochemical processes. Zooplankton, represented mainly by large copepods (for example, *Calanus glacialis*), is the food source for capelin, polar cod, and numerous birds associated with the ice edge.

The trophic food web of the polar seas includes the flora and fauna under the ice (Melnikov 1989; Gulliksen and Lonne 1989). The fauna under ice includes copepods, Mysidacea, isopods and other amphipods, foraminifers, round worms (Nematoda), Polychaeta, arrow worms (Chaetognatha) etc. The main components of the fauna under the ice in the Arctic Seas are amphipods (*Ampherusa glacialis, Gammarus wilkitskii, Onisimus nanseni*) and polar cod. The total biomass (in wet weight) near the lower ice surface is usually 1-10 g / m^2, but sometimes it reaches 20-40 g / m^2.

Both the quantity of individual species and the bioproductivity of the seas is enormous. The processes of pollutant fixation, transformation and accumulation by plankton organisms are controlled by the complex dynamics of plankton productivity.

In the period of the Arctic «warming» (1930s) the biomass of the Atlantic ecosystem of the Barents Sea was the following: phytoplankton - 8 billion tons, zooplankton - 1.2 billion tons, planktivorous fishes - 180 million tons, carnivorous fishes - 28 million tons. In the cold period (end of 1960s - beginning of 1970s) these indices were significantly lower; phytoplankton - 2 billion tons, zooplankton - 300 million tons, carnivorous fishes - 3-4 million tons (Evolution ... 1994). In the cold years the natural delay of the zooplankton development occurs which leads to incomplete utilization of the primary organic substances in surface waters and its sedimentation on the sea bottom.

3.2.2 Radionuclide accumulation by marine plankton

The role of Arctic and subarctic pelagic and cryopelagic biofilters in the primary distribution of radionuclides is not well studied. At present it is possible to imagine the character and scale of the process of radionuclide bioaccumulation and sedimentation only in very general terms. We have used for this purpose MMBI data on the bioproductivity of pelagic and cryopelagic zones. The dynamics of cesium, strontium and plutonium in seawater was reconstructed for the period 1950-1990s.

Phytoplankton, zooplankton and ice fauna are the first marine barrier for the transfer and transformation of radioactive fallout into the biological cycles. The primary distribution of radionuclides determines the dose loads and radiation effects in the marine environment. Plankton organisms filtering the water fix radionuclide compounds removing them from the intensive cycle. The radionuclides incorporated into marine organisms begin migrating through the trophic food web.

In 1956, after a series of atmospheric nuclear explosions by the United States, the radioactivity of plankton in the Pacific Ocean was 7 thousand times higher than of seawater. Japanese scientists determined that the radioactivity of copepods in the Bikini atoll area exceeded the environmental radioactivity levels by more than 1000 times. One must note that the artificial radioactivity of the plankton remained high even several months after the initial impact of the explosions on the marine ecosystem (Gromov et al. 1985).

Before the Chernobyl disaster, Baltic waters contained about 20-45 Bq / m^3 ^{137}Cs. In June 1986 the ^{134}Cs and ^{137}Cs concentrations reached correspondingly 5900 and 1100 Bq / m^3 in the water and 1500 and 2500 Bq / kg (dry weight) in plankton (Ilus et al. 1988a,b, 1993; Ilus and Ikaheimonen 1991; Ikaheimonen and Ilus 1995). In mide-summer 1986, the ^{134}Cs and ^{137}Cs level decreased almost by an order of magnitude, possibly, because of biofiltration. The concentration of ^{134}Cs and ^{137}Cs in seawater was about 260 and 510 Bq / m^3 and in plankton about 110 and 220 Bq / kg (Table 3.3). In the 1990s radiocesium levels in seawater (to 50-90) Bq / m^3) and in plankton had returned to their previous levels.

Table 3.3. Gamma radionuclides in plankton (Bq / kg dry weight) and seawater (Bq / m^3) in the Finnish Bay of the Baltic Sea (1986) (Ilus and Ikaheimonen 1991)

	06.06	11.06		04.07		30.07		26.08	
Nuclide	Plankton	Plankton	Water	Plankton	Water	Plankton	Water	Plankton	Water
^{40}K	39	0	1800	0	1900	320	2300	0	2100
^{95}Zr	0	230	28	0	0	0	0	0	0
^{95}Nb	0	300	100	0	13	0	0	0	21
^{103}Ru	2200	2600	1600	370	230	0	91	180	62
^{106}Ru	910	1200	670	0	200	0	57	0	100
^{110}Ag	200	140	19	170	0	0	0	1500	8.5
^{125}Sb	0	0	250	0	68	0	0	0	130
^{129}Te	2200	2500	5900	0	1200	0	0	0	0
^{131}I	500	530	410	0	0	0	0	0	0
^{134}Cs	1200	1500	560	250	290	110	260	500	410
^{136}Cs	140	0	25	0	0	0	0	0	0
^{137}Cs	2200	2500	1100	570	570	220	510	850	790
^{140}Ba	280	0	6200	0	1200	0	230	0	0
^{140}La	430	0	6000	0	1100	0	210	0	0
^{141}Ce	68	0	68	0	24	0	0	0	0
^{144}Ce	0	0	120	0	0	0	0	0	0

In summer 1999 the plankton of the Azov Sea coastal zone was characterized by small ^{137}Cs concentrations (2.8 Bq / kg wet weight). The activity of this radionuclide was about 11 Bq / m^3 dry weight (Table 3.4).

Table 3.4. ^{137}Cs and ^{40}K plankton and algae from the Azov Sea in 1999

Object	^{137}Cs	^{40}K	^{137}Cs	^{40}K
	Bk/kg wet weight		Bk/kg dry weight	
Plankton	2.79±0.18	103±3	10.9±0.76	405±12
Algae, st. 4.	0.31±0.06	81±2	5.89±1.12	1546±31
st.18	0.69±0.03	61±1	8.59±0.34	768±11
st.2	0.25±0.08	71±2	2.12±0.71	608±20

Few determinations exist of radioactivity in plankton from the polar seas. However, some suppositions can be made by analogy, using the above-mentioned data on the radionuclide content in seawater and plankton contamination levels in the Baltic and North Seas. The maximum ^{137}Cs concentration in the Barents and

Kara Seas plankton in the beginning of the 1960s and in the middle of 1980s was 20-60 Bq / kg (dry weight). In the 1990s the contamination of plankton from the Arctic Seas was very low, evidently, on the order of 5-20 Bq / kg.

Certainly the radioactivity of the Barents and Kara plankton in the area of direct influence of the Novaya Zemlya atmospheric and underwater nuclear explosions must be higher than in the Baltic after the Chernobyl events. We can suppose that the ^{137}Cs, $^{239,240}Pu$, ^{90}Sr concentration in plankton in areas, for example, of Chernaya Bay or Matochkin Shar Strait, could be 2-10 thousand Bq / kg. In any case, the heightened levels of radionuclides in the bottom sediments from the mentioned areas results from past Novaya Zemlya nuclear tests and also from the radioactive discharges from West Siberian rivers to the polar seas (Matishov et al. 1994a,b,c, 1997c).

In June 1986 two sediment cores were sampled in the deep-water (967 and 1426 m) regions to the west of Norway. They contained in the upper one- centimeter layer correspondingly 220 and 330 Bq / m^2 of ^{134}Cs and ^{137}Cs (Erlenkeuser and Balzer 1988). These values are significantly lower than the content (of order 1000 Bq / m^2) which was formed after the Chernobyl disaster in the lakes of Finland and Germany and in the Baltic Seas.

The structure of plankton organisms is characterized by different mechanisms for passive swimming: small, often microscopic, dimensions, the abundance of water in tissues (to 98 %) gaseous and adipose inclusions, and existence of different protrusions on the surface. These features may contribute to the sorption of radioactive particles by organisms. The transport and redistribution of microelements and radionuclides in the marine environment with the participation of biotic particles is an important process that increases the rate of descent for ^{60}Co, ^{106}Ru, ^{109}Cd, ^{237}Pu, ^{241}Am to the seafloor (Fishek et al. 1991). The concentration of radionuclides occurs both as a result of adsorption of different microelements and plankton feeding.

The removal of suspended material from the pelagic zone occurs mainly through feeding. The dimensions of newly suspended particles, consumed by zooplankton, are of microns or several tens of microns, while the excreted material emerging after filtration is hundreds and sometimes thousands of microns (Lisitsyn and Vinogradov 1982). The daily need for nourishment of zooplankton is on average 50-70 % of its own body weight. Accordingly, suspended matter from the entire Ocean (Vinogradov 1967) can be filtered and removed from the pelagic zone in less than one month. Laboratory studies and experiments in natural conditions showed that the rate of removal of fecal pellets, consisting of phyto- and zooplankton remnants, from the photic zone, is approximately several days.

With all the diversity of factors influencing the accumulation of radionuclides by plankton, of key importance are the total stocks of biomass and the structure and dynamics of growth. Evidently, these indices influence also the volumes of surface and intra-tissue accumulation of radionuclides. The extraction of the smallest particles from seawater and their transfer to large pellets, descending on the bottom, goes on especially intensively in the areas with high biomass of planktonic filter feeders which, in turn, is determined by the high primary production. The annual production of phytoplankton and zooplankton in the Barents Sea is correspondingly 2-3 billion and 120-300 million tons (Matishov and Drobysheva 1994).

Copepods, being filter feeders, are the primary source of food for many fish species, sea birds and, through detrital production, for benthic organisms. The consumption of euphausiids provides about 20 % of the biomass increment of cod and 30 % - of sea perch (Marti and Martinsen 1969). The annual consumption of all zooplanktic animals by Barents Sea fish varies from 20 t0 45 million tons. Large biomass quantities of crustaceans is also consumed by sea birds.

Thus, artificial radionuclides are removed from the pelagic zone both as a consequence of water filtering by plankton and the material transfer along food chains to fish, birds etc. The role of pelagic and cryopelagic biofilters in the primary distribution of radionuclides requires additional investigation. It is clear that the formation of fecal pellets accelerates the accumulation of radionuclides on the sea bottom.

3.3 Radionuclides in pelagic ecosystems

Particles of different types are suspended in seawater. During the transport of suspended particles through seawater, some particles will quickly descend to the bottom and others will continue to migrate slowly through the seas, resulting in the cross-border movement of contaminants attached to particles. The pelagic ecosystem is the primary filter of radioactive suspension.

The general picture of the distribution of radionuclides in seawater and also assimilated in planktonic organisms reflects the downward dynamics of radionuclide content in the Arctic Ocean basin (Fig. 3.16-3.18). During 50 years the ^{137}Cs concentration in the upper water layer of the northern seas varied mainly from 5-20 to 30-70 Bq / m^3 (Table 3.5). However, the water masses of the Irish and North Seas near the waste outflows from radiochemical plants were concentrated with radiocesium to 100-130 Bq / m^3.

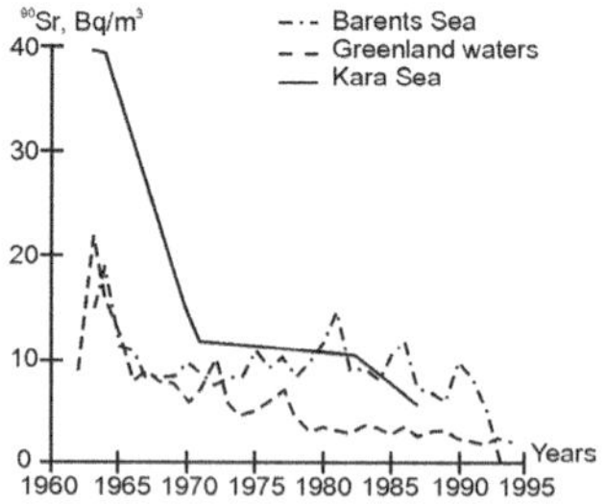

Fig. 3.16. ^{90}Sr concentration in surface coastal waters of the Greenland, Barents and Kara Seas during 1960-1995 (AMAP… 1997)

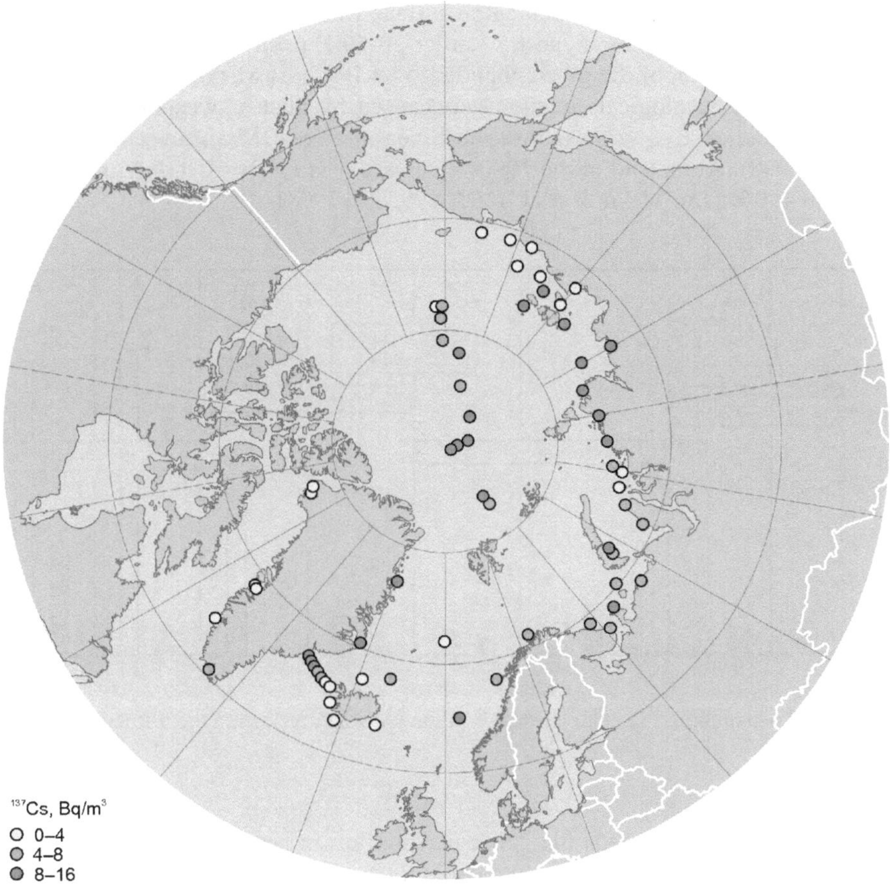

Fig. 3.17. ^{137}Cs concentration in surface waters of the Arctic Ocean and adjacent seas (AMAP... 1997)

The maximum concentration by ^{137}Cs of the Irish and North Seas (to 100-400 Bq / m^3) and the Norwegian, Barents and Kara Seas waters (from 15 to 75-100 Bq / m^3) was noted only in 1980-1984. At the same time the ^{137}Cs content in the Baltic Sea had increased sharply (from 100-650 to 1000 Bq /m^3) at the end of the 1980s owing to the Chernobyl discharges. The waters of polar seas have maintained radiocesium concentrations from 5-15 to 50-90 Bq / m^3 because of the influx from West European sources and Chernobyl fallout.

In the 1990s ^{137}Cs levels were insignificant in all northern seas - from 3-8 to 10-30 Bq / m^3. The Baltic Sea and a number of Novaya Zemlya bays are the exceptions, maintaining relatively high radiocesium concentrations to 100-200 Bq / m^3).

During the 1960s the ^{90}Sr content in surface waters of the Barents, White, Kara and Laptev Seas was the highest, of order 20-50 Bq / m^3 (Table 3.6). The maximum concentration of the given nuclide (to 84 Bq / m^3) was observed in the Kara Sea. This radioecological situation was caused by nuclear weapon tests and releases of radioactive wastes from the Siberian rivers (Matishov et al. 1994a,c; Strand 1998). At the end of the 1990s increased ^{90}Sr levels exist only in Chernaya Bay (to 140 Bq / m^3) (Smith et al. 1993, 1995a,b, 1998).

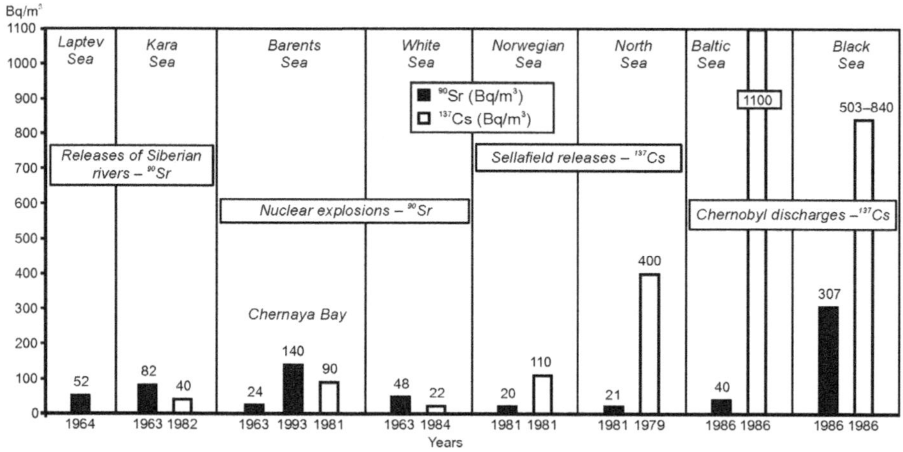

Fig. 3.18. Maximum ^{137}Cs and ^{90}Sr levels in the sea water reflecting the influence of different sources

In the 1980s and 1990s the ^{90}Sr concentration in seawater began to decrease sharply. Its content changed correspondingly from 21 to 2 and from 12 to 1 Bq / m^3. The decrease in ^{90}Sr level occurred as a result of the cessation of atmospheric nuclear tests. The ^{90}Sr concentrations in the Barents and White Seas decreased 3-5 times during the 29-year period of observations (1963-1992) (Izrael 1998).

At the same time these levels remained almost an order of magnitude higher than the ^{90}Sr concentration in the North Atlantic waters which was 2-4 Bq / m^3 (Nies et al. 1998). The Atlantic received this nuclide only from global atmospheric fallout.

$^{239,240}Pu$ levels in the waters of the Kara, Barents, Norwegian and Greenland Seas in 1970-1990s varied from 2 to 18 mBq / m^3. In the 1980s the concentrations were significant only around the British Islands - from 30 to 260 mBq / m^3. The waters of the Kola Bay near Atomflot contained 18-70 mBq / m^3 of $^{239,240}Pu$ in 1998.

The increased concentration of artificial radionuclides in the sea undoubtedly is caused in the first place by proximity to the release points of technological radionuclides and their transfer by major currents. It is necessary also to take into account the noticeable influence of salinity on the concentration of radionuclides.

For example, in the Norwegian Coastal Current having a salinity less than 33 ‰ and a ^{137}Cs concentration from 50 to 62 mBq / l.

In the surface waters of the open central part of the Norwegian-Greenland Basin (between 0.5°E and 10°W) with salinity more than 35 ‰ the quantity of radiocesium was many times lower (3-7 mBq / l). The same condition is observed in 100 m depth waters (Kautsky 1986). The 90Sr level also decreases sharply (from 10-20 to 2-4 mBq / l) from the coastal waters of Norway to the central part of the basin.

Thus, the levels of artificial radionuclides in the ecosystem of the Polar basin were and are an order of magnitude higher than those in the surface waters of the north Atlantic and north Pacific. In 1980-1990s the waters of the Arctic Ocean contained ^{137}Cs (1-15 Bq / m^3) and other radionuclides at levels that were several times higher than those of the North Atlantic.

The above mentioned levels of artificial radionuclides are characterized as rather low and practically safe for the habitation and development of pelagic organisms. They are minimal also in comparison with the standards for drinking water accepted by western radiation authorities.

The level and accumulation rate of radionuclides in phyto- and zoo- plankton organisms depends on the concentration of radionuclides in the environment, i.e. in the surface sea layer. In particular, the contamination level of the Baltic Sea increased significantly after June 1986 as a result of Chernobyl fallout. The 134Cs and 137Cs content in water and plankton of the Finnish Gulf reached correspondingly 6 and 1 thousand Bq / m^3, 1.5 and 2.5 thousand Bq / kg. In the middle of the 1990s the ^{137}Cs levels in the Baltic seawater and plankton returned to their pre-Chernobyl levels.

During periods of maximum discharges from the Sellafield facility, when the ^{137}Cs concentration in the coastal waters of Scotland increased to 200-400 Bq / m^3, the planktonic crustaceans likely accumulate to 70-150 Bq / kg of this nuclide. The Barents Sea plankton likely contain 5-50 Bq / kg of ^{137}Cs in the beginning of the 1960s and approximately 2-10 Bq / kg in the 1990s. The trends of artificial radionuclides from the beginning of nuclear weapons testing until the present time, can be used to forecast the ecological situation in the Russian Seas.

During the period 1960-1990s, the uniform (almost by an order of magnitude) decrease of the contamination level in the Arctic occurred. This was the result of the natural decay of cesium and strontium, reduction of discharges of technologically enhanced radionuclides and self-cleaning of sea basins. At the same time the phenomena of differentiation and sedimentation of radioactive suspension on the seabed and of the accumulation of these particles in organs and tissues of sea animals occurred.

Taking into account that the maximum quantity of the radionuclides was contained in seawater during the 1960s and 1980s, one can expect that precisely these stages must be reflected in the bottom sediments by peaks in radionuclide levels in sediment deposits. Also it is necessary to take into account that the life cycle of many mollusks, bottom fish, seals and whales is from 10 to 30 years. So some tissues and organs of the oldest age groups can reflect the periods of maximum concentrations of radionuclides in the seawater.

Table 3.5. ^{137}Cs in surface waters, Bq / m^3

Region	1960–1964	1965–1969	1970–1974	1975–1979	1980–1984	1985–1989	1990–1994
Azov Sea	–	–	–	–	–	70–250	8–11
Black Sea	–	–	20	16	–	720–840	40
Baltic Sea	–	–	–	20–25	15–20	100–180*	100–150
Northern Sea	–	–	–	60–400	80–220	10–150	6–20
Irish Sea	–	–	–	–	–	40–120	–
Irish Sea, to the west of Scotland	–	–	–	20–130	100–190	15–60	–
Norwegian Sea	–	–	–	20–70	15–100	10–50	3–28
Barents Sea :							
west	–	–	5–10	5–30	10–75	5–90	–
east	–	–	–	–	14–33	–	–
Kola Bay	–	–	–	–	–	–	–
White Sea	–	–	–	–	13–22	–	11–15
Kara Sea	–	–	–	–	3–27	–	2–12**
Laptev Sea	–	–	–	–	–	–	1–13
Arctic Ocean	–	–	–	–	–	1–15	1–15
North Atlantic	–	–	–	–	3–6	–	2–4

*Bothnia and Finnish bays – 500–650 Bq/m^3.
**Bays to 30–210 Bq/m^3.

Table 3.6. ^{90}Sr in surface waters, Bq / m^3

Region	1960–1964	1965–1969	1970–1974	1975–1979	1980–1984	1985–1989	1990–1994	1995–1999
Azov Sea	–	–	–	–	11	30–230	20–30	–
Baltic Sea	–	–	–	–	10–16	15–40	–	–
North Sea	–	–	–	–	6–22	2–20	–	–
Irish Sea, to the west of Scotland	–	–	–	–	–	–	–	–
Norwegian Sea	–	–	–	–	2–20	–	–	–
Barents Sea:								
west	20–23	10–20	8–12	10–14	9–14	7–13	4–7	–
east	–	–	–	–	4–14	–	3–5*	–
Kola Bay	20–54	–	9–15	–	8–12	–	5–9	–
White Sea	48	–	–	–	8–12	8	–	–
Kara Sea	18–84	–	9–28	–	7–21	–	3–12	1–7
Laptev Sea	52	–	–	–	–	–	4–7	–
Arctic Ocean	–	–	–	–	–	6–8	–	–
North Atlantic	–	–	–	–	–	–	1–2	–

*Chernaya Bay – 26–140 Bk/m^3

4 The benthic zone

4.1 Patterns of radionuclide distribution

The ecosystem of the benthos, which is situated on the continental shelf, slope and abyssal plains, consists of bottom sediments, zoobenthos, bottom fish fauna, bottom waters and porewater in sediments. The process of accumulation of suspended and dissolved radionuclides on the shelf, continental slope and further on the Ocean bottom is determined by a number of natural factors. The dependence of radioactive pollution of the sea bottom ecosystem on the flux of pollutants from the coastal and pelagic zones is quite evident. The northern seas with shelf depths to 300-600 m are a large reservoir to which radioactive material of different origin is delivered by numerous pathways.

4.1.1 Barents Sea

The composition and distribution of artificial radionuclides in surface bottom sediments of the Barents, Kara and other northern seas became the focus of investigations mainly during the 1990s (Gurevich 1992; Matishov et al. 1992b,c, 1994a,b,c, 1995b, 1996f, 1998b,c; Saetre et al. 1992; Matishov D. end Rissanen, 1993; Matishov D. end Szczypa 1993; Matishov D. 1993a; Foyn end Semenov 1993; Strand et al. 1993; Strand end Nielsen 1995; Szczypa 1997; Kuznetsov et al. 1997b; Strand 1998). It was established that the migration and accumulation of radionuclides is determined both by the chemical character of the radionuclides and by the properties of unconsolidated marine sediments (Fig. 4.1). All bottom or suspended sediment samples consist of particles of different size (from 1.0 to 0.001 mm and less). Several main types of bottom sediments, depending on bottom relief and sedimentation regime, are distinguished on the Arctic shelves (Klenova 1960; Belov end Lapina 1961; Lisitsyn 1978, 1994; Matishov 1984). We shall consider below some aspects of the topography and geomorphology of Arctic shelves using the Barents Sea as an example. The waters of the Barents Sea traverse a seabed that is highly irregular. The mean water depth of the shelf is 250 m; maximum depths reach 400-500 m (Matishov et al. 1992a). The outer edge of the shelf on the north and west of the Barents Sea exhibits depths from 200-350 m along the banks and 400-550 m along the troughs. The shallow shelf areas are situated mainly in the south-eastern part of the Sea, on the narrow coastal strip of the Novaya Zemlya and Spitsbergen archipelagoes and also on the Bear-Hope ridge. In other areas small areas on the bottom and tops of some banks and ridges are have depths less than 80-100 m. The deep troughs and vast banks control to a great extent the vertical and horizontal water circulation.

Along the coastal shelf of the Northern Norway, Spitsbergen, Novaya Zemlya, Franz-Josef Land the characteristic edge troughs are situated. There is a system of narrow (3-6 nautical miles) and relatively shallow (100-200 m) successively situated valleys and wide (20-40 miles) hollows. In the open part of the sea deep

(300-500 m) Bear, Finmarken and Zudcape Troughs which are several hundred kilometers long and 100-200 km wide, are distinguished.

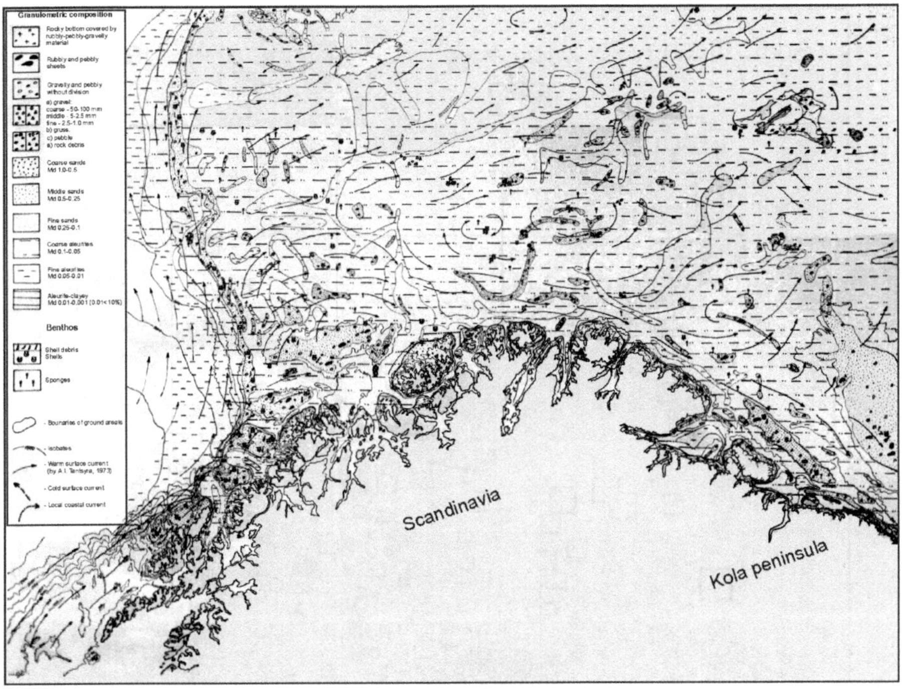

Fig. 4.1. Bottom sediment characteristics in the southwest Barents Sea

In the beginning of the 1980s the ^{137}Cs concentration in the Barents Sea bottom sediments was from 10 to 30 Bq / kg. The West European nuclear reprocessing facilities were the main contributors (Vakulovsky et al. 1985, 1988). In the 1990s the bottom sediments in almost all areas were characterized by low (background) levels of artificial radionuclides (Matishov et al. 1994a,c, 1998b,c, 1999d; Strand et al. 1994; Foyn end Swaren 1995; Strand 1998; Nies et al. 1998; Ivanov 1999). The activity of all artificial radionuclides varied on average from 0.5 to 6 Bq / kg.

^{137}Cs is present everywhere in shelf sediments (Fig. 4.2-4.4). The content of this element in the sediments of the Kola Peninsula and Franz-Josef Land coastal zones is from 0.2-1.0 to 9 Bq / kg. In these areas ^{137}Cs levels in coastal waters range from 4 to 8 Bq / m^3.

The sands and gravelly-pebbled sediments in the shallow areas around the Kanin Peninsula also have insignificant (0.2-0.6 Bq / kg) quantities of this nuclide. From the coastal shelf to the west from Novaya Zemlya and Spitsbergen, the 137Cs content in samples was not more than 3 Bq / kg. An exception is the aleurite and clayey iceberg sediments of Kongs, Horsunn, Nordensheld and other bays. Here the upper reaches of the bays are occupied by discharge glaciers and radioceasium concentrations are 10-17 Bq/kg.

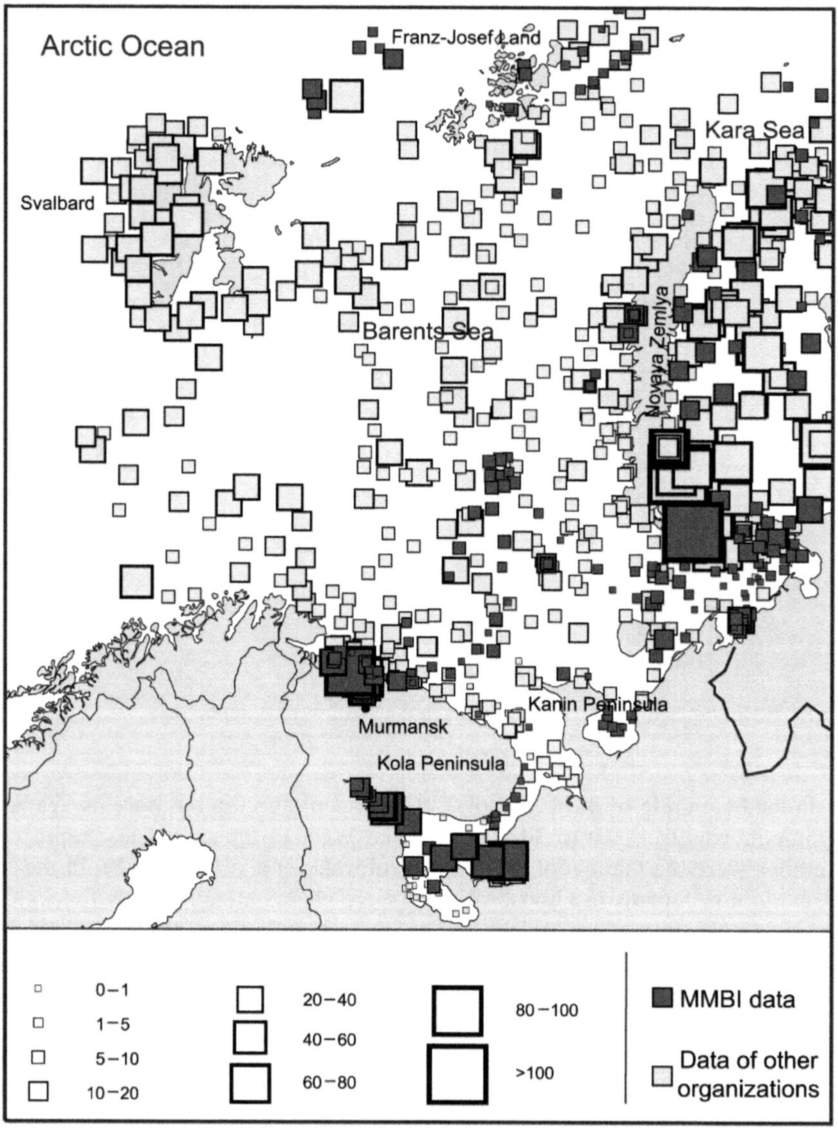

Fig. 4.2. ^{137}Cs content in the surface layer of the Barents and White Seas bottom sediments in 1990s

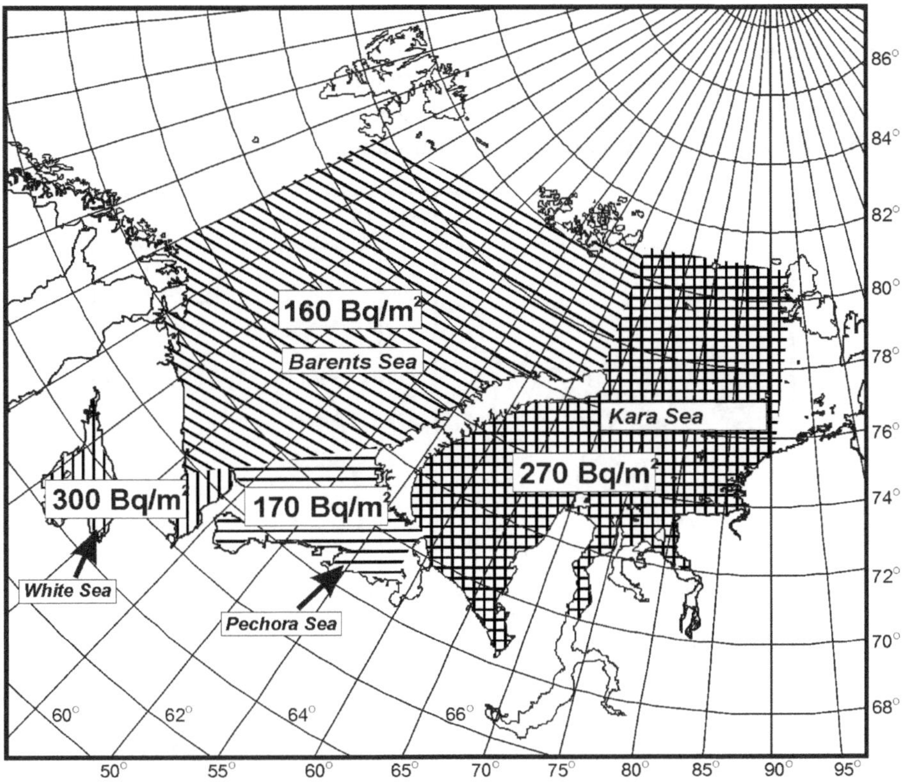

Fig. 4.3. Maximum ^{137}Cs concentrations in bottom sediments of the Barents, White and Kara Seas

^{137}Cs accumulation in the open central and western parts of the Barents Sea varies depending on bottom topography, hydrodynamic regime and geographic conditions. The ridges and banks are usually covered by aleurite and fine sands containing low concentrations of radionuclides. On the large Myrmanskaya, Bear, North Kanin, Gusinaya and other banks where depths range from 50-160 m, ^{137}Cs is present in sediments from 0.2 to 4 Bq / kg. Overall it should be noted that the bank slopes which are exposed to waters of Atlantic origin are contaminated somewhat more.

Higher ^{137}Cs accumulation (5-11 Bq / kg) is typical for all shelf troughs and hollows (Fig. 4.2, 4.3). In particular, the clayey sediments, covering the bottom of the Central Hollow (depth 300-380 m), contains from 5 to 9 Bq / kg of ^{137}Cs. This vast hollow is situated approximately at a distance of 250 km from the Novaya Zemlya nuclear test sites.

The investigation of bottom sediment radioactivity in the region situated to the north-west of the test site is only slightly contaminated. Releases of liquid radioactive wastes were carried out in this area (Matishov et al. 1992c, 1994a, b; Saetre et al. 1992). Here the ^{137}Cs content in sediments varies from 15 to 160 Bq / m^3 reflecting the global fallout levels of this isotope. It can be noted for comparison that on the shelf to the south from Novaya Zemlya the level reaches 910 Bq / m^3 (Strand et al. 1994). The Pechora Sea with depths increasing northward to 70-120 m has its own specific features of ^{137}Cs distribution.

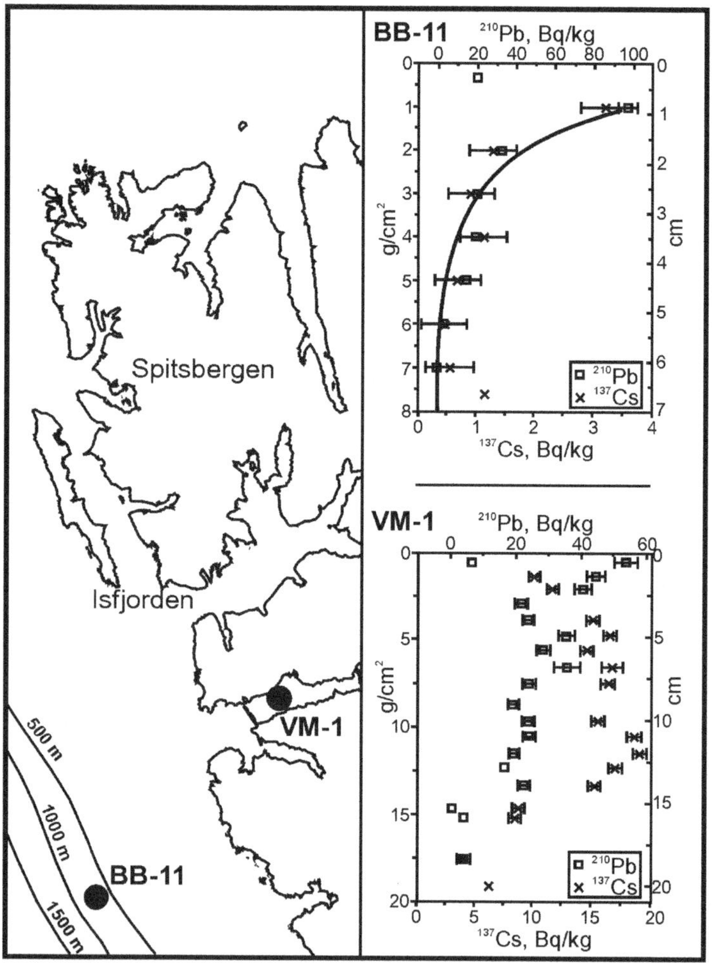

Fig. 4.4. Vertical distribution of ^{137}Cs and ^{210}Pb in bottom sediments of the Spitsbergen shelf

Chernaya Bay serves as a vivid example of a local shelf area where sedimentation process play a dominant role in the migration of ^{137}Cs in seawater. This is the location of the first Russian underwater, atmospheric and underground nuclear tests. The radioactive products of explosions together with particles of materials from the remnants of bombs and disintegrated bottom sediments descended quickly to the seabed in a limited area. One can see the impact from the radionuclide compostion and concentration and spectrum in bottom sediment collected from this area (Fig. 4.5).

During the 1990s the concentrations both of ^{137}Cs and $^{239,240}Pu$ in surface sediments of Chernaya Bay reflected a uniform trend of decreasing with distance from the source. The maximum ^{137}Cs content (44 Bq / kg) in bottom sediments was found in the deep trough near the Novaya Zemlya southern coast, and the minimum (6 Bq / kg) in the Vajgatch Island area. The clayey sediments of the South Novaya Zemlya Trough with depths 170-220 m contain on average from 18 to 27 Bq / kg of ^{137}Cs (Matishov D. 1993a; Smith et al. 1995a,b). The migration of this isotope to 30 cm deep into silts is noted in sediment columns taken from this trough.

In sandy sediments from shallow areas of the Pechora Sea the ^{137}Cs concentration varies from 0.6 to 6.5 Bq / kg. Coarse sands in the Gulyaevsky sandbanks area have very low ^{137}Cs content (0.3-0.7 Bq / kg). The sands in the Pomorsky Trough between The Kolguev Island and Timan Peninsula have from 1 to 3 Bq / kg of radiocesium. Clayey sediments in Pechora Sea bottom depressions contain noticeably more ^{137}Cs (5-15 Bq / kg). The content of ^{137}Cs is also relatively high in the silts of the Pechora Bay (8-16 Bq / kg) (Matishov D. 1993a; Ivanov 1999).

The somewhat increased radiocesium accumulation in the aleurite sediment fraction from shelf depressions located to the east of Kolguev Island (10-15 Bq / kg) and to the west of Vajgach Island ((7-9 Bq / kg) probably are the result of suspended sediment transport from the White Sea and Pechora River respectively. The sedimentary ^{137}Cs activity is significantly correlated with clayey silts (size fraction <0.001 mm) and this material is typically enriched in organic carbon.

Considering the Barents Sea, only small differences are observed in the ^{137}Cs concentrations of bottom sediments. In high salinity waters, ions of Cs^+ tend to remain in the dissolved phase and are weakly sorbed by suspended particulate matter and bottom sediments (Gromov end Spitsyn 1975). The existence of cesium in aqueous solution, mainly in the ionic state, means that hydrologic factors determine the migration pathways for radiocesium in throughout the sea. As a result, small cesium concentrations are characteristic for the majority of the Barents Sea.

^{90}Sr in bottom sediments is investigated only in several areas (Matishov et al. 1994a,b,c, 1996f; Rissanen et al. 1995). Based on analysis of more than forty bottom sediment samples from the Barents shelf, it is clear that the ^{90}Sr concentrations vary from 0.11 to 0.74 Bq / kg (Fig. 4.6). The central and northeastern portions of the Barents Sea are covered by aleurite and fine sands which contain very small amounts of artificial radionuclides. The sediments of the large Murmansk Bank in water depths of 120-180 m contain 0.11-0.43 Bq / kg of ^{90}Sr. Evidently, sediments containing radionuclides are winnowed from higher elevations of the sea bottom and subsequently are transported to shelf depressions. In particular, the vast and deep (300-380) shelf depression in the Barents Sea has somewhat higher levels of sedimentary ^{90}Sr, up to 0.8 Bq / kg.

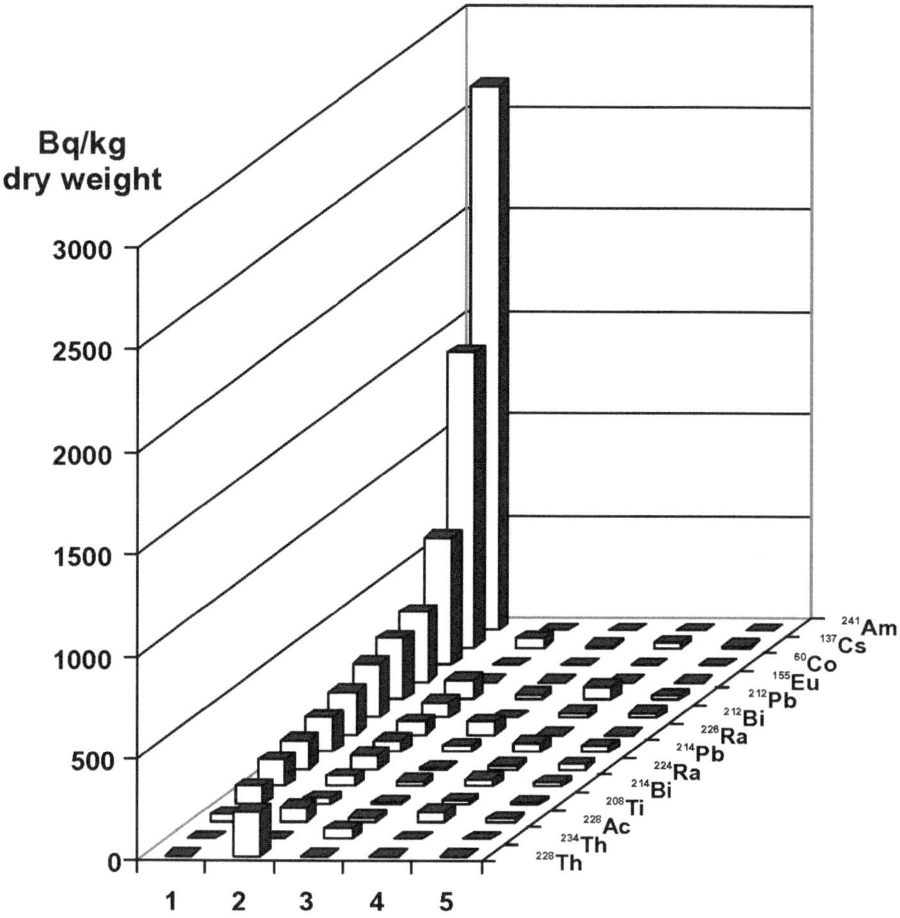

Fig. 4.5. Distribution of radionuclides in surface bottom sediments of the southeastern Barents Sea 1992

The ^{90}Sr content in aleurite sediments found within the coastal zone and to the north from the Kildin Island (depths 150 m) is about 0.4 Bq / kg. Similar concentrations of ^{90}Sr (0.1-0.5 Bq / kg) are present in sediments from along the continental slope around the sunken submarine «Komsomolets» (Table 4.1) (Kuznetsov et al. 1993).

^{239,240}Pu is present everywhere in the Barents shelf sediments. Concentrations vary over a wide range (from 0.1 to 20 Bq / kg) (Fig. 4.7). On an areal basis, ^{239,240}Pu concentrations vary mainly in the range 2.5-25 Bq / m^2 (Matishov et al. 1997c,f; Strand 1998). The uneven distribution of this radionuclide is a result of differences in sediment type, geographic position and proximity to the Novaya Zemlya nuclear test sites. Within the Pechora Sea, the quantity of ^{239,240}Pu in sediments increases to more than 100 Bq / m^2 towards the local source of Chernaya Bay where underwater nuclear explosions were carried out by the Former Soviet Union (Tables 4.2, 4.3).

Table 4.1. ^{90}Sr content in bottom sediments in the vicinity of the sunken atomic submarine «Komsomolets»

Station	Layer, cm	^{90}Sr, Bk/kg
2508	0–2	<0.06
2519		2.6±0.5
2525		<0.06
2542		0.5±0.1
M2-3-11/1	0–5	0.5±0.1
M2-3-11/2	5–10	0.5±0.1
M2-3-11/3	10–15	0.19±0.05
M2-3-11/4	15–20	0.18±0.05

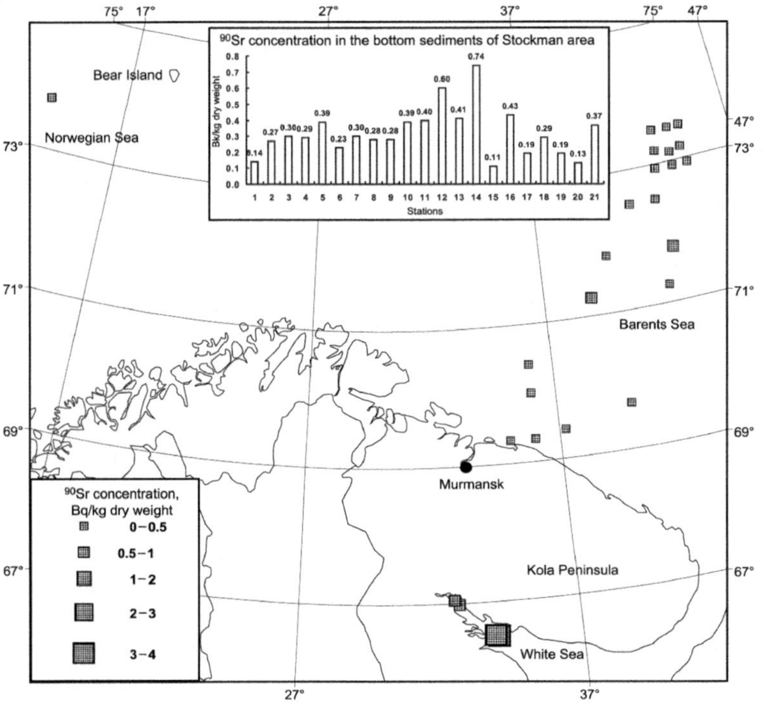

Fig. 4.6. ^{90}Sr content in the surface layer of bottom sediments in the Barents and White Seas. The inset graph displays variations of ^{90}Sr content in sediments depending on the bottom topography in the area of the Stockman gas-condensate complex

Table 4.2. Geochemical data and radionuclide concentrations in the surface layer (0-2 cm) of the Pechora Sea bottom sediments (Smith et al. 1995a), Bq / kg dry weight

Distance from the Chernaya Bay, km	TOC, ppt, convertion	Error, %	^{210}Pb	^{137}Cs	$^{239,240}Pu$	^{238}Pu:$^{239,240}Pu$
0	16.5	79	152.0	158.0	8.47	0.024
21	19.2	84	220.0	18.2	14.20	0.032
36	20.0	91	247.0	27.4	13.20	0.025
47	17.0	86	173.0	14.4	4.08	0.035
54	20.1	81	29.0	18.8	5.20	0.045
72	1.9	12	37.7	1.68	0.45	0.048
100	4.5	26	51.2	3.3	0.75	0.044
106	7.8	40	76.7	5.43	1.65	0.020
125	3.8	9	34.2	3.57	0.32	0.053
161	8.1	57	93.2	7.83	1.13	0.010
163	0.5	2	18.5	0.23	0.07	0.175
165	0.5	2	16.8	0.38	0.08	0.140
169	6.1	32	47.5	10.8	0.65	0.026
190	1.0	5	14.2	0.70	0.33	0.050
194	9.6	75	37.2	11.6	0.30	0.028
250	6.7	51	68.3	9.18	0.77	0.022

When the specific features of the $^{239,240}Pu$ distributions are analyzed, it should be noted that the aleurite sediments on the coastal shelf of the Kola Peninsula and around Franz-Josef Land and Spitsbergen contain 0.3-1.7 Bq / kg of plutonium. The silts on the bottom of the central depression of the Barents Sea have accumulated from 0.9 to 3.2 Bq / kg of $^{239,240}Pu$. The $^{239,240}Pu$ concentration in the sands and coarse aleurites on the Gusinaya Bank and shallow areas (water depths of 70-90 m) in the Pechora Sea are very small (0.1-1.0 Bq / kg). In some bottom depressions and troughs the content of this radionuclide increases to 3.2-4.3 Bq / kg.

The most significant accumulations of $^{239,240}Pu$ are characteristic for the bottom sediments of the Novaya Zemlya shelf where the distribution of this radionuclide displays a mosaic image. $^{239,240}Pu$ concentrations on the order of 0-3 Bq / kg occur in coastal fine-grained sediments as a result of local atmospheric fallout from nuclear tests. The natural transport of contaminated sediments from Chernaya Bay has resulted in increased $^{239,240}Pu$ levels in the south-eastern part of the Barents Sea. In the Novaya Zemlya depression, $^{239,240}Pu$ concentrations reach 13 Bq / kg. The relatively high plutonium activity in samples from all shelf stations located to the south of Chernaya Bay shows that the surface sediments were formed under the influence of $^{239,240}Pu$ influx during and after underwater nuclear tests (Matishov D. 1993b; Smith et al. 1995a,b).

The ratio between ^{238}Pu and $^{239,240}Pu$ activities in sediments from throughout the Barents Sea irange from 0.02-0.05. Such values reflect global fallout from the atmosphere. In the Pechora Sea, in particular to the south from Novaya Zemlya, the ratio ^{238}Pu : $^{239,240}Pu$ in bottom sediments increases to 0.06-0.08 (Smith et al. 1995a,b).

Table 4.3. Concentration of ^{238}Pu and $^{239,240}Pu$ in the surface sediment layer (0-2 cm) of the Barents Sea (Rissanen et al. 1997a), Bq / kg dry weight

Station	Dry substance content, %	Loss when drying the samples, %	^{238}Pu	$^{239,240}Pu$
1	47	7.5	0.037	0.63
2	40	9.6	0.062	1.70
3	40	8.9	0.074	2.00
4	46	6.3	0.050	1.20
5	40	9.4	0.073	1.90
6	42	7.5	0.021	0.87
7	43	8.0	0.053	1.60
8	48	7.2	0.025	1.10
9	42	8.8	0.059	1.60
10	48	6.0	0.033	0.88
11	41	9.5	0.095	1.90
12	41	9.1	0.140	3.20
13	38	10.0	0.110	2.40
14	47	6.5	0.076	1.60
15	73	0.7	<0.020	0.59
17	66	7.8	0.047	1.00
18	70	1.8	0.049	0.94
19	67	2.0	0.052	1.10
20a	64	1.9	0.078	0.95
21	79	1.4	<0.030	0.50
22	76	0.5	<0.020	0.55

It should be noted that bottom sediments in the White Sea have accumulated ^{137}Cs in the range from 1 to 60 Bq / kg (Matishov D. et al. 1995a). Similarly high accumulation levels were not determined in shelf sediments of the Barents Sea. As a result, bottom fauna of the Kandalaksha Bay are thus subjected to a radiation exposure that is 5-6 times higher than for benthic organisms in the Gusiny, Bear and other troughs of the Barents Sea.

During the1990s, waters of the White Sea depressions contained relatively high salinity concentrations (28.0-29.7 ‰) as well as maximum concentrations of ^{137}Cs (16-22 Bq / m^3). We suppose that the global fallout of radioactive substances are the main source of ^{137}Cs received by the White Sea. The radionuclide migration from land to sea occurs mainly through the river system of the Severnaya Dvina. However, in the regional aspect, judging by the presence of ^{134}Cs and ^{60}Co in bottom sediments, there are evidently local sources of contamination that may play a very important role in some areas, for example the Kandalaksha Bay.

From materials collected during a number of MMBI surveys, the ^{137}Cs content in Barents Sea bottom sediments, where the main stream of Atlantic waters flows, was 8.5 Bq / kg (0-2 cm layer). In a region to the south of Franz-Josef Land (water depth 345 m), the cesium concentration in sediments was near to zero. This region is not subject to the influence of Atlantic waters that have traversed the shallow, sediment-rich Barents Sea.

Overall the ^{137}Cs distribution in relatively deep shelf areas (300-400 m) having complex bottom relief depends not only on the lythology of bottom sediments but also on the proximity of sampling sites to major current pathways and nuclear test sites (or nuclear waste burial sites). The ^{137}Cs and $^{239,240}Pu$ contents in the sediments of the south-eastern part of the Barents Sea was noticeably influenced by the proximity to the southern Novaya Zemlya nuclear test site. This is clear from the increased ^{137}Cs content in the sediments of the deep channel along the souther edge of Novaya Zemlya and in the Pechora Sea.

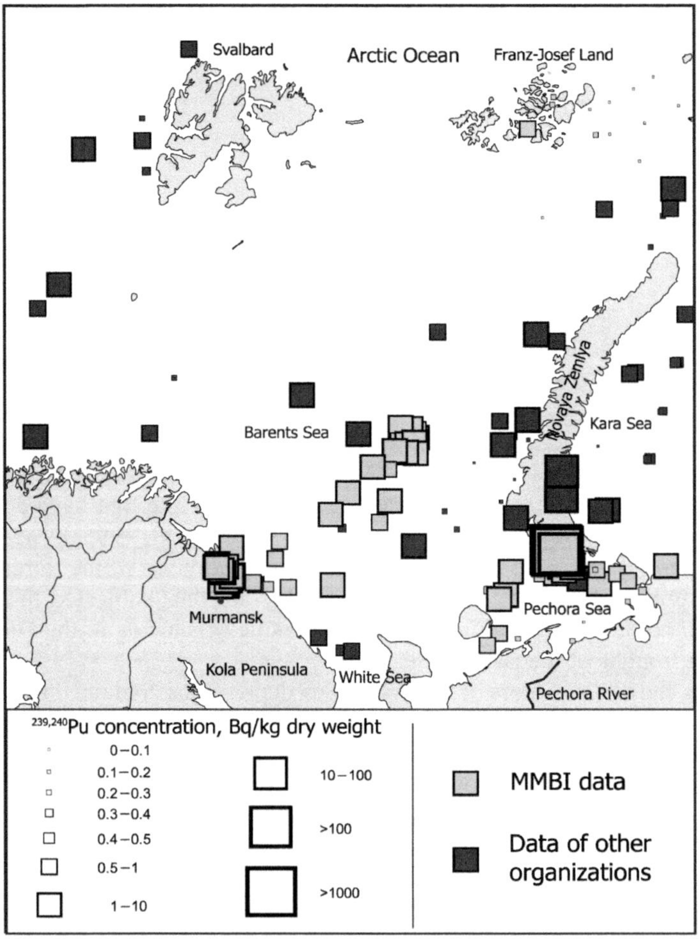

Fig. 4.7. $^{239,240}Pu$ content in the surface layer of the Barents Sea bottom sediments in 1990s

4.1.2 West European and Black Seas

Important results were achieved through the study of radioecological processes during the 1980-1990s following the contamination of seas situated near the region of the Chernobyl accident and the West European nuclear reprocessing facilities. It must be noted that the greatest absorption of artificial radionuclides in pelagic water masses takes place mainly during the productivity period, when radionuclide fluxes from the water column to the benthic zone result in the highest deposition of radionuclides in bottom sediments.

The concentrations of radionuclides in the water and surface sediments of the Irish Sea shelf during the 1960-1990s have always been highest near the Sellafield facility (Huntend Kershaw 1990; Kershaw 1992). While ^{137}Cs concentrations in surface waters reached 200-210 Bq / m^3, in bottom waters the concentrations reached only 13 Bq / m^3. In 1993-1994, bottom sediments near Sellafield contained about 400 Bq / kg of ^{137}Cs and 250 Bq / kg of ^{90}Sr. Contrary to Chernaya Bay, the aquatic system adjacent to Sellafield is characterized by the constant presence of ^{134}Cs. The ^{134}Cs:^{137}Cs ratio in the surface sediments is 0.33.

The important specific feature of the contamination of the newest sedimentary strata of the Irish Sea is the saturation by ^{137}Cs, ^{90}Sr, $^{239,240}Pu$ and ^{241}Am to concentrations of 5500, 2000, 1700 and 1600 Bq / kg respectively within a sediment depth interval of 14-18 cm (Fig. 4.8). These extreme concentrations of anthropogenic radionuclides reflect the period of maximum discharges from Sellafield (Kershaw, 1992). At present the strata of «dirty» silts is source of radionuclides desorbed from sediments to the bottom waters. The mobile ions of ^{90}Sr, $^{239,240}Pu$ and other radionuclides migrate during the remobilization process across the Irish Sea boundaries.

The radionuclide content in the North, Norwegian and Greenland Seas in 1992-1995 was relatively small. ^{137}Cs levels in North Sea sediments varied from 4 to 14 Bq / kg (Strand 1998). Sediments within the 10-12 cm layer contained up to 31 Bq / kg of ^{137}Cs. In the Norwegian Sea, bottom sediments accumulated 25-45 Bq / kg of ^{137}Cs. Sediments on the shelf and continental slope of the Greenland Sea accumulated 2-9 Bq / kg of ^{137}Cs. This radioactive contamination of bottom sediments in the various northern seas is partially related to Sellafield because the transport pathway from sources in the Irish and Baltic Seas follow athe underwater continental boundary (Hunt, Kershaw 1990; Nies et al. 1998).

The ^{137}Cs content in the surface aleurite sediments on the continental slope near the sunken atomic submarine «Komsomolets», varies from 4 to 16 Bq / kg. Bottom waters contain concentrations in a range of 0.5-3.0 Bq / m^3. The short-lived radionuclide ^{134}Cs is absent from sediments near the sunken submarine and also from sediments on the Barents Sea shelf.

Surface and bottom waters from the continental slope (in the area of the submarine «Komsomolets», water depths 1200-1800 m) contain extremely low concentrations of $^{239,240}Pu$ - 0.01-0.05 Bq / m^3 (Table 4.4) (Kuznetsov et al. 1993). More recently deposited bottom sediments have also accumulated low $^{239,240}Pu$ concentrations (0.1-1.65 Bq/kg). However there are areas where plutonium is distributed in sediments to depths of 10-15 cm and concentrations of about 1 Bq / kg. It is highly unlikely that the migration of these artificial radionuclides occurred during the last 3-4 years.

In areas free of sunken nuclear objects, in particular the North and Norwegian Seas, ^{239,240}Pu concentrations are noticeably lower. For example, the ^{239,240}Pu level in coastal waters and bottom sediments of Denmark and Norway vary correspondingly within the limits 9-15 mBq / m^3 and 0.07-2.8 mBq / kg. Consequently, local releases of radionuclides near the «Komsomolets» submarine accident have been minimal.

A wide spectrum of radionuclides is observed in the closed Baltic Sea. Radionuclide concentrations increased sharply during the summer of 1986 as a result of the Chernobyl accident (Ilus et al. 1993; Ikaheimonen et al. 1995) (Tables 4.5, 4.6; Fig. 4.9). ^{137}Cs distributions mirror the bottom geomorphology and penetrate up to depths of 30 cm in bottom sediments. In 1982, prior to the Chernobyl accident, marine sediments contained ^{137}Cs, ^{90}Sr and ^{239,240}Pu in quantities of 20-120, 1-3 and 1-4 Bq/kg respectively (Study ... 1983).

Of course, very high ^{137}Cs concentrations in both seawater and sediments are connected primarily to the radioactive Chernobyl plume that covered several times the area of the Baltic catchment area. At the same time, the Baltic Sea is characterized by high river runoff (472 km^3 of fresh water annually) and impeded water exchange with the open Atlantic Ocean. As a result, the Baltic contains low salinity and high ^{137}Cs concentrations (to 10-21 mg / l) in the sea surface layer (Emelyanov et al. 1999).

After the accident, the maximum ^{137}Cs concentrations were observed in the shallow (50-120 m) and freshened (1-7 ‰) Finnish and Bothnia Bays. In 1986-1987 the ^{137}Cs concentration in aleurite sediments varied from 60 to 1400 Bq / kg with some locations exhibiting a concentration range of 200-3600 Bq / kg. The value of contamination per unit of area also was rather high: from 200-800 to 2000-6000 Bq / m^2 and in some areas between 14000-22000 Bq / m^2. By the beginning of the 1990s, ^{137}Cs levels had decreased significantly, ranging from 50 to 470 Bq / kg (Ilus et al. 1993).

In the open part of the Baltic Sea, where the salinity of water masses is 8-13 ‰, sediment ^{137}Cs concentrations were considerably lower. The ^{137}Cs content in the fine aleurites and clayey silts of the shelf depressions (150-250 m) during 1986-1987 varied from 50 to 400 Bq / kg; the corresponding concentrations per unit of area were 100-1000 Bq / m^2.

The accumulation of ^{239,240}Pu during 1986-1990 provides an entirely different picture. The level of this radionuclide in bottom sediments of the Finnish, Bothnia and other bays varied from 2 to 5.5 Bq / kg (30-170 Bq / m^2) (Ilus end Ikaheimonen 1991; Ilus et al. 1993; Rissanen et al. 1997a). Clayey sediments in the open part of the Baltic Sea contained still less ^{239,240}Pu (2-4 Bq / kg and 9-40 Bq / m^2). The ratio of ^{238}Pu : ^{239,240}Pu was typical of the global fallout signal (0.031-0.047).

Table 4.4. Plutonium in bottom sediments at the site of the sunken submarine «Komsomolets»

Station	Layer, cm	$^{239,240}Pu$	
		Bk/kg	Bk/m^2
1	2	3	4
	1991		
M2-1-01/1	0–5	0.22±0.04	3.9
M2-1-02/1		0.13±0.03	2.2
M2-2-03/1		1.47±0.17	47.6
M2-2-04/1		1.21±0.15	27.6
M2-1-05/1		0.28±0.09	8.0
M2-2-06/1		1.65±0.24	53.3
M2-2-08/1		1.03±0.10	31.2
M2-2-08/2	5–10	0.98±0.15	30.8
M2-2-13/1	0–5	0.24±0.03	6.2
M1-2-11/1		0.97±0.14	35.7
M1-2-14/1		0.22±0.05	5.8
M1-2-15/1		0.72±0.11	16.9
M1-2-17/1		0.90±0.12	24.8
	1992		
M2-2-WH1/1	0–5	0.79±0.10	31.0
M2-2-WH1/2	5–10	1.08±0.20	
M2-2-WH1/3	10–15	0.55±0.08	
M2-2-WH2/1	0–5	0.74±0.13	33.7
M2-2-WH2/2	5–10	0.63±0.09	
M2-2-WH2/3	10–15	1.13±0.53	
M2-3-WH1/1	0–5	0.75±0.09	27.1
M2-3-WH1/2	5–10	0.52±0.09	
M2-3-WH1/3	10–15	0.02±0.02	
M2-3-WH2/1	0–5	0.82±0.14	27.0
M2-3-WH2/2	5–10	0.33±0.09	
M2-3-WH2/3	10–15	0.15±0.05	
M2-2-RI1/1	0–5	0.72±0.09	20.0
M2-2-RI1/2	5–10	0.56±0.10	
M2-2-RI1/3	10–15	0.21±0.02	
M2-2-RI2/1	0–5	0.41±0.06	19.0
M2-2-RI2/2	5–10	1.52±0.59	
M2-2RI2/3	10–15	0.06±0.01	
M2-3-RI1/1	0–5	0.91±0.14	20.1
M2-3-RI1/2	5–10	0.02±0.01	
M2-3-RI1/3	10–15	0.06±0.01	
M2-2-9/1	0–5	0.66±0.10	26.3
M2-11/1		0.60±0.07	16.1
M2-2-12/1		0.13±0.01	5.6
M2-2-12/2	5–10	0.04±0.01	
M2-2-12/3	10–15	0.14±0.02	
M2-2-15/1	0–5	0.06±0.01	2.5

Table 4.4. (Cont.)

1	2	3	4
M2-2-15/2	5–10	0.10±0.01	
M2-2-16/1	0–5	1.01±0.16	40.9
M2-2-16/2	5–10	0.14±0.02	
M2-2-16/3	10–15	0.04±0.01	
M2-2-17/1	0–5	0.12±0.02	3.7

Table 4.5. Radionuclide content in surface sediments of the Baltic Sea 1982 (Ilus end Ikaheimonen 1991), Bq / kg

Station	^{90}Sr	^{137}Cs	$^{239,240}Pu$
C-V1	2.3	24	3.1
EB-1	1.2	47	1.5
XV-1	3.0	120	6.8
LL-3f	1.1	38	2.3
Teili-1	1.8	46	4.0

The accumulation of ^{60}Co and ^{90}Sr after the Chernobyl accident was insignificant. However in the Finnish Bay, the ^{90}Sr level in sediments was up to 24 Bq / kg and of ^{60}Co up to 52 Bq / kg. It is entirely possible that some of this ^{60}Co originated from a local source.

An enormous quantity of ^{137}Cs, ^{134}Cs and other artificial radionuclides were received by the Baltic basin during the period of Chernobyl discharges as a result of both atmospheric and river fluxes. The ^{137}Cs level in surface seawater increased by an order of magnitdue from 50 to 600 Bq / m^3 beginning during the summer period of 1986-1987. The ^{90}Sr and ^{60}Co content in the sea increased very little. For example, before the accident the ^{90}Sr level in seawater was 14-26 Bq / m^3 while after the accident the ^{90}Sr level was 14-35 Bq / m^3.

In the second half of the 1980s and during the 1990s Baltic Sea sediments were saturated by radiocesium. Soon after the accident the ^{137}Cs concentration in sediments reached 1-3.6 thousand Bq / kg. For comparison, bottom sediments in a similarly closed, but more saline system, the White Sea, contained ^{137}Cs concentrations from 1 to 67 Bq / kg. The role of Chernobyl discharges in the contamination of the White Sea basin was much less dramatic.

In the Black Sea, the accumulationof radioactive elements originating from Chernobyl fallout and transported by the Danube and Dnieper Rivers is quite evident (Fig. 4.10). ^{137}Cs in quantities from 20 to 40 Bq / kg was traced in the upper part of sediment columns (to 40-50 cm) on the north-western underwater edge of the Black Sea (Gulin et al. 1997). Several peaks of radiocesium concentration (of order 40-45 Bq / kg) are observed in these sediment profiles (Fig. 4.11).

Table 4.6. Bottom sediment distribution of g-emitting radionuclides in the Baltic Sea 1986 (Ilus end Ikaheimonen 1991), Bq / kg dry weight

Station	Date of sampling	Depth of sampling,m	Depth layer, cm	^{40}K	^{95}Zr	^{95}Nb	^{103}Ru	^{106}Ru	^{110}Ag	^{125}Sb	^{134}Cs	^{137}Cs	^{144}Ce
C VI	02.09	70	0–5	720	0	0	0	0	0	0	0	36.0	0
			5–10	830	0	0	0	0	0	0	0	4.6	0
EB 1	27.08	128	0–5	1000	45	610	140	180	12	17	220.0	500.0	41.0
			5–10	1000	37	0	0	0	0	0	25.0	110.0	0
			10–15	1100	0	0	0	0	0	0	0	14.0	0
Ga 63	26.10	139	0–5	1000	0	0	0	0	0	0	6.4	110.0	0
			5–10	1000	0	0	0	0	0	0	0	71.0	0
			10–15	1200	0	0	0	0	0	0	0	16.0	0
			15–20	1100	0	0	0	0	0	0	0	1.6	0
XV 1	28.10	55	0–5	830	0	0	0	16	0	5.50	30.0	170.0	9.6
			5–10	900	0	0	0	0	0	0.44	4.1	110.0	0
			10–15	900	0	0	0	0	0	0	0	110.0	0
			15–20	900	0	0	0	0	0	0	0	110.0	0
LL 3a	20.08	65	0–5	1000	0	0	0	0	0	0	0	67.0	0
			5–10	1000	0	0	0	0	0	0	2.9	18.0	0
F–8	14.06	84	0–1	–	–	–	–	–	–	–	400.0	890.0	–
			1–2	–	–	–	–	–	–	–	92.0	330.0	–
			2–3	–	–	–	–	–	–	–	66.0	550.0	–
			3–4	–	–	–	–	–	–	–	0	110.0	–
			4–5	–	–	–	–	–	–	–	0	50.0	–
			5–10	–	–	–	–	–	–	–	0	7.0	–

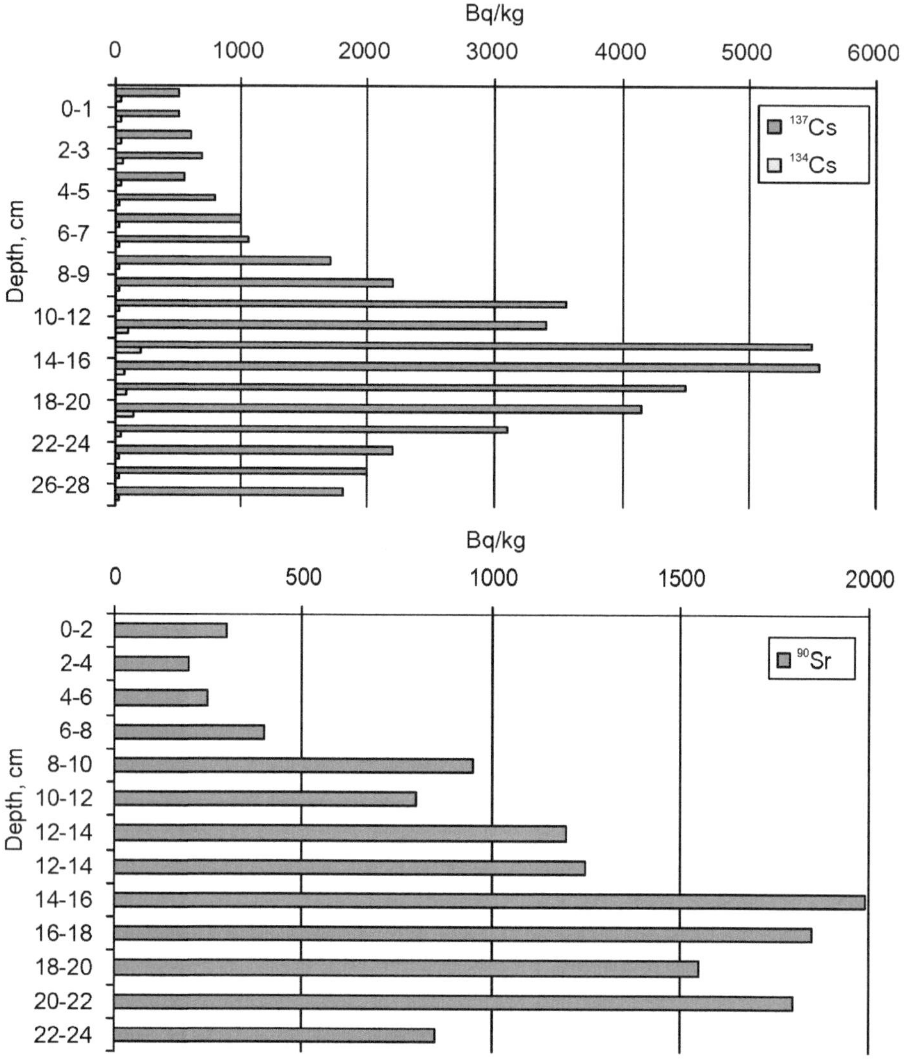

Fig. 4.8. Vertical distribution of ^{134}Cs, ^{137}Cs and ^{90}Sr in the Irish Sea sediments near Sellafield in 1993 and 1994 (Kershaw, Baxter et al., 1995)

Three phases of bottom sediment contamination are noted in the Black Sea. The first two correspond to periods of active nuclear weapons testing (1945-1959 and 1960-1963) and the third - to the accident at the Chernobyl power station in 1986. During the period of atmospheric nuclear testing, the maximum levels of ^{137}Cs in sediments from different parts of the Black Sea were identical. After the

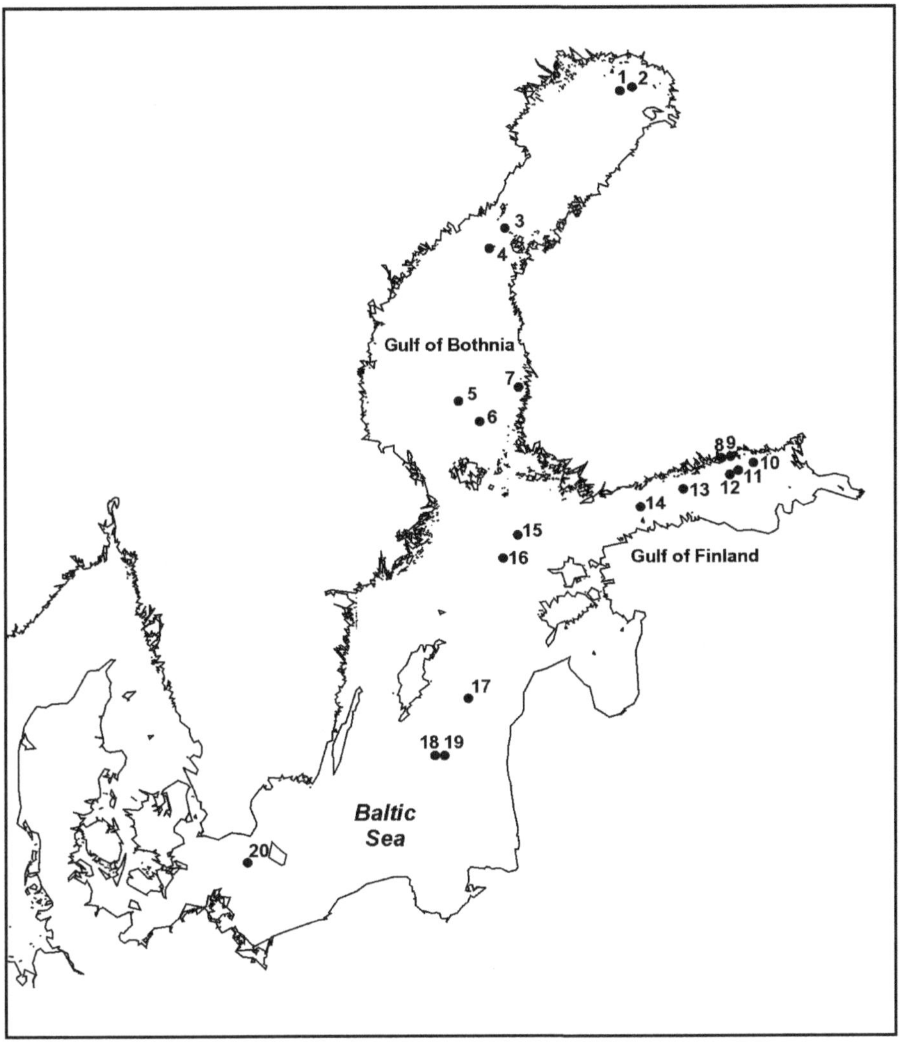

Fig. 4.9. Water, bottom sediments and biota sampling locations in the Baltic Sea (Ilus et al. 1993)

Chernobyl accident, the contamination of Danube River sediment far exceeded the contamination levels in sediments from the continental slope and deep part of the sea. This corresponds to the more uniform ^{137}Cs distribution over the sea surface in the beginning of 1960s as a result of the global character of radioactive fallout from nuclear testing. The input of radioactive substances via the Danube River after Chernobyl was much larger than the supply from the Dnieper because of the nonuniform atmospheric distribution of radionuclides throughout the Black Sea region (Gulin et al. 1997).

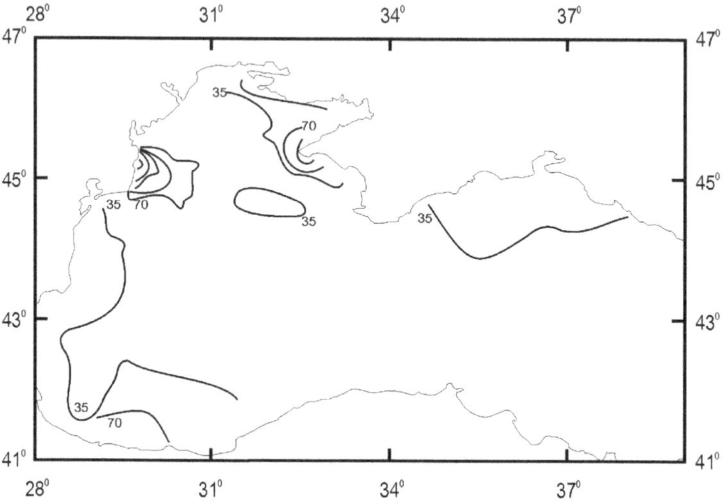

Fig. 4.10. Distribution of ^{137}Cs content in the upper 5 cm layer of the Black Sea bottom sediments averaged for the period 1986-1992 (radioactivity levels as referenced to May 1, 1986) (Gulin et al. 1997)

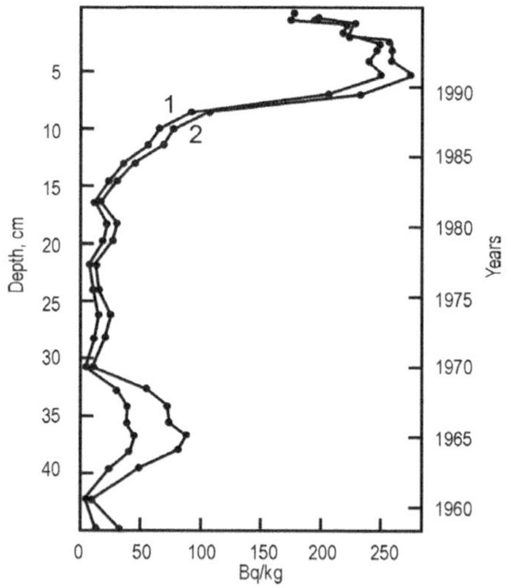

Fig. 4.11. Vertical distribution of the ^{137}Cs content in bottom sediments of the Danube region of the Black Sea, Station 5237: **1** -*July 1994*; **2** - *initial radioactivity, normalized to the age of strata* (Gulin et al. 1997)

The nonuniform distribution of artificial radionuclides in bottom sediments of the West European and Black Seas resulted from (i) differences in the transport pathways for various radioactive elements from their primary sources (e.g. atmospheric and sea transport), (ii) the degree of wash out from catchment areas, (iii) water exchange between neighbouring sea basins, and (iv) conditions of sedimentation on the shelf. It is evident that the large quantities of artificial radionuclides were added to the North Sea basin from the Baltic Sea after the Chernobyl accident. One cannot exclude therefore, that in the beginning of the 1990s, Chernobyl ^{137}Cs originating from the Baltic Sea arrived to the Barents Sea.

4.1.3 Kara Sea shelf

The composition of radionuclides in Kara Sea sediments is similar in many respects to sediments on the Barents Sea shelf. It is quite evident that the process of sedimentation of radioactive substances is controlled by characteristics of the water circulation and lithology of modern sediments, the proximity or remoteness of the sources and by other related environmental factors. The ice regime is the most important factor determining to a great extent the conditions of modern sedimentation on the shelf.

Considering the importance of bottom topography, it is important to note that about half of the sea area to the south-east from the line Vajgach Island - October Revolution Island (Severnaya Zemlya) is occupied by vast areas containing sandy sediments. These areas, known as shallows, are located in water depths of 80-100 m and contain sands of different grading and sizes. In the south-east Kara Sea there are many narrow depressions of relief covered by coarse aleurite sediments.

The organic carbon content in these sands does not exceed 0.5 % (Kuznetsov 1976). This supports the viewpoint that the lighter (and finer) sediment fractions are present mainly during transport through the region without the possibility of long-term deposition. Correspondently, bottom fauna in the area is represented by well developed communities of bivalve mollusc-filterers *Astarte borealis* and *A. crenata* which are prominantly associated with sea bottom areas containing coarse aleurite sediments (Kuznetsov 1976).

In deeper water depths (to 250-450 m and more) located in the western Kara Sea, fine aleurite sands predominate. The bottom of all large troughs (East Novaya Zemlya, Saint Anna, Voronin) is covered by aleurite-clayey and clayey silts. In these areas, sediments contain the highest organic content (to 2.0 % of C_{org} and more), the bottom fauna communities consist of detritus feeders (*Portlandia arctica, Macoma calcarea, Spiochaetopterus typicus, Ophiocten serriceum, Ophiopleura borealis*) and unselective sediment feeders (*Elpidia glacialis*) are developed (Kuznetsov 1976).

Flux investigations of sediment-related substances is of special interest in the Arctic. The large Siberian rivers and shelf seas are the main sources not only of terrigenous (mineral) and organic substances but also of radioactive elements. The study of marine suspensions using sediment traps supports the following view: in the western part of the Kara Sea within the zone of weak river runoff (surface salinity 21.39 %), the daily flux of sedimented substance is (in mg m^{-2}) 18.7 (water depth 33 m) and 60.2 (water depth of 43 m). For C_{org}, fluxes were 4.28 and 7.29 mg C m^{-2} respectively (Lisitsyn et al. 1994). The main component of the sedimentation flux is not related to fecal pellet formation but to the diatom remnants.

For the Yenisey, sedimentation fluxes vary from minimums of 9.4 mg m^{-2} per day in the oligotrophic area of the open sea to the north of Dikson Island to 158.5 mg / m^{-2} per day within the outer boundary of the Yenisey estuary. However, large sediment slides or avalanches can result in sediment fluxes of 22156 mg m^{-2} per day as observed at the freshwater – seawater interface (Lisitsyn et al. 1994). The mentioned values indirectly indicate why it is not surprising to find differences in the rates and volumes of radionuclide fluxes on the Kara shelf.

^{137}Cs concentrations in surface (0-3 cm) bottom sediments are within the limits 0.5-2.0 Bq / kg (Fig. 4.12). These limits are substantially lower than observed in sediments of the Baltic, Irish and North Seas (Matishov et al. 1993c, 1994c, 1995b, 1997a,d, 1998b,c; Rissanen et al. 1995, 1997b; Nies et al. 1998).

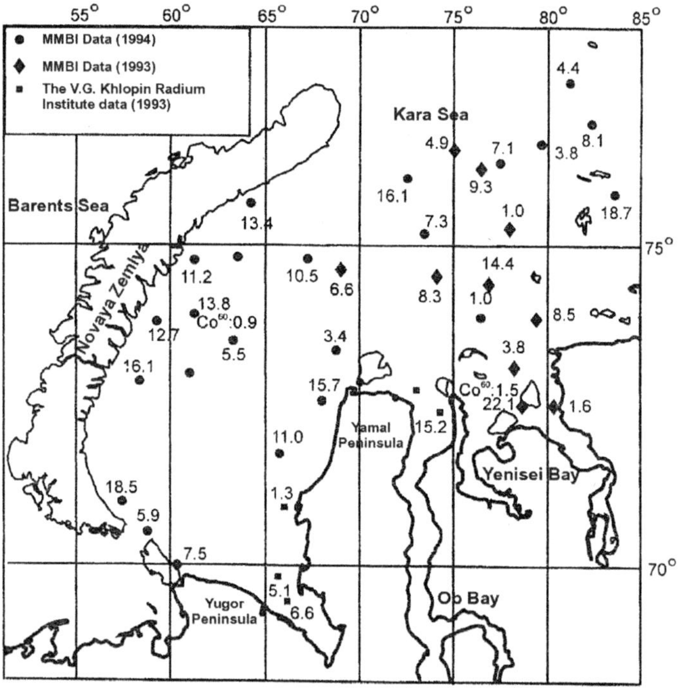

Fig. 4.12. ^{137}Cs concentration in Kara Sea bottom sediments (0-3 cm layer) in 1993-1994, Bq / kg

In sandy and gravelly-pebbled shelf sediments little ^{137}Cs accumulates (1-8 Bq / kg). The quantity of cesium in the sandy and coarse sediments on the coastal shelf to the east of Novaya Zemlya varies from 3 to 10 Bq / kg. However, some closed depressions in the bays of Novaya Zemlya, for example, Stepovoy Bay, contain clayey sediments with ^{137}Cs up to levels of 90 Bq / kg (Foyn end Semenov, 1993; Nikitin et al. 1995).

The ^{137}Cs concentration of sandy sediments of the Bajdaratskaya Bay, and within the shallows of the Sharapovy Koshki area range from 0.8 to 6.2 Bq / kg (Fig. 4.13). However, levels of 12-16 Bq / kg have been noted for a few locations. These distribution characteristics are normal for coastal marine environments located in the vicinity of major river inflows to the sea. In water depths of 95 m located to the north of Yugorsky Peninsula, aleurite sediments have accumulated 27-31 Bq / kg of ^{137}Cs and about 0.5 Bq / kg of ^{60}Co (Rissanen et al. 1995) (Table 4.7). These values appear to be somewhat anomalous. Additional observations are needed in this situation.

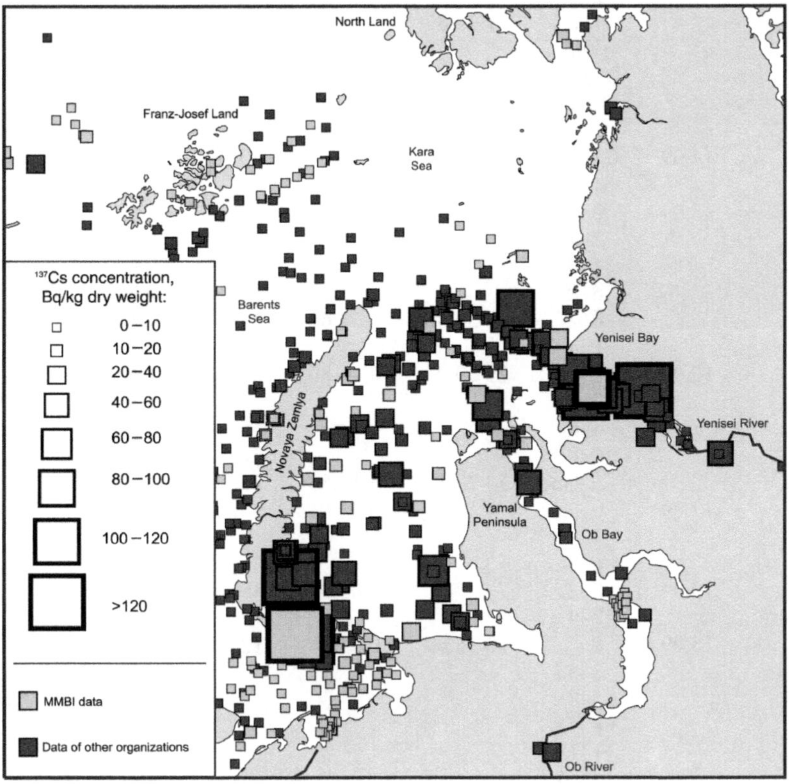

Fig. 4.13. ^{137}Cs content in the surface layer of the Kara Sea bottom sediments in 1990s

Table 4.7. γ-emitting radionuclides in bottom sediments of the Kara Sea 1993, (Bq / kg dry weight)

Station	Date	Layer thickness, cm	Coordinates N	Coordinates E	Depth, m	^{137}Cs	^{60}Co	^{41}K
1	12.09	0–3	69°59.72′	61°47.01′	95	27.5	0.5	21.4
		3–5				31.4	–	20.9
		5–10				27.8	–	22.2
2	13.09	0–2	60 39.00	65 36.00	31	5.4	–	15.1
		2–4				6.1		15.2
3		0–2	69 19.81	66 14.30	22	5.6	–	15.9
		2–4				6.2	–	16.4
4a		0–1, sand	71 02.82	65 54.61	20	1.3	–	14.1
4b		0–2–3, sand	70 00.06	66 00.59	32	12.5	–	18.5
		2–4				16.0	–	18.8
5	14.09	0–2–3, sand	71 02.89	65 54.61	20	0.8	–	11.6
6	15.09	0–2	72 30.34	74 25.62	15	14.6	–	16.4
		2–4				14.1	–	17.0
		4–6				13.8	–	18.8
		6–8				14.3	–	19.1
7		0–2	72 59.86	72 58.62	29	10.7	–	13.5
		2–4				11.9	–	18.2
9	16.09	0–2	73 59.03	73 16.80	30	24.2	0.3	17.5
		2–4				28.5	–	19.6
		4–6				27.6	–	19.2
		6–8				22.8	–	19.1
10a		0–2, sand	74 27.11	73 17.84	25	2.3	–	13.0
11	17.09	0–2	74 18.64	78 38.81	29	5.7	–	12.8
		2–4				7.6	–	13.8
		4–8				6.3	–	14.9
12	19.09	0–2	73 37.00	80 05.00	45	32.3	0.6	17.3
		2–4				28.7	–	17.3
13	18.09	0–2	72 35.47	80 31.11	12	81.5	5.8	15.7
		2–5				75.0	5.1	15.1
		5–8				76.0	4.0	16.1
		8–11				70.0	2.3	17
		11–14				67.3	1.8	16.4
		14–17				75.2	1.8	16.1
14	17.09	0–2	73 54.57	81 03.48	41	36.0	–	19.8
		2–4				32.5	–	19.3
		4–8				24.7	18.9	18.9

^{137}Cs concentrations in sandy sediments deposited on the shallows situated between the Tamal Peninsula and Severnaya Zemlya archipelago vary from 4 to 10 Bq / kg. In this same area, radiocesium levels increase to 15-18 Bq / kg in depressions filled by aleurites (Matishov et al. 1994b,c, 1997c; Champ et al. 1994;

Krosshavn 1995; Nies et al. 1997). Higher ^{137}Cs accumulations (to 15-30 Bq / kg and more) are found in the clayey silts that are accumulating in depressions located opposite to the Siberian rivers (Table 4.8). These increased contamination levels are traced from the sea coast to 75°N. Within a column of sediment, the ^{137}Cs level sharply decreases with depth to 0.5-0.1 Bq / kg and less, beginning from a depth of 7-10 cm. Evidently, during the time period between 1950-1990s, radioactive discharges from Russian nuclear facilities situated in the Ob and Yenisey basins exited the rivers and subsequently deposited in the depressions located on the coastal shelf.

Table 4.8. Radionuclides in surface sediment layers of the Kara Sea, Bq / kg (Kuznetsov et al. 1994)

Station	Coordinates		^{90}Sr	^{137}Cs	^{239,240}Pu
	N	E			
1	2	3	4	5	6
		Yenisey Bay			
3	69°41′	84°04′	0.5±0.3	3.0±0.7	0.2±0.03
5	70 32	83 28	3.5±0.2	0.6	0.5±0.11
6	71 03	83 13	2.1±1.0	2.5±1.0	0.7±0.14
7б	71 23	82 59	1.3±0.4	21.4±1.9	1.0±0.19
9	71 47	82 46	1.4±0.1	19.0±1.7	0.4±0.18
12	72 07	81 36	3.5±0.2	52.6±3.2	1.0±0.20
13	72 08	80 56	1.7±0.2	53.6±2.7	1.0±0.32
14	72 23	80 48	2.4±0.2	73.6±1.8	1.3±0.25
15	72 50	78 32	0.6±0.2	22.3±2.8	1.3±0.22
16	73 10	80 22	1.9±0.3	63.6±3.8	3.1±0.50
22	72 43	80 38	–	61.2±1.9	1.6±0.30
23	71 41	83 25	1.1±0.3	36.7±2.6	0.8±0.20
26	72 21	80 08	1.7±0.3	26.3±2.1	1.2±0.20
27	72 41	78 19	0.6±0.3	17.6±1.6	0.8±0.06
28	72 44	78 39	1.3±0.4	20.3±1.4	5.7±1.20
29	72 58	78 12	0.5±0.1	4.8±0.9	0.5±0.16
30	73 17	77 55	<0.2	2.1±0.8	4.8±0.4
31	73 13	78 36	0.4±0.2	7.9±0.9	0.3±0.08
4413	72 12	81 27	–	51.5±5.6	–
4403	73 00	81 27	–	58.3±8.8	–
4402	73 33	79 51	–	25.2±4.0	–
4409	71 19	82 59	–	42.7±3.4	–
		Sea			
17	74 20	78 45	–	21.6±2.4	–
19	74 45	74 18	–	6.0±1.1	0.31±0.12
20	74 04	73 06	–	10.8±1.3	–
30	73 17	77 55	–	2.1±0.8	4.83±0.04
31	73 13	78 36	–	7.9±0.9	0.27±0.08

Table 4.8. (Cont.)

1	2	3	4	5	6
34	73 44	79 39	–	31.5±3.8	2.16±0.43
35	74 00	79 18	–	6.3±1.3	0.71±0.22
36	74 14	78 48	–	12.0±1.7	–
37	74 27	78 31	–	25.4±2.3	1.58±0.52
38	74 43	78 11	–	17.5±2.4	–
39	74 56	77 50	–	11.4±1.4	–
40	75 12	77 35	–	<0.9	0.45±0.13
41	75°28'	77°09'	–	3.0±1.4	–
42	75 43	76 44	–	8.4±1.3	–
43	75 58	76 18	–	6.5±1.2	0.48±0.18
44	76 15	75 59	–	5.6±1.3	0.41±0.12
45	76 05	75 00	–	8.6±1.8	–
47	76 43	74 42	–	8.7±1.6	–
48	75 28	75 03	–	9.0±1.6	–
49	75 12	75 28	–	5.8±1.5	–
50	74 56	75 54	–	13.6±1.8	–
51	74 37	76 20	–	11.8±1.3	0.65±0.21
52	74 22	76 40	–	<1.3	–
53	74 14	74 59	–	4.8±1.1	–
54	74 30	74 35	–	4.9±1.3	–
55	74 46	74 13	–	3.7±1.1	–
56	75 02	73 52	–	<2	–
76	74 56	70 25	–	<2	–
77	74 52	69 15	–	6.3±1.1	0.62±0.19
4376	70 08	55 02	–	22.0±5.5	–
4380	72 37	59 55	–	13.0±2.9	–
4382	74 33	64 04	–	11.0±1.9	–
4383	74 20	64 37	–	12.0±2.5	–
4386	72 41	64 34	–	9.6±2.4	–
4388	71 01	64 37	–	<15	–
4391	69 46	64 51	–	19.0±5.1	–
4393	70 20	63 08	–	24.0±4.6	–
4415	72 47	73 25	–	<2	–

In fine aleurite and clayey silts within the Novaya Zemlya Trough (water depth = 300-400 m), ^{137}Cs concentrations range from 10-30 Bq / kg (Matishov et al. 1994c; Strand et al. 1994; Strand 1998). ^{137}Cs in the silty sediments of the deepest areas (to 500-600 m) and at the entrance to the Central Arctic basin through the Saint Anna Trough are comparatively small (from 2 to 10 Bq/ kg).

The determinations of ^{137}Cs concentrations in bottom sediment samples from the south-western part of the Kara Sea (RV «Professor Stockman» cruise, 1994) showed comparatively high contamination levels. The maximum values of cesium in sediments (to 150-240 Bq / kg) were found in water depths of 110-120 m on the shelf to the west of the Yamal Peninsula (Ajbulatov et al. 1994). The most likely explanation as put for by the authors of the work is that the sediments were collected near leaking radioactive waste containers. Additional sampling at these locations is desirable to exclude any doubt with regard to instrumentation problems during the determination of these uncharacteristically high radioactivity levels.

It is almost impossible now to analyze the dynamics of ^{137}Cs during the period 1970-1990 because of the few radioactivity data available for the Russian Arctic from this time period. We can note only that the bottom sediments on the south-west of the Kara Sea in 1982 had ^{137}Cs activity of order 4-20 Bq / kg (Strand et al. 1994). By the results of a number of researches (JRNEG 1993; Strand et al. 1994, 1997; Kuznetsov 1994), the radiocesium concentration in sediment per unit of area on the south of the sea in 1971, 1982 and 1992 varied correspondingly in the ranges 100-800, 5-100 and 100-500 Bq / m^2 (Fig. 4.14). In principle, all these levels are not high for the open shelf and correspond to the global fallout signal.

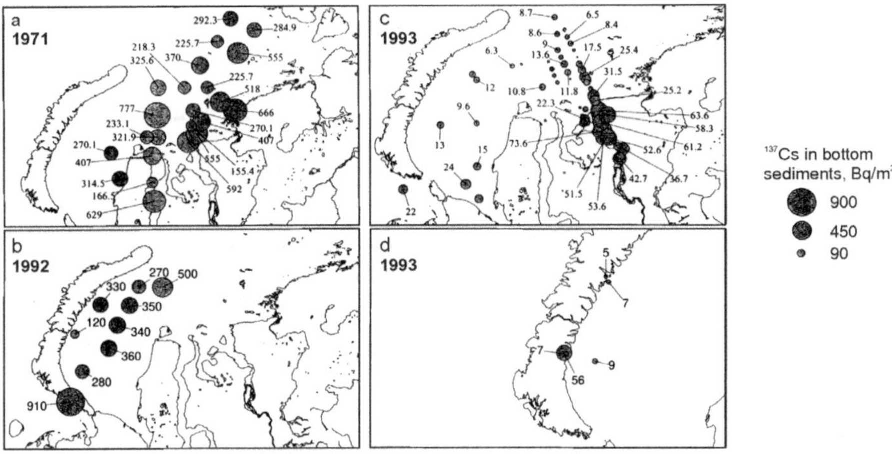

Fig. 4.14. ^{137}Cs in the Kara Sea bottom sediments (Strand et al. 1997): **a** - *1971 (AMAP 1997)*; **b** - *1992, to 0-10 cm depth interval (JRNEG 1993)*; **c** - 1993 (Kuznetsov et al. 1994); **d** - *1993 (JRNEG 1993)*

Overall during the 1990s, the ^{137}Cs content in the bottom sediments of the Kara Sea was within a range of 1 to 40 Bq / kg. Areas with the highest levels of radiocesium are connected with the zones of avalanche sedimentation of material from the Siberian rivers and also to areas in close proximity to radioactive waste dumping sites on the shelf.

^{60}Co was present in bottom sediments of a number of coastal shelf areas during the 1990s (Matishov et al. 1994b,c; Nies et al. 1997). This short-lived radionuclide was found in the north of the West Siberian rivers Ob and Yenisey. ^{60}Co was present throughout the length of a 20 cm sediment core taken at the mouth of the Yenisey Bay at levels ranging from 1 to 6 Bq / kg. In some clayey interlayers the

^{60}Co activity reached as high as 19 Bq / kg. This radionuclide is also observed in small quantities in the silts of the depression located opposite to the Ob. In 1993 the ^{60}Co was found (0.5 Bq / kg) in the coarse aleurites to the west from Yamal in the area of Sharapovy Koshki on the depth about 100 m.

As noted earlier, ^{60}Co is present in sediment deposits near containers of nuclear waste. For example, ^{60}Co (3-5 Bq / kg) was detected in bottom sediments of the Stepovoy Bay. This radionuclide was also detected in the silts of one depression within the East Novaya Zemlya Trough. Probably, this is an indication of the existence of a local source of bottom sediment contamination.

$^{239,240}Pu$ is present in Kara Sea bottom sediments in insignificant amounts. During the 1990s, $^{239,240}Pu$ concentrations in the upper sediment layers of shelf sediment deposits varied mainly from 0.1 to 3.5 Bq / kg (Fig. 4.15) (Matishov et al. 1994c; Strand et al. 1994, 1997; Champ et al. 1994; Rissanen et al. 1997a; Stepanets et al. 1999). Plutonium particles was detected in sediment cores to depths of 7-12 cm. $^{239,240}Pu$ concentrations on a per unit area basis was 2-30 Bq / m^2 and in some places - to 120 Bq / m^2. ^{238}Pu and ^{241}Pu was detected at levels of 0.02-0.11 and 4.6-11.6 Bq / kg in sediments (Table 4.9) (Strand et al. 1994; Rissanen et al. 1997a).

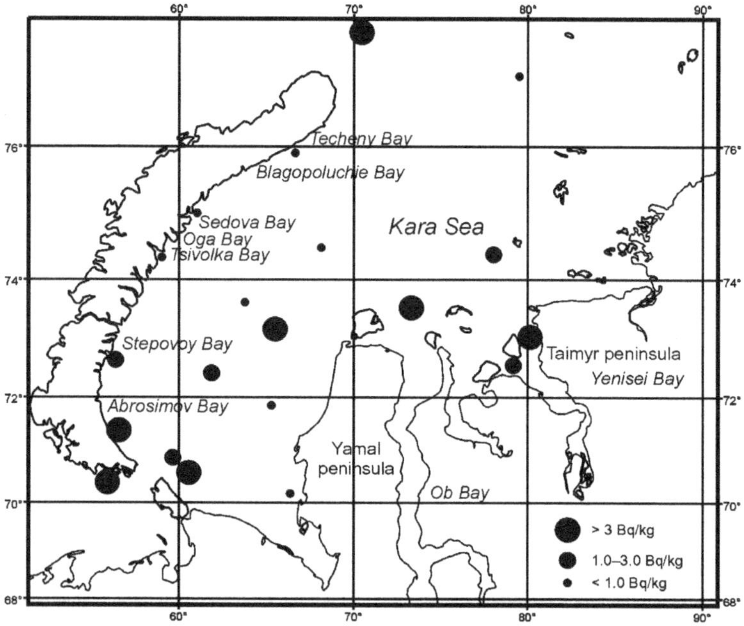

Fig. 4.15. $^{239,240}Pu$ content in the Kara Sea bottom sediments (0-2 cm) (Stepanets et al. 1999)

Table 4.9. ^{238}Pu and $^{239,240}Pu$ concentration in Kara Sea bottom sediments 1993 (Rissanen et al. 1997a), Bq / kg dry weight

Station	Depth of column by layers, cm	Date	^{238}Pu	$^{239,240}Pu$
1	0–3	12.09	0.100	2.20
	3–5		0.100	2.40
	5–10		0.110	3.30
Total, Bk/m^2			3.700	100.00
6	0–2	15.09	0.020	0.51
	2–4		<0.020	0.46
	4–6		<0.020	0.66
	6–8		0.028	0.88
Total, Bk/m^2			–	24.00
7	0–2	15.09	<0.020	0.55
	2–4		<0.020	0.46
9	0–2	16.09	0.034	1.10
	2–4		0.030	1.10
	4–6		0.047	1.20
	6–8		0.048	1.20
Total, Bk/m^2			1.300	37.00
10	0–2	16.09	<0.020	0.18
11	0–2	17.09	<0.020	0.30
13	0–2	18.09	0.026	0.74
	2–5		0.029	0.65
	5–8		0.042	1.10
	8–11		0.040	1.10
	11–14		0.041	1.20
	14–17		0.032	0.72
Total, Bk/m^2			2.500	67.00

Bottom sediments located to the north of the West Siberian rivers Ob and Yenisey are comparatively more contaminated by transuranic elements. The maximum (0.8-3.5 Bq / kg) $^{239,240}Pu$ content is noted within the coastal zone where river and sea water mix. At some locations, this radionuclide is found at levels up to 4-6 Bq / kg (Johannessn et al. 1996).

Beyond the coastal zone, in water depths ranging from 60-100 m, the influence of river suspended sediments on radionuclide accumulation on the bottom is as important as near the estuaries. Fine sand and coarse aleurite sediments contain from 0.2 to 1.5 Bq / kg of $^{239,240}Pu$. However $^{239,240}Pu$ levels increase to 1.6-3.5 Bq / kg in the aleurite sediments located in bottom sloping areas of seabed depressions (Fig. 4.16).

To the southwest of the Kara shelf $^{239,240}Pu$ concentrations in surface sediments vary from 0.25 to 1.4 Bq / kg (Strand et al. 1994). In different sands of the Bajdaratskaya Bay, the concentration of this radionuclide is insignificant (0.4-0.8 Bq / kg). However, aleurite sediments deposited to the east from Amderma (depths about 100 m) contain on the order of 2.2 Bq / kg of $^{239,240}Pu$ Its concentration increase with depth in sediment cores to 3.3 Bq / kg within the 5-10 cm layer of sediment (Rissanen et al., 1997a) (Fig. 4.17). On a per unit area basis, the concentration of $^{239,240}Pu$ is 100-120 Bq / m^2.

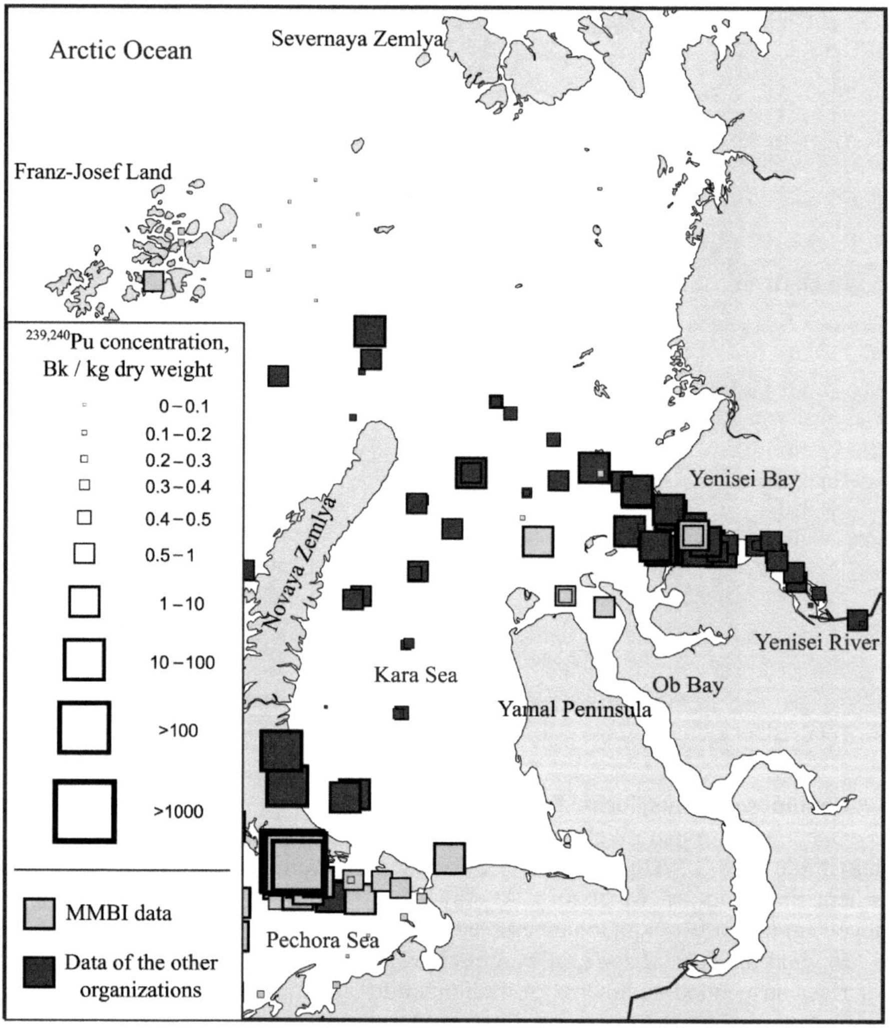

Fig. 4.16. $^{239,240}Pu$ content in the surface layer of Kara Sea bottom sediments in the 1990s

Within the East Novaya Zemlya Trough, the upper layer of sediment contains 0.8-3.2 Bq / kg of $^{239,240}Pu$ or 6-20 Bq / m^2.

Sediments from the Santa Anna Trough have low $^{239,240}Pu$ concentrations (0.05-0.5 Bq / kg) (Rissanen eyt al. 1997a). Only to the north-east of Zhelania Cape (Novaya Zemlya) are higher concentrations of $^{239,240}Pu$ detected, up to 3 Bq / kg.

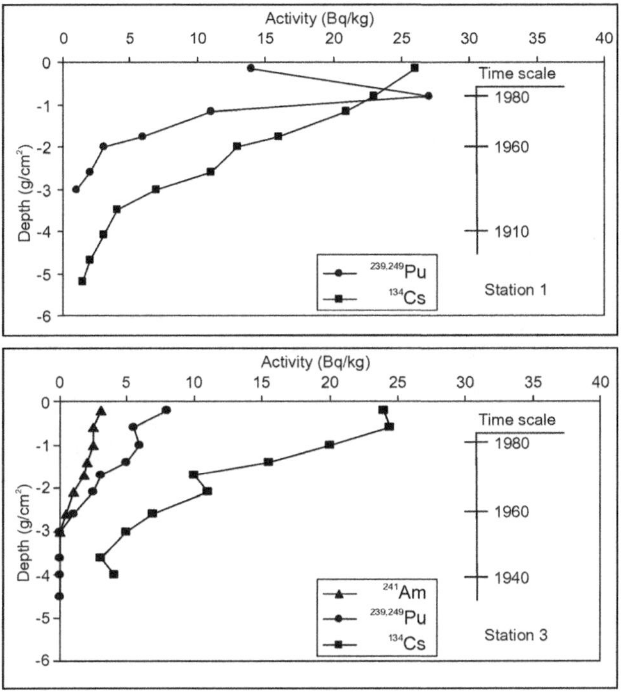

Fig. 4.17. ^{137}Cs, ^{239,240}Pu and ^{241}Am distribution in the bottom sediment columns from the Kara Sea (age determination by ^{210}Pb geochronolgy) (Hamilton et al. 1994)

Considering activity ratios, the ratio of ^{239,240}Pu to 137Cs varies from 0.021-0.074 and for ^{238}Pu and ^{239,240}Pu activity ratios vary from 0.01 to 0.065 (Strand et al. 1994; Champ et al. 1994; Rissanen et al. 1997a). All of these values indicate global fallout from the atmosphere as the primary source of radionuclides to sediment deposits in the investigated areas. This ratio increases sharply in areas where local sources exist today or in the past. Thus, in Chernaya Bay, where underwater nuclear tests were carried out, this ratio is equal to 53.6 in the epicentre of the test site (MMBI expedition on RV «Dalnie Zelentsy», 1992) and gradually decreases in the Pechora Sea with increasing distance from the test point: at a distance 21 km - 0.78; 36 km - 0.480; 100 km - 0.230; 125 km - 0.090; 161 km - 0.144; 250 km - 0.084 (Smith et al. 1995a).

^{90}Sr concentrations in bottom sediments has been investigated only in a few locations in the southern Kara Sea (Fig. 4.18). During the 1990s ^{90}Sr concentrations in bottom sediments varied from 0.4 to 6 Bq / kg (Kuznetsov et al. 1994; Strand 1998; Stepanets et al. 1999). In contrast, the content of this radionuclide in sediments from the Yenisey and Ob Bays reaches 10 and 20 Bq / kg respectively. Between 1970 and 1977 a trend of decreasing dissolved conentrations of ^{90}Sr with increasing distance toward the mouth of the rivers. Near the town of Salekhard of the Ob Bay, ^{90}Sr concentrations were close to the global fallout level. Evidently dissolved ^{90}Sr concentrations decrease over time due to ^{90}Sr removal from the catchment areas, the processes of self-purification of the river water and the dilution processes (Kuznetsov et al. 1994).

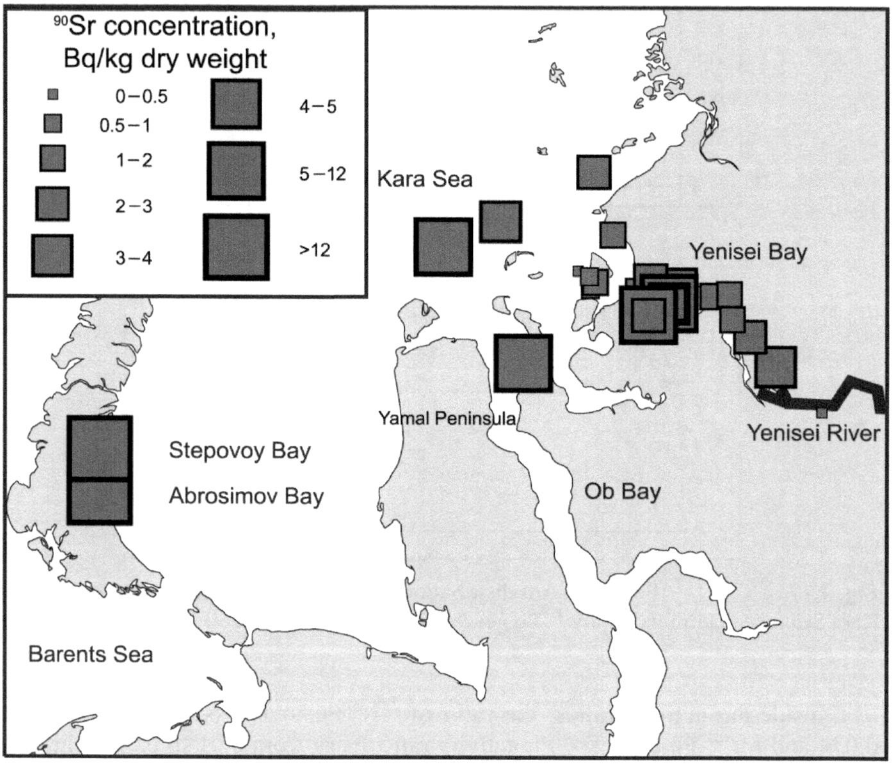

Fig. 4.18. ^{90}Sr content in the surface layer of the Kara Sea bottom sediments in 1990s

Determinations of ^{90}Sr content in sediments indicates that fine aleurites and clayey silts from depressions in the Novaya Zemlya Trough have accumulated on the order of 1-4 Bq / kg of ^{90}Sr. Sediments to the north of the Yenisey and Ob estuaries contain from 0.5 to 5 Bq / kg of ^{90}Sr.

S.M. Vakulovsky and others (1985, 1994a) estimate that the quantity of ^{90}Sr carried by the Techa and Iset rivers to the Ob and further to the Kara Sea was 4700 Ki for the period 1961-1990. At the same time, the total quantity of ^{90}Sr transported by the Ob during this period was, according to these calculations, 17500 Ki. Moreover, 75 % of this quantity is thought to result from the removal of global radioactive fallout through washout of the large catchment area of this river and its tributaries (2.990 mln km^2).

In general, when considering artificial radionuclide accumulation in the Kara Sea, the observed variations of ^{137}Cs and $^{239,240}Pu$ in sediments (0.6-100 and 0.2-10 Bq / kg respectively) are similar to both the Yenisey and Ob coastal areas as well as for the Baltic Sea. It is important to note however that results for the Baltic Sea were obtained in 1982, i.e. before the Chernobyl accident (Ilus et al. 1988a,b). The source of pre-Chernobyl radionuclide concentrations in the closed Baltic basin can be explained by their input from the global sources.

There is not sufficient evidence to connect definitively the formation of radioactive contamination of the Kara Sea, which is open to the Arctic Ocean, with global fallout. The great influence of the Siberian radiochemical plants on cesium, strontium, and plutonium accumulation in bottom sediments is quite evident. A thorough investigation of the mechanisms of radionuclide transfer from northern rivers to bottom sediments of the Kara Sea is necessary for identifying the exact contributions of various sources to the observed environmental levels.

4.1.4 Laptev Sea

Radionuclides in Laptev Sea bottom sediments have been investigated to a limited degree (Matishov et al. 1995b; Pavlov et al. 1999). The large Lena River, with an annual water runoff of more than 500 km^3 (AMAP, 1997), flows into this Arctic Basin. The average annual suspended sediment concentration of the Lena is 40 mg / l with maximum concentrations of 400 mg / l (Kuptsov et al. 1999). These suspended sediments, which also contain pollutants, have been traced distances up to 500 km from the Lena mouth.

The upper turbid surface water layer extends from the outer edge of the Lena delta northward, almost to the shelf boundary. Its thickness is 5-7 m, i.e. it is situated in the layer of ice formation. In autumn this giant layer of turbid water transforms into sea ice which is transported throughout the Arctic basin via the Transpolar drift system (Lisitsyn 1978).

The suspended sediment content in some samples of the dirty ice reaches 300 mg / l, more frequently it is within the limits of 5-100 mg / l, i.e. is similar to concencentrations in rivers near the sea boundary. The main part of this suspension (30-60 %) is represented by pelite material (fraction finer than 2 μm).

The Laptev Sea shelf area is shallow (50-100 m). There are no apparent sources of artificial radionuclides in this area. Lithological sections carried out in some sea areas has been used to characterize ^{137}Cs levels in bottom sediments from water depths of 120 to 3100 m. The surface sediments in all areas of the continental underwater border and abyssal plains contain radiocesium at background levels. The 40K concentration in bottom sediments varied from 630 to 880 Bq / kg. These radiocesium concentrations conform to expectations based on bottom topography and water circulation patterns.

The coarse aleurites of the continental slope to the north from the Kotelny Island (Novosibirskye Islands) have accumulated from 7 to 10 Bq / kg of ^{137}Cs. Aleurite sediments to the north of the Lena River taken on a section from the shelf edge to the continental shelf, contain from 1.9 to 8.8 Bq / kg of ^{137}Cs (Fig. 4.19). On the continental shelf to the north-east from the Tajmyr Peninsula the radiocesium content in fine sands and aleurites ranges from 3.8-6.8 Bq / kg. The bottom sediments in the Vilkitsky Strait in water depths of 120-160 m have somewhat increased concentrations of ^{137}Cs (5.7-13 Bq / kg) (Matishov et al. 1995a, 1997c).

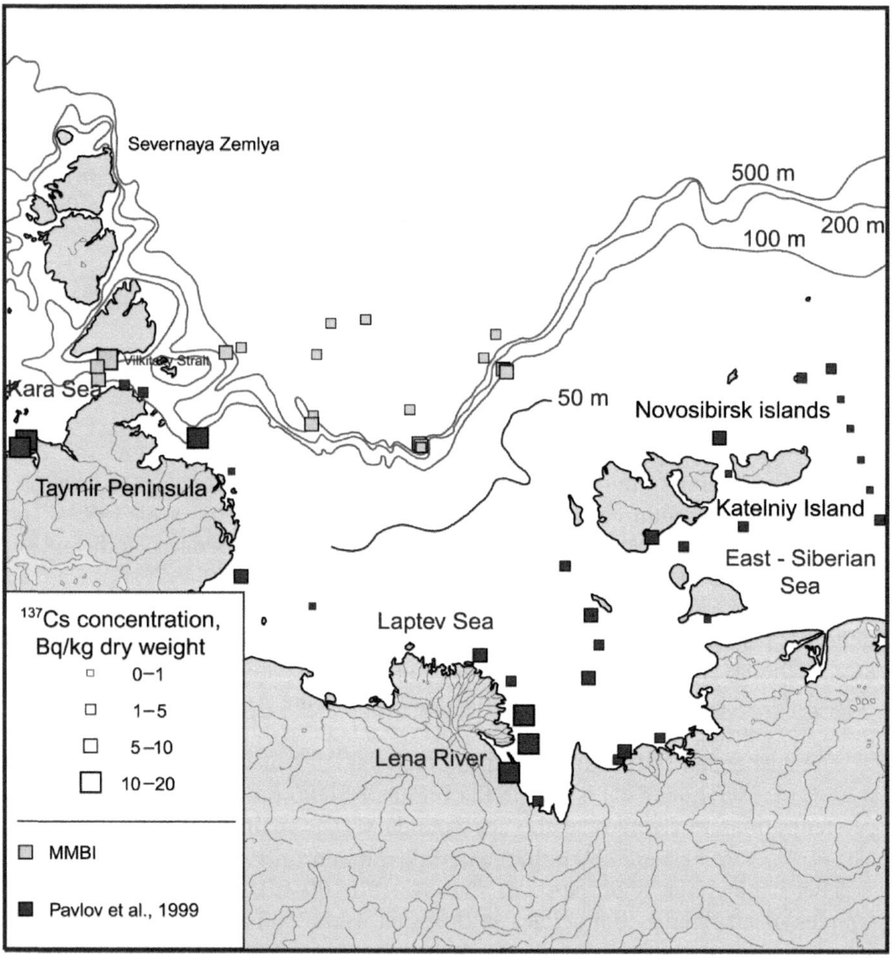

Fig. 4.19. ^{137}Cs content in the surface layer of Laptev Sea bottom sediments in the 1990s (Matishov et al. 1995; Pavlov et al. 1999)

In the southern part of the Laptev shelf adjoining the Lena delta, ^{137}Cs activity in sediments varied from 48 to 500 Bq / m^2 (Pavlov et al. 1999). However, radiocesium concentrations in a number of areas was higher - 100-1900 Bq / m^2.

These high values are probably the result of radionuclide releases from river suspended sediments, sedimenting particles from melting sea ice and global fallout.

The lowest ^{137}Cs levels (1.6-2.2 Bq / kg) were noted in clayey silts from the abyssal plains in water depths of more than 3000-3200 m. Unconsolidated sediments in the upper part of the continental slope were distinguished by relatively high radiocesium concentrations (6.8-9.8 Bq / kg). These studies did not reveal any anomalous ^{137}Cs concentrations in bottom sediments from the Laptev Sea. However, radiocesium levels of 8-13 Bq / kg are higher than observations from many areas of the Barents shelf (Matishov et al. 1994c, 1995b).

The characteristics of radionuclide distributions in Laptev Sea bottom sediments are not detailed enough to identify the exact sources of ^{137}Cs to bottom sediments. One can suppose that the origin is connected with transport from the vast territory of East Siberia by the large Lena River system. Radioactive particles from the Chernobyl accident could reach the Lena catchment area and also from Chinese nuclear tests carried out until 1994. ^{137}Cs from northern Europes nuclear reprocessing facilities transported by warm Atlantic waters of the Gulf Stream system may also enter the continental slope region of the Laptev Sea. Overall, the ^{137}Cs content in Laptev Sea bottom sediments varied from 1.6 to 13.0 Bq / kg. These low levels characterize global radioactive fallout. The process of radionuclide accumulation on the sea bottom is caused by the noticeable release of radioactive particles from the land by Siberian rivers (Lena, Yenisey). The ^{137}Cs influx with Atlantic waters to the Laptev Sea continental slope is also possible.

4.2 Sedimentation of radioactive substances on the shelf

An analysis of the available data on artificial radionuclides indicates that during the 1990s radionuclide levels in the bottom sediments of bays and shelves of the Norwegian, Barents, White, Kara, Laptev and other northern seas reflected the background signal of global fallout from atmospheric nuclear tests. From the beginning of the 1950s, the increase of ^{137}Cs and ^{90}Sr that resulted from atmospheric and underwater nuclear tests and later by the Chernobyl accident, was counterbalanced by a decrease in activity resulting from radioactive decay (^{137}Cs half-life = 30.2 years; ^{90}Sr half-life = 27.7 years). Shelf sediments exhibit a small concentration range of ^{137}Cs (1-45 Bq / kg) (Matishov et al. 1994a,c, 1996f, 1997c,f, 1998b,c). The very low activity of ^{90}Sr (from 0.2 to 6 Bq / kg) is characteristic for marine sediments.

The incorporation of radioactive substances that enter the ocean from the atmosphere into seabed sediments can be envisioned to occur analogous to marine sedimentogenesis (Leontjev 1968, 1974; Lisitsyn 1978). The following four stages of accumulation of radionuclide-containing particles can be distinguished: 1) the initial sorption of radioactivity onto particles; 2) particle transport over the sea or ocean area; 3) particle differentiation or sorting; and 4) sedimentation on the seabed.

The existence of large, un-sorted material in the surface sediments of the Arctic Seas testifies to the importance of glacial and iceberg transport. The following parameters influence glacial sedimentation:

- the period of ice cover on the surface of the basin;
- coastal zone morphology and pathways of the seasonal ice drift;
- the ratio between the rate of input of total terrigenous debris to the accumulation of fine material.
- The glacial and iceberg characteristics play a noticeable role in the accumulation, transport and ultimate burial of radioactive particles in bottom sediments.
- As a rule, the composition and origin of bottom sediments are closely interconnected in space and time. With respect to radionuclide accumulation, the role of the main sources, the spectrum of their discharges, the half-life of individual radionuclides, and radionuclide speciation are important.

Overall, radionuclide concentrations in the environment of the Arctic Seas are very low in comparison with the shelf basins of the Baltic, Irish, North and also of the Azov Seas. Sediments on the bottom of Chernaya Bay, where the levels of ^{137}Cs, ^{60}Co, ^{155}Eu and ^{241}Am were correspondingly 1440, 618, 344 and 2862 Bq / kg, are an obvious exception. Such high concentrations of artificial radionuclides are characteristic only for the few places where nuclear tests were carried out or at nuclear wastes burial sites where leakage have occurred. Low levels of ^{60}Co indicate no noticeable leakages of radionuclides from reactors, buried on the shelf, in 1990s.

The local anomalies of ^{137}Cs (60-120 Bq / kg), ^{90}Sr (10-50 Bq / kg) and ^{60}Co 5-80 Bq / kg) in bottom sediments are found only in the bays near the objects of the nuclear fleet and in the estuaries of the Siberian river where radioactive wastes were released. The Ob-Yenisey coastal zone is the area of intensive discharge of river suspension, acting as a large-scale marginal filter for the Kara Sea. During the downstream transport of river suspended material there is a large loss (more than 90 %) to the bottom (Lisitsyn, 1994). Dissolved elements which sorb to particles during transport downstream and through the saltwater-freshwater salinity gradient zone are also lost to the bottom with suspended material.

The concentration of ^{137}Cs in the sediments of the relatively deep Barents Sea, with its complex system of currents and incised seabed, is influenced only by the proximity of the sampling places to the main current pathways and nuclear test sites. The highest ^{137}Cs concentrations are found in the Southeastern Barents Sea. The sediments of the central and Northeastern areas of the sea contain lower cesium concentrations. Within the Pechora Sea, ^{137}Cs concentrations in sediments are noticeably influenced by proximity to the Southern Novaya Zemlya test site. This is apparent from the ^{137}Cs content in sediments taken from the deep-water channel along the southern coast of Novaya Zemlya.

The $^{239,240}Pu$ accumulation in the longitude belt 60-70°N that resulted from atmospheric nuclear weapons testing is estimated to be aproximately 60 Bq / m^2 (Hardy et al., 1973). In the shelf sediments of the polar seas, the $^{239,240}Pu$ concentration varies from 0.1 to 3.0 Bq / kg. This range of concentrations for sediments does not exceed the levels expected from global radioactive fallout. However, nearer to the old Novaya Zemlya test sites, the $^{239,240}Pu$ content in sediments increases to 5-15 Bq / kg. Additional excpetions include coastal military bases of the nuclear fleet (Olenjya, Zapadnaya Litsa), underwater dumping sites (Abrosimov, Stepovoy) and nuclear reprocessing facilities (e.g. Kola Bay, Yenisey River). Sedimentary $^{239,240}Pu$ concentrations at these more contaminated sites varies from 6 to 20 Bq / kg.

The plutonium content in sediments from the test-site in Chernaya Bay, Novaya Zemlya, where nuclear tests were carried out, is as high as 5000 Bq / kg. These concentrations will not diminish appreciably for many thousands of years because the half-life of this isotope exceeds 24000 years. However for the Barents-Kara shelf as a whole, $^{239,240}Pu$ concentrations in bottom sediments vary from 0.1 to 20 Bq / kg. These sediment concentrations are approximately one order of magnitude higher than in the Azov Sea (Table 4.10) where the origin of both $^{239,240}Pu$ and ^{90}Sr is connected only with global fallout from nuclear weapons tests and the Chernobyl accident.

Table 4.10. Plutonium concentrations in surficial sediments from major seas (Bq / kg dry weight)

Region	$^{239,240}Pu$
Norwegian Sea	0.7–1.6
Kara Sea (open part)	0.3–1.7
Yenisey Bay	3–3
Baltic Sea (open part)	1.5–6.8

The clear dependence of all anthropogenic radionuclides on sediment size distributions and material composition is quite evident. The size spectrum of sandy, aleurite and clayey particles is standardized, in mm into the following classes: more than 1.6, 1.6-1.0, 1.0-0.5, 0.5-0.25, 0.25-0.125, 0.125-0.063, 0.063-0.032, 0.032-0.02 and less than 0.02.

The distribution of seabed sediments exhibits some general patterns. As a rule, sandy sediments (0.25-1.0 mm) are deposited on plateaus, typically in water depths up to 130 m. Coarse and medium-grained sands are characteristic for water depths of 50-80 m and fine grained sediments are typically found in water depths of 80 to 130 m. Sandy sediments are found along the coastal shelf, within the Murmanskaya (Central), on the Gusinaya, North Kanin and other banks, and on the Kanin and Pechora shallows. The degree of sorting and size distribtuion of the sandy fractions are determined by the influences of waves and the velocity of stationary currents.

Clayey sediments (coarse and fine pelite fractions) are associated mainly with low areas within deep troughs and shelf depressions. The fine-grained pelite fraction (0.001-0.01 mm) is deposited in water depths of 300-500 m and more, and characteristically contains high water contents (70-90%) (Matishov end Pavlova 1990). This fine, widely dispersed, sedimentary material is exposed to large quantities of seawater and has strong sorptive properties. These characteristics result in the efficient adsorption of radionuclides onto clayey silts. The highest rate of sediment accumulation for modern terrigenous and organic sedimentary material occurs in troughs where sediment thicknesses are (5-30 cm). The analysis of granulometry together with radioactivity for bottom sediments has shown a strong negative relationship between ^{137}Cs, $^{239,240}Pu$ and ^{90}Sr concentrations and grain-size within the size fraction of <0.063 mm.

Aleurite sediments of the Arctic shelf are mainly associated with deposits in water depths of 130-300 m. These sediments represent a transitional phase in sediment size sorting. They contain a prevailing size group of sediment fractions, representing 50 to 75 % of the total sedimentary material and an additional size fraction that makes up the remaining 25 to 50 % of material. Aleurite sands and sandy aleurites cover the slopes of all heights and banks of the Barents, Kara, White and other northern seas.

It is well established that the dispersement of radionuclides is largely a function of their chemical behavior and sorption properties on shelf sediments. Measurement of ^{137}Cs and ^{90}Sr concentrations in the bottom sediments of different granulometric types indicates clearly that there is an increase of ^{137}Cs content with decreasing grain-sizes from sands to clay and silty sediments. This is probably mainly the result of differences in the organic composition of the sediments, which is many times higher in silts, and also with the hydrochemical properties of sea water. Plutonium is mainly associated with organic complexes and sesquioxides (Cook et al. 1997).

A significant correlation between organic carbon and fine silts is also characteristic of sedimentary basins located in the depressions and troughs of the shelf and bays. In these areas, the content of the aleurite-silty sediment fraction is up to 70-95 %. There is a direct dependence between the amount of the aleurite-clayey sediment fraction and ^{137}Cs and $^{239,240}Pu$ levels (Matishov et al. 1998c). For radiocesium, the linear dependence is expressed by a high correlation coefficient 0.9 between the calculated and measured values of surface sediment ^{137}Cs activity and the aleurite-pelite fraction in the less than 0.063 mm sediment size class ($r = 0.91$, $P < 0.001$).

The accumulation of artificial radionuclides in surface shelf sediments depends on the large number of environmental factors. While radionuclide accumulation in deep closed depressions and troughs is determined in part by the distance from local release points and by global fallout, it is also influenced by the rate of sediment accumulation which in turn depends on the surface area of the bank and angle of the slopes from which sediments are washed away by bottom currents. The maximum volume and rate of sedimentation of clayey radioactive particles is noted in the deepest places of troughs (Table 4.11). The aleurite-clayey sediments which contain very large amounts of silt are usually found here (Matishov end Pavlova, 1990).

Table 4.11. Radioactive contamination of shelf troughs and depressions during the 1990s (water – Bq/m^3, bottom sediments – Bq/kg)

Area	Depth, m	^{137}Cs		^{60}Co		^{90}Sr		$^{239,240}Pu$	
		water	ground	water	ground	water	ground	water	ground
Bear Trough	300–500	6–10	4–7	–	No	4–7	–	–	–
Zudcape Trough	300–400	2–7	6–9	–	No	–	–	–	–
Central Hollow	300–390	2–6	5–9	–	No	1–4	0.3–0.8	–	0.9–3.2
Hollow of the Kola Bay	200–350	7–9	1–24	–	0.5–5	2–5	–	0.02–0.07	0.2–5
Gusiny Trough	200–300	–	5–11	–	No	–	–	–	–
Southern Novaya Zemlya Hollow	150–220	4–8	15–44	–	–	3–4	–	0.05–0.1	4–14
Saint Anna trough	400–600	1–4	3–6	–	No	1–2	–	–	0.01–0.5
Eastern Novaya Zemlya Hollow	250–420	4–14	7–30	–	0.5–0.9	2–3	–	0.01–0.02	0.6–1.5
Underwater Ob Valley	10–20	3–6	4–40	–	0.3–0.6	3–11	2–16	–	2–5
Underwater Yenisey valley on the shelf	10–20	3–6	60–100	–	2–6	–	3–11	–	0.6–6
Continental slope in the area of the sunken submarine «Komsomolets»	1700–1800	2–6	4–16	–	–	0.5–4	0.1–0.5	0.01–0.05	0.1–1.7

The natural porosity of sediments influences diagenesis processes in seabed sediment deposits (Matishov end Pavlova 1990). The quantity of such porewaters increases from sandy (10-25 %) to clayey (40-60 % and more) sediments. Obviously, the high saturation of sediments by water directly influences the convective and diffusive transfer of radionuclides. One can envision a mass transfer resulting from the downward flux of porewater and further moderated by the granulometric composition of the bottom sediments.

The majority of biologically available elements that migrate through sediment porewaters (including radionuclides) is concentrated in the liquid phase of sediments in association with soluble organic substances. Undoubtedly, the study of the content and dynamics of radionuclides in porewater is necessary for understanding the processes of artificial radionuclide geochemistry in the biologically-active sediment zone. An understanding of bioturbation phenomena and the resulting transfer of contamination is needed to obtain a complete picture of radionuclide migration.

Radionuclides are fixed in the upper layers of loose fine-grained sediments (to 10-40 cm). It is quite evident that radionuclides detected in the deepest sediment layers were supplied from water masses from 1950-1980s (Fig. 4.20).

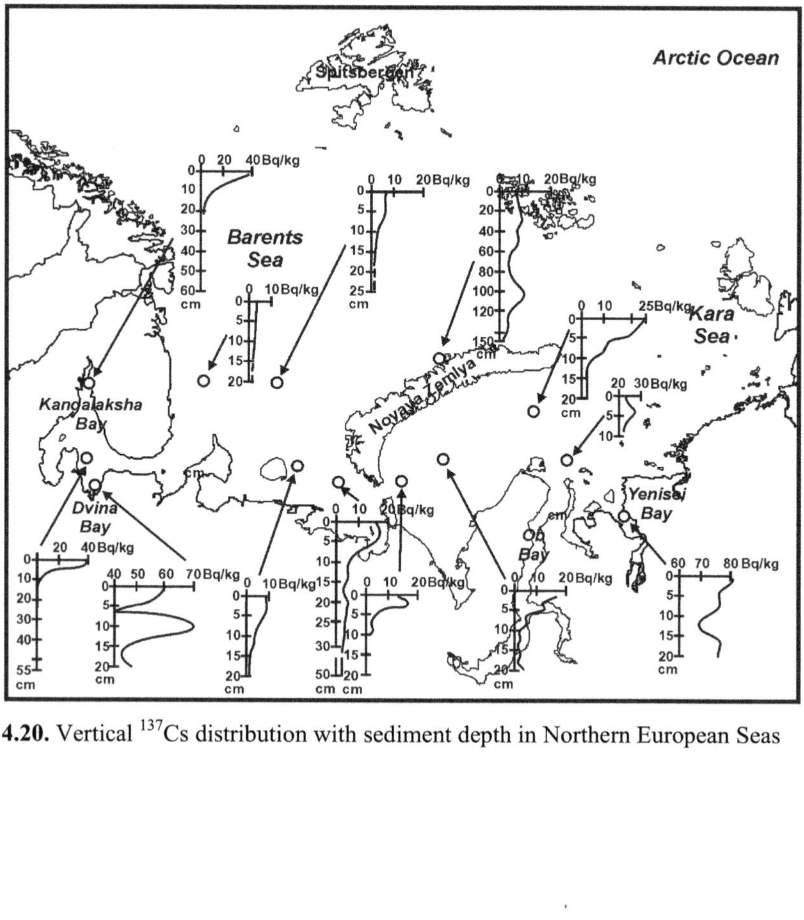

Fig. 4.20. Vertical ^{137}Cs distribution with sediment depth in Northern European Seas

In unmixed sediments the vertical distribution of radionuclides is heterogeneous. Typically, maximum radionuclide levels are associated with sediment depths from 3 to 25 cm. For example, the ^{137}Cs concentration in the vertical section of the Andreev bay bottom sediments varies in the 0-5 cm layer from 53-114 Bq / kg. It then decreases within the 6 to 12 cm sediment depth interval from 8.3-35.4 Bq / kg. ^{60}Co concentrations are less than the detection limit at all depths except within the 6-7 cm layer where it is 5.8 Bq / kg.

In the open part of the Barents Sea the ^{137}Cs exhibits a nearly uniform depth distribution to a depth horizon of 9 cm with ^{137}Cs concentrations varyinf from 8.7-4.6 Bq / kg. In the deeper layers ^{137}Cs concentrations decrease to below the detection limit. A number of deeper layers exhibit radionuclide concentrations that are higher than at the surface sediment layer. This indicates that the sedimentation was not greatly disturbed by bioturbation or other sediment mixing processes.

One exception is the heterogeneous distribution of radiocesium in a 1.5 m gravity column taken in the Nordensheld bay. This column consists entirely of iceberg silt. Iceberg silt has the highest ability to absorb ^{137}Cs in the series: iceberg silt → foraminiferal silt → deep-water red clay. The maximum ^{137}Cs concentration (12.7 Bq / kg) was observed at 1 m depth in the sediment column (Fig. 4.20).

The long-term variation of radioactive contamination in the upper layers of Holocene sediments results in decreasing concentrations of radiocesium and radiostrontium as a result of radioactive decay. At the same time it is obvious that the activity level of the long-lived radionuclide $^{239,240}Pu$ will increase within all components of marine ecosystems. The accumulation of this isotope during the 50 year period of nuclear tests is observable in sediment deposits today.

Radioactive fallout from nuclear tests is the main cause of plutonium accumulation in the Barents and Kara Seas. In this connection, it is worthwhile to investigate the vertical distribution of radionuclides in bottom sediments using the ^{210}Pb geochronology method. This method is used to reconstruct the dynamics of radioactive contamination by assigning time periods to individual sediment depth horizons.

In a number of sediment columns, radionuclide distributions in the lower depth intervals are distributed between many microscopic zones. This feature is commonly observed in silty sediments containing polichaeta (worms) and other silt eating benthic organisms. It indicates a highly localized nature of mixing that results in nuclide migration over vertically connected microzones. As a result of biologically-mediated migration through localized mixing, particles can be transported without destruction (Matishov et al. 1994c; Smith et al. 1995a).

Thus, the sedimentation of radioactive substances in the bays and on the shelf of Arctic and Sub-Arctic Seas depends on a combination of many phenomena and processes. In a number of bays the role of local and regional sources of contamination is important. The ecosystems of such bays are influenced by the emission of technologically-derived radionuclides. Overall during the 1990s, ^{137}Cs and $^{239,240}Pu$ were the only biologically significant radionuclides in shelf sediments. Radiocesium concentrations in shelf sediments are a thousand times lower than in Chernaya Bay. However, the widespread distribution of ^{137}Cs in shelf sediments (average level 2-10 Bq / kg) makes it the more significant dose-forming radionuclide in these regions.

The role of organisms is not fully appreciated among the factors influencing the accumulation of radioactive contamination on the seafloor. The role of biofiltration in the transfer and transformation of artificial radionuclides in marginal seas and in the open Ocean should be more intensively investigated because different environmental conditions result in different levels of radioactive contamination in benthic organisms and, correspondingly, different dose loads to organisms inhabiting the sea floor.

4.3 Ecological characteristics of benthic organisms

Radioactivity levels in benthic organisms of the northern seas are connected with the behavior of organisms living on the sea bottom. The benthic zone is the main area where suspended mineral and organic substances undergo transformation and accumulation. Particulate matter is transferrred from the pelagic zone to the seabed where microorganisms, together with benthic animals transformation the sedimented materials.

Sediment deposits are the main accumulators of cesium, plutonium and other radionuclides and serve as the habit for zoobenthic organisms. The species composition and behavior of bottom fauna are determined by the characteristics of shelf topography, the texture of bottom sediments, bottom currents and the hydrochemical regime. These factors taken as a whole determine the complicated interconnection of macrophytes, benthos, bottom fish and, through trophic relationships - of birds and seals. The comparability of benthic and zooplankton biomass is indirect evidence of the importance of organic sedimentation processes on the shelf.

The existence of species on the shelf, continental slope, abyssal plains, and mid-ocean ridges depends directly or indirectly on the vertical flux of particles to the bottom. In the course of sedimentation from surface waters these particles influence the chemical composition, forms of chemical elements and composition of sea water and, further, of bottom sediments and sediment porewater. Plankton plays a primary role in the formation of suspended matter in the Ocean and in the transportation of sediment containing radioactive substances to the bottom (Lisitsyn end Vinogradov 1982; Izrael end Tsyban 1989; Lisitsyn 1994; Matishov et al. 1994c). It must be added that the transport of organic substance to the sea bottom is the primary source of nourishment and energy for bottom dwelling animals.

The total biomass of the Barents Sea benthos is estimated at 150 million tons (Zenkevich 1963; Life and environment ... 1989). The biomass of benthos per unit of area varies mainly from 10-15 to 100-400 g / m^2. However, in invertebrate colonies, of for example the Icelandic comb-bearer *Chlamus islandica*, the population density reaches 2 kg / m^2 (Denisenko 1989). The species composition of zoobenthos groups will be characterized below in order to represent the conditions leading to pollutant accumulation in the organs and tissues of bottom invertebrates.

The benthos composition of Arctic and Sub-Arctic Seas includes, depending on the basin, from one to four thousand animal species. The largest species groups among the bottom invertebrates of the Barents, Norwegian, Kara and other seas are Echinodermata, Polichaeta, Bryozoa, molluscs, sponges, Hydroides and crustaceans (Galkin 1985; Denisenko 1989; Denisenko et al. 1992; Anisimova end Luppova 1995). Significant differences in benthic species composition is observed between the coastal and open sea areas. Molluscs, with biomass to 30 kg / m^2 prevail in the coastal zone, for example, in the beds of blue mussels. For the transition from rocky to sandy and silty sediments the quantity of bottom animals usually increases while the average biomass decreases because of a reduction of epifauna (Zenkevich 1963).

Polichaeta, bivalve molluscs, Echinodermata and crustaceans prevail in the Barents Sea benthos. These groups are distributed on the shelf unevenly. In the Pechora Sea and on the Novaya Zemlya shelf Bivalvia represent more than 75 % of all benthic species. The small forms prevail - Portlandia, Leda, Yoldia, and some species of Astarte. An increase in biomass for bivalve molluscs is also noted on the Hope-Bear Island Bank. Both small species (Astarte) and large ones (Pecten, *Macoma calcarea*) are widespread on the shallows. In deep depressions and troughs, the quantity and biomass of bivalve molluscs decreases sharply because of the absence of sufficient quantities of food. Polychaeta also prefer soft clayey sediments.

Echinoderma represent 50-60 % of all zoobenthos in the northern deep-water areas and also in the south-western and central parts of the Barents Sea. The *Ctenodiscus crispatus*, *Brisaster fragilis*, *Ophiura sarsi* and others are among the most numerous species. The prevalence of areas with depths of 150-350 m in the south-west are occupied by species of sponges among which the large species *Geodia barretti* prevails, and is noted for having a biomass of as much as 20 kg. Sponges constitute 95 % of all benthos in these areas. Their biomass reaches 5-6 kg / m^2.

Bottom crustaceans, the shrimp Pandalus borealis and Kamchatka crab Paralithodes camchatica occupy a unique niche in the near-bottom zone. Shrimps have adapted to hide themselves in the silty sediment. Many crabs and shrimps migrate long distances, up to 100-200 km from the open parts of the sea to the coastal zone.

The life cycles of benthic organisms average 4-9 years (Zenkevich 1963). These inhabitants of the seabed extract both the nutritive substances and radioactive particles from bottom waters and sediments during their life spans.

Bottom invertebrates are divided into three trophic groups depending on their feeding behaviors. The first group includes filter-feeders (sestonofags) for example, bivalve mollluscs, which consume detritus suspended in the near-bottom water layer. Many organisms seize the material from suspension and thereby facilitate sedimentation. Mussels, for example, consume on average 1.5 liters of water per hour, filtering almost all particles from the water. Mussels, which are found in populations of high density, process as much as 150 tons of water hourly per 1 m^2 of the mussel colony surface. As a result, the water becomes completely clean from particles which then are deposited on the seafloor (Klenova 1948).

The second group consists of detritofags, ophiuroids for example, which gather and sort suspended organic particles both from the bottom water layer and from the upper layer of sediments. detritofags are found more often in areas of intensive sedimentation in shelf depressions.

The third group includes the unselective feeders - Polychaeta, sea cucumbers and sea urchins. These organisms consume sediment, penetrating to sediment depths of 5-10 and sometimes to 30 cm. Benthic organisms, nourished by organic matter contained in soft sediments (50-90 %) usually inhabit shelf troughs and depressions. As mentioned previously, clayey silts contain about 4 times more detritus than sands. This type of sediment contains the largest quantity of radioactive contamination.

All trophic groups are distinctly connected with the sedimentation regime. The distribution of detritus on the seafloor is largely controlled by bottom relief. In shallow areas water circulation intensifies, and sedimentation is hindered. In bottom depressions and on flat bottom areas detritus is accumulated as seabed sediments.

The character of sedimentation and trophic groups inhabiting the benthic zone depends to a great extent on the bioproductivity of the pelagic zone. Zooplancton, which filter organic and radioactive particles from surface waters, contributes to their sedimentation on the sea bottom. Artificial radionuclides accumulating in the marine ecosystem are passed from phyto- to zooplankton. Radionuclides are transferred to benthic animals with fecal pellets produced by planktonic crustaceans. Benthic organisms receive the bulk of their radioactive contamination from bottom sediments.

4.4 Radionuclide accumulation by benthic organisms

There are thousands of species of benthic organisms. For the present the contamination of about 60 species has been analyzed comprising the principal taxonomic groups (Bivalvia, Gastropoda, Echinodermata, Crustacea, Decapoda, Annelida, Bryozoa) of the Barents, Pechora, White and Kara Seas (Strand et al. 1993; Kuznetsov et al. 1993; Matishov et al. 1994b,c, 1995a, 1996c,f; Matishov D. et al. 1995a; Sickel et al. 1995; Rissanen et al. 1996, 1997a,b, 1999a; Kuznetsov et al. 1996, 1997a,b). About 300 specimens of benthic fauna were studied in 1991-1999 from approximately forty areas of the polar shelf (Matishov et al. 1994b,c: Matishov D. et al. 1995a). In all specimens, the ^{137}Cs concentrations were low. In particular, they were below the limit of detection (0.1 Bq / kg wet weight) in 160 samples. Similarly, shrimp taken from the Barents Sea were found to have ^{137}Cs

concentrations below the detection limit (0.2 Bq / kg wet weight) in all but one case (Sickel et al., 1995).

During the 1990s invertebrate organisms contained from 1 to 9.5 Bq / kg of ^{137}Cs, ^{90}Sr and $^{239,240}Pu$ (Tables 4.12, 4.13). Northern shrimp contained from 1.6 to 7.9 Bq / kg, and sea urchin - 1.8 Bq / kg of cesium. Artificial radionuclide concentrations were not investigated in bottom organisms during the period of maximum radioactive fallout during the atmospheric and underwater nuclear weapon tests in 1960s.

Table 4.12. Radionuclides in marine organisms (data of the Murmansk District Sanitary and epidemiological Station)

Species	Date	^{137}Cs, Bk/kg	Total β-activity, Bk/kg
Sea comb-bearer	19.07.91	1.0	119.2
		1.2	42.5
	27.07.92	3.2	–
		6.4	–
	11.08.92	2.0	–
		1.5	–
		9.5	–
Sea urchin	15.06.92	1.8	11.6
		0.0	10.3
Shrimp	03.07.92	2.5	49.0
		3.5	88.2
	23.07.92	7.9	–
		1.6	–
Crab	11.09.92	4.5	–

The phenomenon of higher ^{137}**Cs** absorption by silt-eating worms (Annelida) living in Arctic Seas has been documented (Matishov et al. 1996c,f; Rissanen et al. 1997a) (Fig. 4.21, Table 4.14). Large polychaeta (Sipunculidae, Golfingia) contain from 4.7 to 6.2 Bq / kg of radiocesium. However, the highest concentrations (to 17-20 Bq / kg) of this nuclide are noted in the chitinous tubes of polychaeta Pectinaria, Nepthys. A positive correlation exists between radiocesium in worms and the perrcentage of fine aleurite and clayey silts in the sediments in which the worms live.

Benthic organisms collected between 1991-1995 on the continental slope of the West Barents Sea edge (the area of sunken atomic submarine «Komsomolets»), for example, amphipoda, decapoda, gastropoda, foraminifera, contained low radiocesium concentrations (1-10 Bq / kg) (Kuznetsov et al. 1997) (Table 4.15). Organisms such as Ampeliscidea contained relatively higher concentrations of radiocesium (13-15 Bq / kg). The amount of radioactive contamination found in benthic organisms increased with decreasing distance toward the hull of submarine.

The influence of Chernobyl discharges on the benthos can be ascertained by looking at the radionuclide content of mollluscs (*Macoma baltica*) and crayfishes (*Mesidothea entomon*) in the Bothnia and Finnish Bays (Table 4.16) (Ilus et al.

1993). In summer and autumn of 1986 the ^{137}Cs and ^{134}Cs concentrations in these fauna species varied correspondingly from 20 to 150 Bq / kg and from 11 to 74 Bq / kg.

Table 4.13. ^{137}Cs, ^{40}K, ^{99}Tc content in benthic organisms from 1996-1998 (by: Bryngot et al. 1999a) and 1999, Bq / kg wet weight

Species	Coordinates, area	^{137}Cs	^{99}Tc	^{40}K
	Barents Sea, 1996			
Starfishes	69°13'N 36°11' E	<0.4		–
Neptunea sp.+Actinaria sp.		<1		–
Starfishes	69°17'N 35°47' E	<1		–
Paralithodes camtschatica		<0.2		–
Starfishes	69°19'N 35°49'E.	<0.4		–
Echinus sp.	71°29'N 48°39'E	<0.5		–
Sponges + Ophiuroidea	69°19'N 35°49'E	0.16±0.05		–
Holoturoidea		0.4±0.2		–
Hyas sp.+Pandalus borealis		0.9±0.3		–
Ehinus sp.		<0.6		–
	Norwegian Sea, 1997			
Homarus vulgaris	Bergen	0.8–1.0	42–35	–
		Not found	–	–
	Oslo Fjord	–	11.1–14.6	–
Mytilus edulis	Hyllesoy, Troms	<0.08	0.5	–
	Oslo Fjord	–	0.7	–
Pandalus borealis	Oslo Fjord	–	0.3–0.8	–
	Norwegian Sea, 1998			
Homarus vulgaris	Oslo Fjord	0.13–0.16	–	–
Cancer pagurus	Sunnhordaland	0.14–0.80	–	–
		0.70	–	–
Mytilus edulis	Hyllesoy, Troms	<0.08	–	–
	Oslo Fjord	0.20		–
Pandalus borealis	Norwegian, North Seas	0.12	–	–
	North Seas	0.14	–	–
	Oslo Fjord	0.18	–	–
	Barents Sea, 1999			
Pandalus borealis	St. 10	<0.59	–	67±8
	St. 1	<1.41	–	71±19
	St. 12	<0.20	–	48±8
	St. 9	<0.36	–	46±3
	St. 3	<0.49	–	56±6
	St. 1, 2	<0.22	–	59±3

The order of decreasing concentrations of ^{137}Cs (Bq / kg) is as follows: Annelida (worms, polychaeta)-4.7-6.2, chitinous tubes-10-20→Bivalvia-1-9→Decapoda (decapods, shrimps)-0.3-8→Amphipoda (sand hoppers, amphipods)-2-3→Echinodermata-0.3-2.2→Gastropoda-0.3-1. Polychaeta, shrimps and other animals that dig into the seabed and ingest silty sediments enriched by radionuclides contain the highest radiocesium concentrations (Fig. 4.22, 4.23).

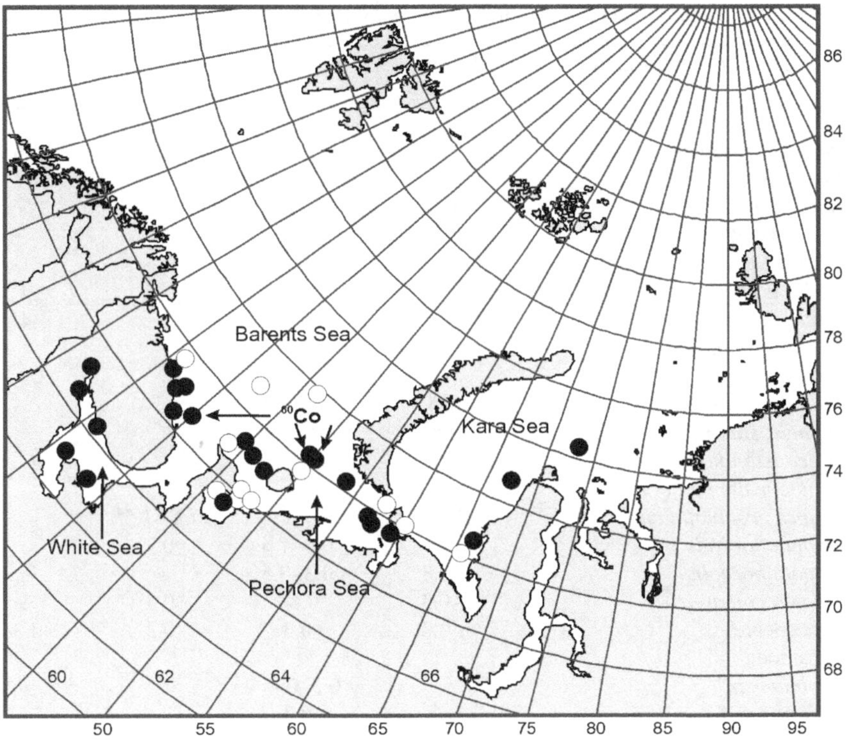

Fig. 4.21. Benthos sampling. The stations where ^{137}Cs concentrations are greater than 0.3 Bq / kg dry weight are designated by black circles. Traces of ^{60}Co were detected at three sites only

^{60}Co is found in very small quantities in benthic organisms (0.1-1.5 Bq / kg) (Strand et al., 1993; Strand, 1998; Matishov et al., 1994c; Rissanen et al., 1996, 1997a) (Table 4.17). The shells of *Mutilus edulis*, *Modiolus modiolus* and polyps on modiolus species collected in the Barents Sea to the north from Kolguev Island contained about 0.1-0.2 Bq / kg of ^{60}Co. Some macrophytes also contained this nuclide. Higher concentrations of ^{60}Co were detected in polyps removed from the shells of *M. modiolus* (to 4 Bq / kg).

Gammarida, Ophiura and crabs (*Hyas* sp.) collected from the Novaya Zemlya bays near dumping areas for nuclear waste have accumulated insignificant amounts of ^{60}Co (0.7-1.3 Bq / kg) (Table 4.18) (Foyn, Semenov, 1993; Strand et al., 1993). Ophiures (*Ophiocten gracialis* Sars) and starfishes (*Pontaster tenuispinus*), inhabiting the bottom near the submarine «Komsomolets», contained similarly low levels of ^{60}Co (Kuznetsov et al., 1996, 1997a). Isopods *Mesidothea entomon* from the Finnish Bay also had approximately the same level of ^{60}Co contamination, about 2.7 Bq / kg after the Chernobyl accident (Ilus et al. 1993).

Table 4.14. ^{137}Cs concentration in water (Bq / m^3), biota and bottom sediments (Bq / kg) of the Barents, Pechora, Kara and White Seas in 1993-1996

Object of investigation	Barents Sea	Pechora Sea	Kara Sea	White Sea
1	2	3	4	5
Upper water layer	6	7	7	12.6
Macrophytes				
Fucus vesiculosus	0.8	1.5	–	3.7
Benthic organisms				
Bivalvia				
Mytilus edulis[a]	<0.3–2.2	<0.3		1.4
Mytilus edulis[b]	0.4	<0.3		1.4
Modiolus modiolus[b]	<0.3–0.5	<0.3		–
Musculus niger[b]	–	–		3.2
Yoldia hyperborea[b]	1.5	<0.3–1.1		–
Chlamys islandica	0.8[a]	0.3[b]		–
Arctica islandica	–	–		1.6[a]
Ciliatocardium ciliatum[a]	1.6	1.6	<0.3	–
Serripes groenlandicus	<0.3[a, b]	<0.3[b]	<0.3[b]–1.5[a]	–
Tridonta borealis[a]	–	<0.3–1.4	<0.3	–
Tridonta borealis[b]	<0.3	<0.3–0.5	–	–
Macoma calcarea[b]	<0.3	0.8	<0.3	1.1
Other species	<0.3	<0.3	<0.3	<0.3
Gastropoda				
Buccinum sp.[a]	–	0.5–0.8	–	–
Buccinum sp.[b]	–[a]	<0.3	–	–
Cryptonatica sp.[b]	–	0.4–0.5	–	–
Littorina saxatilis[b]	<0.3	–	–	<0.3
Other species	<0.3	<0.3	–	–
Echinodermata				
Asterias rubens	<0.3	<0.3–0.6	<0.3	1.1
Stegophiura nodosa	<0.3	<0.3–0.5	–	–
Strongylocentrotus sp.	0.8–0.9	<0.3	–	–
Cucumaria frondosa	0.8–1.4	0.9	–	–
Myriotrochus rinkii	–	–	9.5	–
Decapoda				
Pagurus pubescens	<0.3	<0.3–0.8	–	–
Hyas araneus	<0.3	<0.3–0.7	–	–
Balanus crenatus	0.3–0.4	–	–	–
Amphipoda				
Gammarus sp.	2.2	–	–	–

Table 4.14. (Cont.)

1	2	3	4	5
Bryozoa				
Alcyonidium disciforme	–	1.8	–	–
Isopoda				
Mesidothea entomon	–	–	3.1	–
Annelidae				
Golfingia sp.	6.2	4.7	5.9	–
Polychaeta				
without tubes	1.2	–	2.2–4.8	–
tubes	6–12	–	17	–
tubes Pectinaria	1.7	–	12	–
tubes Nephtys	–	–	20	–
tubes Spiochaetopterus	–	–	6	–
Bottom sediments	0.1–11	0.5–8.8	1–32	2–40

FOOTNOTE: a - organic part; b - whole organism

Table 4.15. Natural and anthropogenic radioactivity in benthic organisms from different trophic groups and of bottom sediments in the area of the sunken nuclear submarine «Komsomolets» (by: Kuznetsov et al. 1997a), Bq / kg

Station	Trophic groups, species	^{40}K	^{137}Cs	^{90}Sr
1	2	3	4	5
	Immobile (fastened) filtering sestonofags			
3177	Spongia	<800	<30	9.5±2.5
	Alcyonaria			
3495	*Eunephthya* sp.	<800	<30	<10
	Ascidiacea			
3500	*Styella* sp.	550±200	26±10	3.4±2.0
	Average	550±200	<30	6.4±3.0
	Mobile filtering sestonofags			
	Amphipoda, Ampeliscidae			
3153	*Haploops setosa* Sars	340±300	<3.0	1.4±0.7
3500		350±150	<20	9.6±4.0
3500		<800	67±30	2.6±2.0
3555		<800	<30	<3
3500	Shelters *H. Setosa*	580±200	29±15	6.9±0.3
	Average	460±200	<30	5.0±2.5
	Gathering detritfags			
	Foraminifera			
3482	*Cribrostomoides subglobosum*	470±150	<2.5	0.6±0.7
	Polychaeta			
3153	*Myriochele danielsseni* (Hansen)	320±300	14±15	1.4±0.7
3177		330±300	15±16	1.4±0.7
	Ophiuroidea			
3177	*Ophiocten gracilis* (G.O.Sars)	210±200	1.5±0.7	1.4±0.7

Table 4.15. (Cont.)

1	2	3	4	5
3153		210±200	1.5±0.7	1.4±0.7
3482		800	<30	<5
3544		750±300	<50	9.0±6.0
	Elpidiidae			
3495	*Elpidia glacialis* (Theel)	600	<28	<3
		800	<30	<10
	Average	380±150	7.0±5.0	2.5±1.0
	Carnivores (predators and carcass eaters)			
	Coelenterata			
3495	Actiniaria	<600	<20	<5
	Polychaeta			
3153	*Augeneria albidentata sadko* Annenkova	340±300	<3	1.4±0.7
	Gastropoda			
3495	*Natica (?)* sp.	220±100	4±5	3.7±3.0
	Buccinidae	<600	<20	<10
		<800	<30	<10
	Isopoda			
3153	*Mesidothea* sp.	340±300	<3	2.5±2.0
	Decapoda			
3555	*Bythocaris leucopis* Sars	340±300	<3	1.4±0.7
		<800	<30	2.5±2
	Asteroidea			
3482	*Pontaster tenuispinus* (Duben et Koren)	280±200	2.3±3	1.4±0.7
3495	*Bathybiaster vexillifer* (Wyville Thonson)	550±200	<10	2.4±1.5
3500		<850	33±20	<3
		<300	<14	3.3±2.0
		<800	<20	<10
	Average	340±150	<20	2.5±1.5
	Bottom sediments			
3499	Aleurite-pelite silt	390±150	16±6	3
3537		440	20±15	5.0±6
3543		770±300	14±0.8	5.2±0.8
3549		190±100	11±4	3
3554		660±300	11	3
3594		860±400	11	3
	Average	570±200	13.8±6	3.7±1.2

FOOTNOTE. The average values being calculated, upper limits were not taken into account

^{90}Sr is present in some species of bottom invertebrates in small quantities (0.01-0.4 Bq / kg). ^{90}Sr accumulation in foraminifera, ophiures and other benthic species on the continental slope near the submarine «Kursk» is also very low - 0.01-0.34 Bq / kg (Kuznetsov et al. 1996, 1997a).

Table 4.16. Radionuclides in Baltic Sea benthic animals in 1986, Bq / kg dry weight

Nuclide	*Macoma baltica*			*Mesidothea entomon*		
	1		2	1		2
	61°14' N 21°24' E, h=12–14 m, 13.08	60°50' N 20°24' E, h=53–58 m, 26.10	60°22' N 26°23' E, h=15–17 m, 15.05	63°00' N 20°30' E, h=76 m, 25.10	60°50' N 20°24' E, h=53–58 m, 26.10	60°09' N 26°35' E, h=50 m, 28.10
^{40}K	70	61	52	220	300	260
^{54}Mn	0	0	0	0	8.8	1.6
^{60}Co	0	0	2.7	0	0	0
^{89}Sr	26	–	34	–	–	–
^{90}Sr	27	–	8.2	–	–	–
^{95}Zr	0	0	18	0	0	0
^{95}Nb	0	0	75	0	0	0
^{103}Ru	34	8.1	140	0	5.2	6.0
^{106}Ru	46	26	51	0	22	14
^{110m}Ag	86	56	12	49	180	59
^{125}Sb	0	0	5.2	0	0	0
^{129m}Te	140	0	170	0	0	0
^{134}Cs	11	16	58	26	74	25
^{137}Cs	22	31	99	49	150	58
^{141}Ce	0	0	21	0	0	0
^{144}Ce	0	0	20	0	0	0

FOOTNOTE. 1 - Bothnia Bay, 2 - Finnish Bay

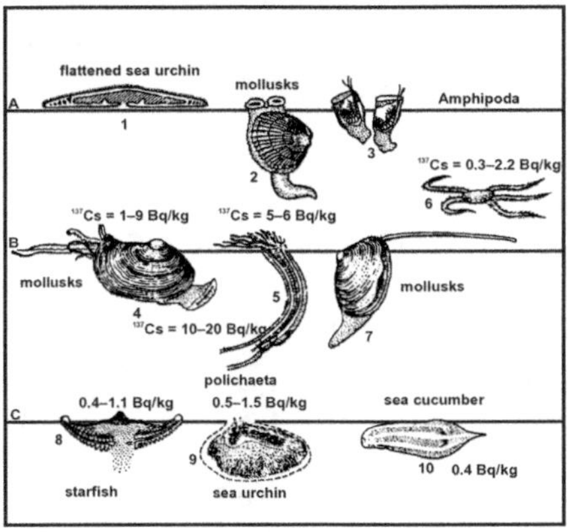

Fig. 4.22. Zoobenthic species in Northern European Seas and representative concentrations of ^{137}Cs: **A** - mobile forms feeding by seston; **B** - detritus feeders; **C** - sediment feeding: **1** - *Echinarachnus parma*, **2** - *Cardium ciliatum*, **3** - Ampheliscidae, **4** - *Yoldia thracialformis*, **5** - Terebellidae, **6** - Ophiura, **7** - *Macoma calcarea*, **8** - *Ctenodiscus crispatus*, **9** - *Brisaster latifrons*, **10** - Malpadiidae

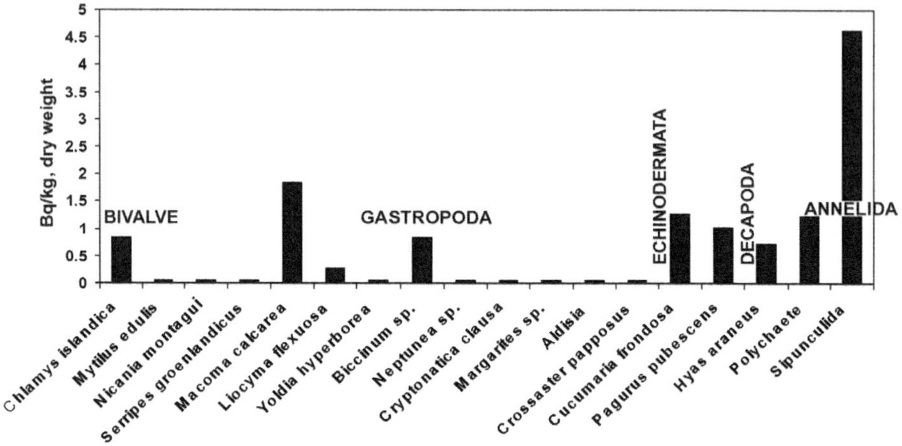

Fig. 4.23. ^{137}Cs concentrations in benthic samples collected in the Pechora Sea

Table 4.17. ^{60}Co concentration in benthic organisms (Bq / kg dry weight)

Date	Coordinates		Sample	^{60}Co
	lat. N	long. E		
13.07.93 г.	65°51’	50°01’	Shells *Mytilus edulis*	0.12
27.07.93 г.	69 50	50 43	Shells *Modiolus modiolus*	0.05
27.07.93 г.	69 50	50 43	Polyps on shells *M. modiolus*	4.00
28.07.94 г.	68 17	39 32	Parts of shells	0.07

Table 4.18. ^{60}Co concentration in benthic organisms (Bq / kg dry weight) ^{137}Cs, ^{60}Co, ^{239,240}Pu concentration in different biota from the Kara Sea in 1992 (Strand et al. 1993), (Bq / kg dry weight)

Station	Species	^{137}Cs	^{60}Co	^{239,240}Pu
1	Gammarida	1.4	<0.4	0.010
2		1.3	<0.5	0.012
6		1.8	<0.7	0.007
Average		1.5±0.3		0.011±0.003
4	Ophiura	0.7	<0.8	0.114
5		<1.2	<1.3	0.160
7		<0.6	<0.6	0.089
11		<0.6	<0.6	0.105
Average				0.118±0.031
12	*Hyas* sp.	0.44	0.4	0.024

The influence of radioactive Chernobyl fallout on ^{90}Sr in the Baltic and Azov Sea benthos did not exceed the levels expected from nuclear weapons testing alone. In the middle of May 1986 the *Macoma baltica* and *Mesidothea entomon* accumulated up to 27 and 2.7 Bq / kg of ^{90}Sr respectively (Ilus et al., 1993). The ^{137}Cs content in mussels (*Mutilus galloprovincialis)* in the Azov Sea was 0.4-2.3 Bq / kg wet weight in hard parts and 1.4-1.7 Bq / kg wet weight in soft parts. The corresponding values for ^{90}Sr were 0.4-12.1 and 1.4 Bq / kg wet weight. The content in Decapoda crayfish was 1.8-6.3 Bq / kg wet weight (^{137}Cs) and 0.3-0.4 Bq / kg wet weight (^{90}Sr).

^{239,240}Pu concentrations in benthic organisms in northern seas is currently extremely low (0.01-0.2 Bq / kg) (Table 4.18). These levels are observed in gammaruses, ophiures and crabs from East Novaya Zemlya Bays near nuclear dumping sites ((Strand et al., 1993; Strand, 1998). Benthic organisms collected during 1992 on the continental slope of the Barents Sea western margin also contained negligible quantities of ^{239,240}Pu (0.03-0.05 Bq / kg). Seawater, in contact with bottom dwelling invertebrates contains levels of ^{239,240}Pu expected from nuclear weapons testing.

The concentration of artificial radionuclides in living benthic organisms does not exceed the levels associated with nuclear weapons testing. The migration and distribution of radionuclides in benthic ecosystems is largely controlled by processes of sorption and filtration by the organisms, burrowing activity, and the advection and diffusion of porewater through the upper layers of bottom sediments.

Some polychaetes from areas of the Barents and Kara Sea shelves accumulate ^{137}Cs up to concentrations of 17-20 Bq / kg. Evidently, this is caused by the feeding behavior of polychaeta. These organisms swallow silty (clayey) sediments which contain relatively high concentrations of radionuclides.

Benthic organisms from the White Sea inhabit sediments containing ^{137}Cs concentrations from 1 to 70 Bq / kg. These levels are considerably higher than for sediments of the Barents Sea shelf. As a consequence, bottom fauna of Kandalaksha Bay are subjected to insignificant irradiation, but these irradiation levels are still several times more than for benthic organisms living in the Gusiny, Bear and other troughs of the Barents Sea.

The inhabitants of the sea bottom in Chernaya Bay are in contact with bottom sediments where the concentration of ^{137}Cs, ^{60}Co, ^{239,240}Pu exceeds concentrations in the Barents Sea by hundreds and thousands times (Matishov D. 1993a; Matishov et al. 1994c). An analysis of benthic species composition conducted during the 1920s and 1990s does not show radical differences. Some taxonomic changes are natural and are not the consequence of nuclear weapons tests carried out in the Bay area during the 1950-1960s (Matishov end Denisenko 1996).

Overall radioactivity levels in benthic species from the Arctic and sub-Arctic seas has been poorly investigated. Benthic organisms inhabiting the majority of bays and parts of the shelf have not been investigated at all. It is necessary to study such commonly found species such as sponges, sea cucumbers, sea urchins and starfishes. The analysis of contamination of several tens of species does not provide a sufficient understanding of radionuclide accumulation in the remaining thousands of species of bottom fauna.

Finally, the high species diversity and abundance of the bottom fauna at least, in some areas, helps to support the high productivity of valuable marketable fish species - seacat, halibut, perch and other bottom fishes. Thus far benthic species do not contain dangerous levels of radioactive contamination. About one third of the population of benthic organisms consists of forms not consumed by fishes, for example, sponges, large mollluscs, Echinodermata and crustaceans.

4.5 Radionuclides in the ecosystem of the Murmansk Bank (site of the submarine "Kursk" accident)

The radioecological situation in the Barents Sea intensified sharply during 1989 and 2000 because of the sinking of two nuclear submarines «Komsomolets» and «Kursk». The misgivings are caused by a number of circumstances: the uncertainty of the technical state of submarine nuclear reactors; the possibility of radiation contamination of cod, halibut, shrimp, scallops, Kamchatka crab and other fauna species important for the world fishery. The Murmansk Bank, where the submarine «Kursk» landed (69° 37.5′ N, 37° 34.6′ E, depth 108 m) is one of the main areas of the native marine fishery. The submarine with two reactors and about 300 kg of fuel was located a distance of 80 km from the Murmansk Marine Biological Institute base in the Dalnie Zelentsy settlement.

The radioactive contamination was very low for all elements of the Barents Sea ecosystem before the «Kursk» accident. In the Barents Sea fauna, ^{137}Cs, ^{90}Sr and ^{239,240}Pu concentrations in different species of benthic organisms varied from 0.3 to 6.2, from 0.01 to 0.4 and from 0.01 to 0.05 Bq / kg respectively (Matishov et al. 1994b,c, 1996f, 1998b). Pelagic fish species (polar cod, herring), benthos-eating fish (haddock, catfish) and carnivorous fish (cod, saithe, halibut) contained very low concentrations of ^{137}Cs (0.2-2.5 Bq / kg) and ^{90}Sr (0.01-0.9 Bq / kg).

During the 1990s, the upper layer of seawater (to 100 m) contained, depending on the sea area, 2-9, 1-7 and 0.0001-0.01 Bq / m^3 of ^{137}Cs, ^{90}Sr and ^{239,240}Pu (Matishov et al. 1997c,f; Nies et al. 1997; Strand 1998).

These concentrations were the result of the influx of Chernobyl contamination from the Baltic Sea and by Sellafield releases. A small quantity of artificial radionuclides accumulated in shelf bottom sediments. Normally, there is very limited amounts of ^{137}Cs, ^{90}Sr and ^{239,240}Pu in the sandy and coarse aleurite sediments on the continental shelf to water depths of 130-150 m - 0.2-4.0, 0.1-0.5 and 0.1-1.0 Bq / kg respectively (Matishov et al. 1993b, 1994b,c, 1997c,f, 1999d). Higher concentrations of radionuclides (^{137}Cs - 5-11 Bq / kg) accumulated in clay and fine aleurite sediments in troughs.

The Murmansk Bank is located in water depths of 90-160 m. It stretches for 260 km in a south-east direction and is bordered with troughs 200-250 m deep. The nuclear submarine «Kursk» sank in the upper part of the bank named «Murmansk shallows». The slopes of the bank are exposed to water from the Murmansk (Nordcape) current which in 1995contained 3-4 Bq / m^3 of ^{137}Cs (Nies et al. 1997). ^{137}Cs and ^{90}Sr were present in small quantities (0.2-4.2 and 0.1-0.43 Bq / kg) in sands and fine aleurites (Matishov et al. 1994c). The ^{137}Cs concentration in cod during the 1990s varied from 0.3 to 0.9 Bq / kg.

During the period September 19-24, 2000, MMBI carried out an expedition to investigate radioactivity in the marine environment and biota of the Murmansk Bank (Fig. 4.24). Water, sediment and samples of local fauna were collected including: zoobenthos (*Cucumaria frondosa*, *Urasterias linckii*, *Spongia* gen. spp., *Hydroidea* gen. spp.) bottom fishes (seacat, ray and others). The radioecological condition of the Murmansk Bank at the end of September 2000 did not differ from the situation before the «Kursk» accident. Seawater on the surface and in the water thickness to the bottom of the bank concentrations of ^{137}Cs were between the limits of 0.5 to 3.5 Bq / m^3. The ^{90}Sr concentration in water was also low for all depth intervals (0.8-5.4 Bq / m^3). The same low level of ^{137}Cs observed in the coastal branch of the Murmansk Current was detected at the control station in the area of Kildin Island (2.1-2.9 Bq / m^3).

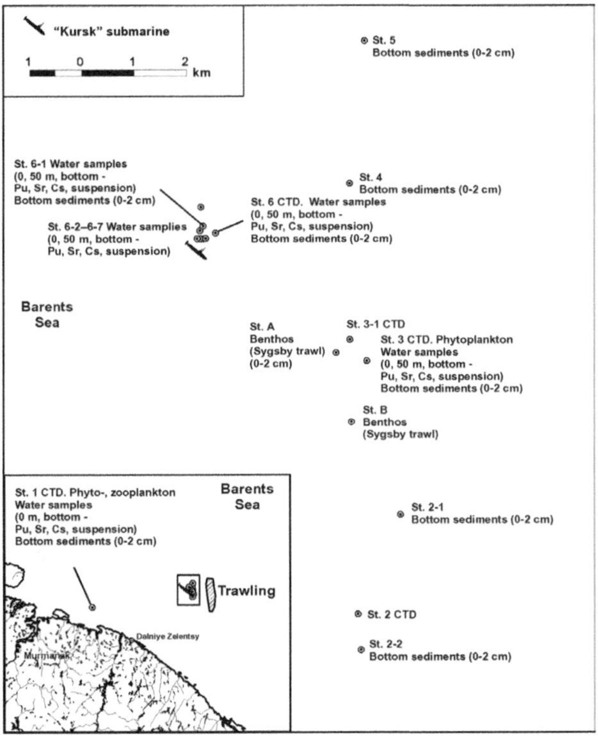

Fig. 4.24. Sampling stations near the site of the sunken submarine «Kursk» September 19-24, 2000) (Matishov et al. 2002)

The radioactive contamination of the seabed was also found to be insignificant. ^{137}Cs accumulations in sediment of the Murmansk Bank southern slope, was 0.6-3.3 Bq / kg dry weight (Fig. 4.25). The activity of ^{137}Cs in zoobenthos (sponges, hydroides, Echinodermata, Bryozoa) also was very small - 0.1-1.0 Bq / kg wet weight. The highest ^{137}Cs activity (26 Bq / kg wet weight) was noted in sea anemones (Actiniaria) (Fig. 4.26). Obviously, this relatively high level is related to the bottom substrate feeding behavior of these bottom animals.

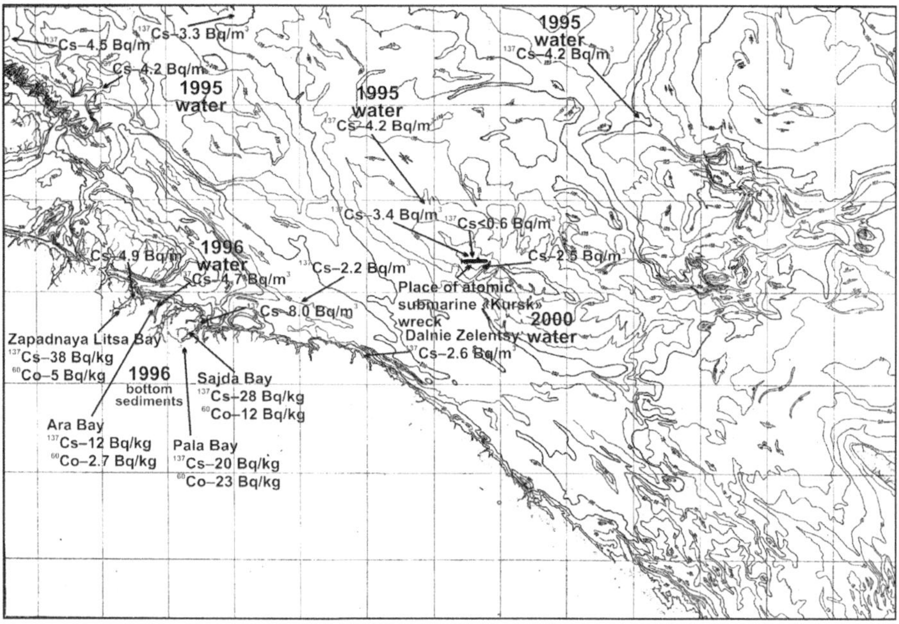

Fig. 4.25. ^{137}Cs content in the Barents Sea water and bottom sediments in the area of nuclear submarine «Kursk» accident in 1995, 1996 and 2000

The 137Cs activity in tissues and organs of marketable fish species varied from 0.2 to 0.8 Bq / kg wet weight (Fig. 4.27). This level of cesium bioaccumulation in pelagic (herring) and bottom (American plaice, catfish, cod, starry ray) corresponds to global radioactive fallout in the northern seas.

The analysis of samples of seawater, bottom sediments, zoobenthos, pelagic and bottom fish species by gamma spectrometry produced no traces of the short-lived radionuclides (^{134}Cs, ^{140}La, ^{131}I, ^{140}Ba, ^{141}Se, ^{95}Ni, ^{103}Ru, ^{95}Zr) which are associated with irradiated nuclear fuel from the «Kursk» reactors. ^{137}Cs and ^{90}Sr in all ecosystem components from the investigated area are within the expected limits. Thus, the submarine nuclear reactors did not leak any radioactive substances during the period of observations.

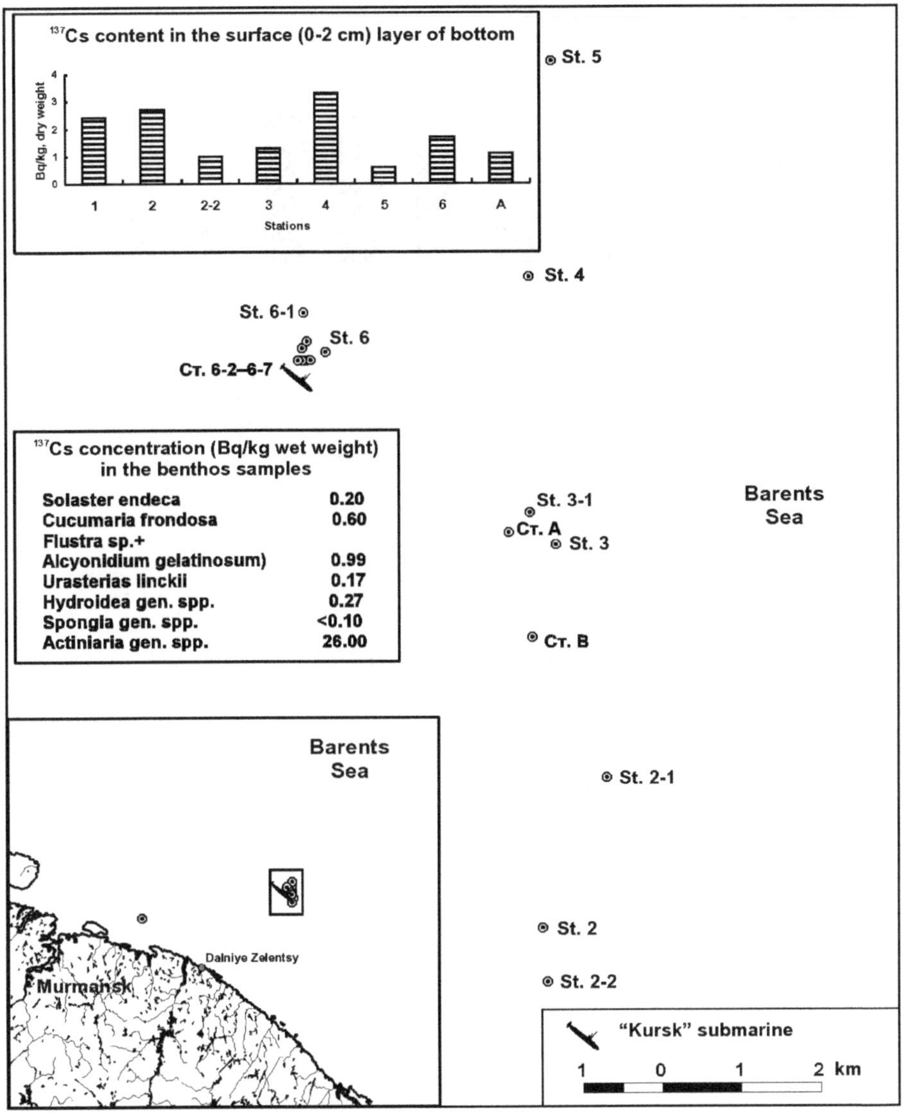

Fig. 4.26. ^{137}Cs content in surficial bottom sediments and benthic organisms near the sunken submarine «Kursk» (Matishov et al. 2002)

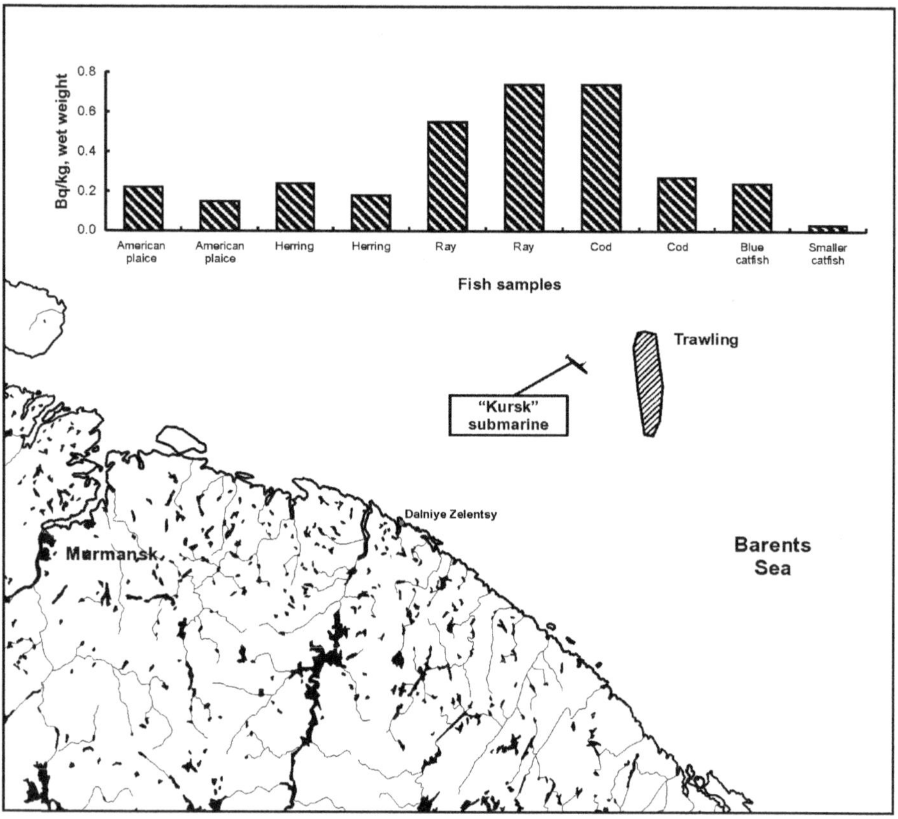

Fig. 4.27. ^{137}Cs in Barents Sea fish near the sunken submarine «Kursk» (Matishov et al. 2002)

5 The coastal zone

5.1 Biogeographical characteristics

The accumulation of artificial radionuclides by coastal ecosystems of the Arctic and Sub-Arctic is influenced to a great extent both by the coastal currents and the influx of radionuclides from land with river runoff, melted snow, rain water and wind. The coastal zone extends over the relatively narrow (hundreds of meters to tens of kilometers) and shallow (mainly to depths 10-90 m) shelf adjacent to land margin. A considerable part of the coastal ecosystem is located in the tidal and storm surge zones and also in the pathway of coastal currents.

The nearshore ecosystem, which occupies a relatively small coastal area, plays an important filtration and transformation role as a result of the biogeochemical processes in operation here. In the coastal zone the highest productivity per unit area is known to occur. The specific biological communities are formed with their complex system of interactions between algae, bacteria, detritus, dissolved organic matter, crustaceans, mollusks and fish owing to the rich species composition of plants and animals, their behavior and habitat characteristics. The interaction of nearshore ecosystem components with the adjacent land ecosystem is also unique relative to other parts of the ocean. The coastal habitats of seals and the nesting places of colonial sea birds are particularly important in this connection (Matishov end Pavlova 1990; Savchuk end Matishov D. 1990).

Coastal bays are an important component of continental margins (Matishov 1984, 1987). They intersect the land by tens to hundreds of kilometers. Estuaries, which adjoin the mouths of large rivers, are also important for the polar shelves. The oceanographic regime of estuaries is characterized by tidal oscillations of sea level and related currents, sea waves, wind currents, river runoff, and density stratification of the waters. The transition between freshwater and marine biological habitats takes place here (Klenova 1948; Zenkevich 1963; Leontjev 1974).

The coastal biological communities are of great interest in radioecology because they are impacted by radionuclide inputs from large rivers, nuclear bases situated in bays, nuclear test sites (e.g. Novaya Zemlya bays), dumped objects containing nuclear wastes. Radioactive particles not trapped in semi-closed coastal basins, partially descend to the bottom and also are transported to the open sea (Matishov et al. 1994a,b,c, 1997b,c,f, 1998b,c).

During recent years MMBI carried out a number of radioecological surveys in Chernaya, Pechenga, Kola, Motosky and Kandalaksha Bays (Fig. 5.1; Matishov et al. 1995a,b, 1996f,g, 1997b,c,f, 1998b,c, 1999d). These areas are characterized by rocky coasts, shallow banks and ridges alternating with relatively deep (150-350 m) hollows. The bottom of these hollows is covered with fluidized aleurite and clayey sediments based on sampling in different geomorphological parts of the bays in water depths from 30-70 to 120-300m.

As a result of interactions between river and sea water, the distinctly pronounced surface halocline and thermocline (to 10-15 m depth) halocline and thermocline is formed in bays along the Kola Peninsula, Novaya Zemlya, Northern Norway, and Spitsbergen. Such water stratification is particularly pronounced during the flood period lasting 2-4 months. A zone of freshened waters (0-20 ‰)

is situated where rivers first flow into the bays. Then the rate of freshening decreases, and the salinity in the next zone is 20-30 ‰. The salinity at the mouth of the bay usually is 30-34.5 ‰.

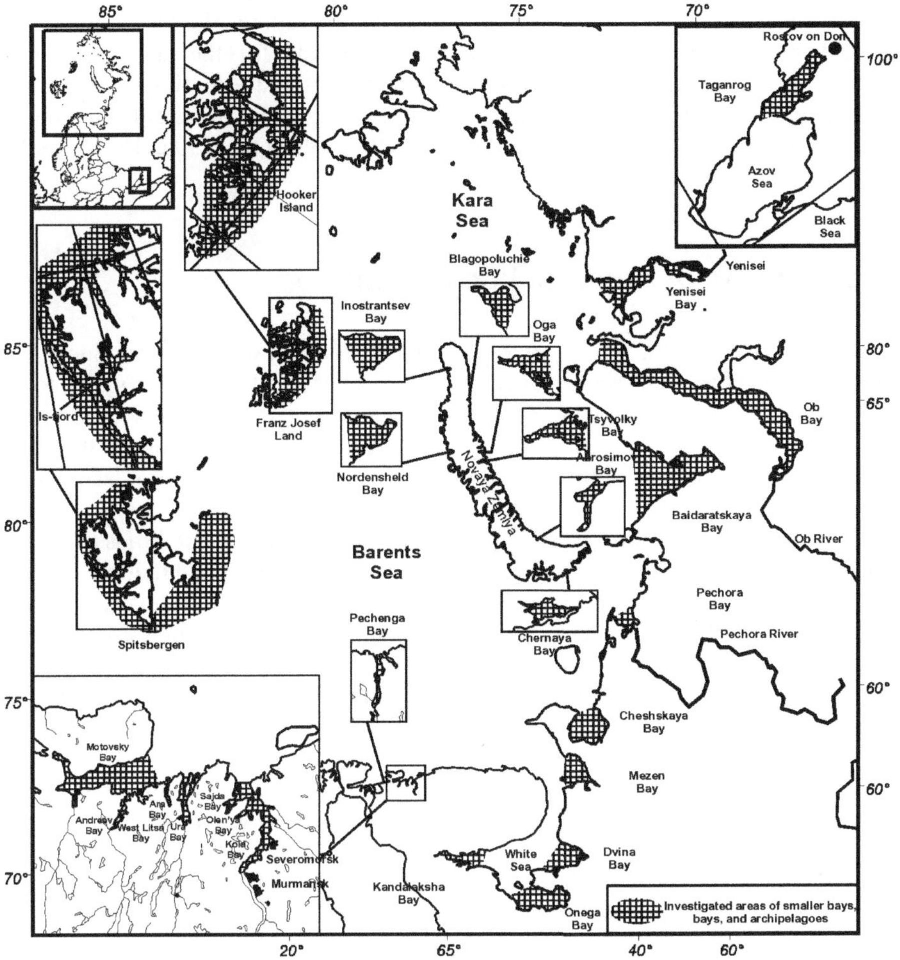

Fig. 5.1. Areas of radionuclide investigations during 1991-2000 by MMBI

Special attention has been given to the study of ecosystems in numerous small inlets dividing Kola and Motovsky Bays. In particular, the bays or inlets Big Motka, Kutovaya, Titovka, Zapadnaya Litsa, Ara, Ura in the Motovsky Bay and Sajda, Olenjya, Pala, Ekaterininskaya Gavan, Vaenga, and Varlamova in the Kola Bay were investigated. All sampling stations are within 1-5 km of the nearest nuclear base.

The artificial radionuclide content in seawater in some of these bays, was considered earlier. Here the accumulation of radioactive contamination in some key components of the coastal ecosystem, including bottom sediments and macrophytes, will be analyzed. The accumulation of radionuclides in the land ecosystem of the coastal zone is presented in brief to make the radioecological picture more complete. The oceanographic conditions responsible for the accumulation of sediments and associated radionuclides in some bays and estuaries are described. The data on hydrology, hydrochemistry and bottom geology were collected during MMBI expeditions in the 1990s (Matishov et al. 1992a,b, 1993a, 1996c, 1997f; Matishov D. 1994, 1997, 1998).

5.2 Patterns of radionuclide distribution

The soil cover of the polar areas is limited. As a result, radionuclides from atmospheric fallout are not retained by soil instead and being transported to the Ocean coastal zone. One can estimate the quantity of radionuclides precipitated from the atmosphere by their concentrations in the lichens growing on the coast. Radioactive substances in both dry and wet atmospheric precipitation accumulate in lichens.

Knowledge of the concentration of radionuclides in lichens and mosses is necessary in order to identify pathways of radionuclide bioaccumulation in birds inhabiting bird bazaars. Many sea birds use lichens for nest building. During the nesting period the birds often consume food with particles of moss, lichen, and soil attached (Golovkin 1982).

In 1991-1994 a number of investigations were carried out on the archipelagoes Franz-Josef Land, Novaya Zemlya, and the islands of Vajgach, Kolguev and others situated in the Pechora Sea, south of the Kara Sea and also on the Kola Peninsula (Matishov et al. 1993b, 1994a,b,c, 1995a,b). The ^{137}Cs content in lichen specimens was 300-500 Bq / kg, in bottom sediment (0-3 cm layer) - 0-12 Bq/kg, in soil - 150-540 Bq/kg. In 1991 the ^{134}Cs concentration in lichens from the southern islands of Franz-Josef Land and the west coast of the northern island of Novaya Zemlya varied from 0.5 to 20.6 Bq/kg dry weight. In these areas the ^{137}Cs concentration varied from 5.6 to 380 Bq / kg dry weight (Fig. 5.2).

The decay corrected ratio ^{137}Cs:^{134}Cs shows that the Chernobyl signal represents less than 30% of the ^{137}Cs activity. The concentration of ^{241}Am, ^{60}Co, ^{152}Eu and ^{155}Eu in moss from Chernaya Bay was correspondingly 70.4, 26.1, 37.8 and 18.8 Bq / kg dry weight. The $^{239,240}Pu$ concentration in lichens from Maybell Island (Franz-Josef Land), Kharlov Island, from the Dalnie Zelentsy and Porchnikha areas on the Kola coast and also on the coast of the Nordensheld Bay (Novaya Zemlya) is 0.9-5.7 Bq / kg (Rissanen et al. 1996, 1999a).

Data from Novaya Zemlya archipelago indicates that the ^{137}Cs content in ecosystem components are as follows: lake and sea coastal waters (to 10-20 Bq / kg), bottom sediments (from 1 to 20-50 Bq / kg), mosses and lichens (10-540 Bq /

kg), Arctic char, birds (geese) and northern reindeer (to 10-20) Bq / kg). Exceptions to these levels are found in bottom sediments and soils of Chernaya Bay where the radiocesium concentration in the environment and biota can reach 5000 Bq / kg or more. The ^{137}Cs content in soils from the northern island of Novaya Zemlya in 1991 ranged from 6 to 107 Bq / kg.

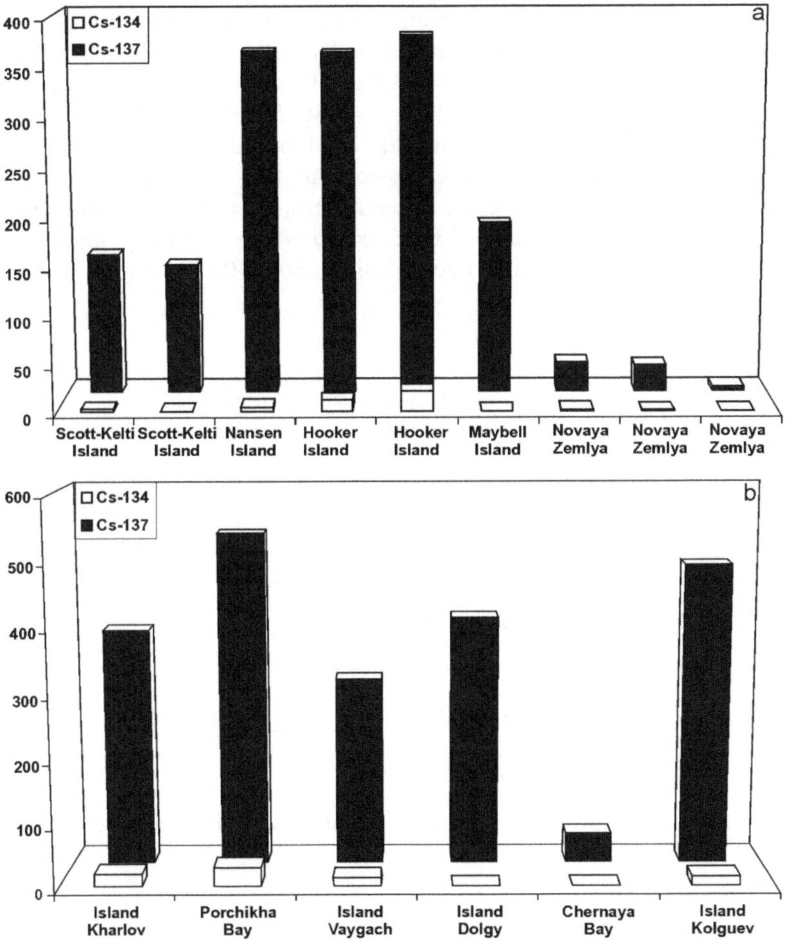

Fig. 5.2. ^{134}Cs and ^{137}Cs concentration in lichens from (**a**) Franz-Josef Land and Novaya Zemlya in 1991 and (**b**) islands in the southwest Barents Sea in 1992, Bq/kg dry weight

^{137}Cs investigations in hundreds of samples of lichens, mosses, mushrooms and berries, has provided a clear view of the radionuclide content on the Kola Peninsula and also on the Pechora Sea coasts. The characteristic feature of anthropogenic ^{137}Cs is that it varies over several tens of kilometers from 5 to 900 Bq / kg (Matishov et al., 1992c, 1993b, 1994c). A number of high contamination areas (800-900 Bq / kg) are apparent when compared to the majority of sites having background concentrations from 80 to 450 Bq / kg. These are the Pyalitsa and Chavanga regions on the Tersky coast, an area north of Verkhnetulomskoye Reservoir (Paust River), and also south of the Pereval meteorological station. Consequently, in 1990-1995 the atmospheric in the polar regions periodically contained significant concentrations of artificial radionuclides.

The comparative gamma-spectrometric analysis of lichen samples taken in Khibiny during the 1990s with samples collected by the Polar Alpine Botanical Garden specialists in 1968 indicate that the ^{137}Cs activity during the earlier period exceeded the present level by 20-25 times (Fig. 5.3). The high concentrations of artificial radionuclides (1-6 thousand Bq / kg) in reindeer moss are a consequence of the Novaya Zemlya nuclear explosions in the atmosphere up until 1963. Current concentrations in different lichen species are as follows: *Alectoria ochroleuka* - 120, *Cladonia alpestris* - 230, *Cetraria nivalis* - 340 Bq / kg. The total activity of anthropogenic radionuclides in soil is about 400 Bq / kg.

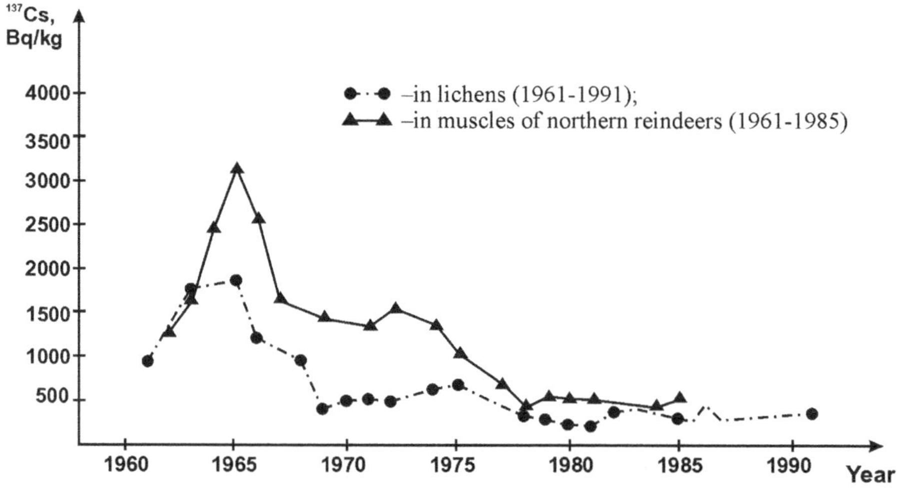

Fig. 5.3. Average ^{137}Cs activity in lichens and muscles of northern reindeers of the Kola Peninsula in 1960-1990

In the beginning of the 1990s the ^{134}Cs and ^{137}Cs concentration on the Kola Peninsula coast and on the southeast islands of the Barents Sea were noticeably higher than the previously mentioned observations on the high latitude archipelagoes. Thus, concentrations changed from 0 to 32 Bq / kg dry weight for ^{134}Cs and to 530 Bq / kg dry weight for ^{137}Cs. The average ^{137}Cs concentration in moss was 335 Bq / kg. The ^{137}Cs content in lichen, soil and loose sediment samples of the Kola Bay coast ranged from 1100-1800 Bq / m^2 (Matishov et al. 1996f). One can note for comparison that after Chernobyl the ^{137}Cs level in lichens on the south of Finland reached 60 thousand Bq / kg.

To study the characteristics of the distribution and concentration of radioactive contamination on the Kola Peninsula in 1992, we have determined the ^{137}Cs content in lichens along the 200-km profile Murmansk - Tumanny - Dalnie Zelentsy. The lichen fields on topographic highs (e.g. plateaus) contained much more radiocesium than on lake coasts and in river valleys. The important regularity of the concentration of cesium in lichens increases by 5-10 times towards the Barents Sea coast reaching 600-700 Bq / kg. This feature is caused by the different accumulation potentials of lichen species and changes in the aeration of the coast (Matishov D. 1994).

5.2.1 Barents Sea coastal zone

The bays Pechenga, Teriberka and Yarnyshnaya are all short fjords (to 12-20 km) that connect to the Barents Sea. These bays are characterized by the absence of ^{60}Co and low concentrations of ^{137}Cs in bottom sediments (1 - 13 Bq / kg) (Matishov et al., 1996f, 1997c, 1998b,c).

Strong currents bring terrigenous materials into deep areas of the bays and to the adjacent shelf. As a result, clayey sediments found in deep areas of the Pechenga Bay (depths to 160 m) are somewhat more contaminated by radiocesium, to 26 Bq / kg. Another feature is seen in aleurite sediment deposits of the Teriberka Bay. The ^{137}Cs activity increases (from 3 to 15 Bq / kg) with deep in seabed sediments. The radiocesium influxes to the bays are mainly from atmospheric fallout and from the marine transport of radiocesium within the Gulf Stream system.

5.2.2 Kola Bay

The Kola Bay is a large fjord valley. Its length is about 70 km, and the width changes from 1-2 to 6-8 km. At the entrance to the Kola Bay, the cross-sectional influx of Atlantic water is 5-6 thousand m^2 with typical salinity values of about 34.5 ‰ in the entering water. The distribution of anthropogenic radionuclides depends on the oceanographic characteristics of the underwater valley.

The hydrodynamics of the Kola Bay is determined by the regime of tidal currents which drives the constant water exchange with the Barents Sea (Hydrometeorology ... 1990). The influence of tidal currents is apparent throughout the entire area of the fjord. Tidal currents are strongest in surface waters of the bay and in the Tuloma River mouth. Current velocities range from 0.7 to 1.4 knots. Current velocities in middle and northern sections of the fjord are somewhat lower than 0.7 knots (Matishov et al. 1997f).

Wind and runoff currents also play an important role in the water dynamics of the upper layer (to 20 m). The velocity of these currents can reach 0.5-3.0 knots in the southern section of the bay. The velocity of runoff currents falls sharply in the middle section (to 0.1-0.2 knots) and is almost undetectable in the northern section. The runoff currents are caused by the fresh water influx from the Kola and Tuloma Rivers to the Kola Bay. The spring river flood period is four months. The total annual runoff of Tuloma and Kola reaches 10 km^3.

The temperature regime of the Kola Bay is determined by the heat advection with the Atlantic waters and also by seasonal climate variations (Fig. 5.4). In March - April cooled waters of Atlantic origin (1-2°) dominate at all water depths. In summer these relatively cold waters fill only the deepest areas of the fjord (200-380 m). In July - August surface waters warm to 9-11 °C (Hydrometeorology ... 1990). In spring bay waters are stratified into four levels. The upper layer (from surface to 10-15 m) is formed by freshened (1-20 ‰) and low salinity (20-30 ‰) waters. The water temperature changes from 3 to 6 °C. Below this layer is a thin intermediate layer of weekly freshened waters (26-33.8 ‰) with temperature from 2.5 to 3.7 °C. The main water layer (depths from 15-25 to 150-180 m) is of Atlantic origin. The salinity in this layer is 34.1-34.4 ‰ and temperatures of 2-2.8 °C. The bottom water layer (200-350 m) is relatively cold (1.5-1.9 °C) and saline (34.4-34.5 ‰). Probably the bottom waters are formed in the bay during winter cold periods.

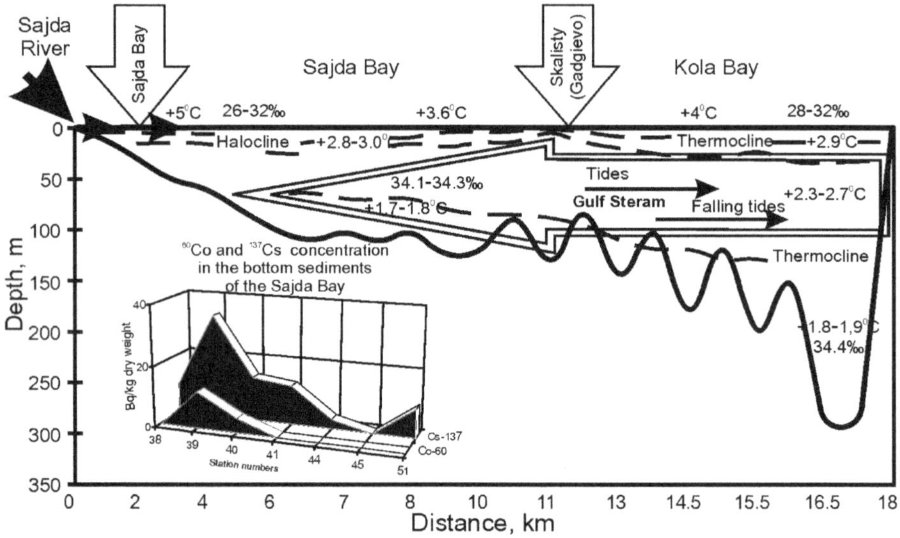

Fig. 5.4. Temperature regime of water masses and ^{60}Co and ^{137}Cs concentration in the bottom sediments of Sajda Bay (Kola Bay) (MMBI, 1990)

Water depths in narrow side bays (Sajda, Olenjya, Pala, Ekaterininskaya Gavan) reach 100-140 m and in some cases to 190 m. The bays contain water of Atlantic origin (Matishov et al. 1997f). Salinities are 34.1-34.4 ‰ and temperatures are 2.2-2.8 °C (Fig. 5.5). During spring floods, the thin layer (5-10 m) of freshened water (26.0-31.1 ‰) is formed in the bays themselves from runoff and falling tide currents which bring some of the melt water from the side bays to the Kola Bay. The temperature of these waters ranges from 3 to 6 °C.

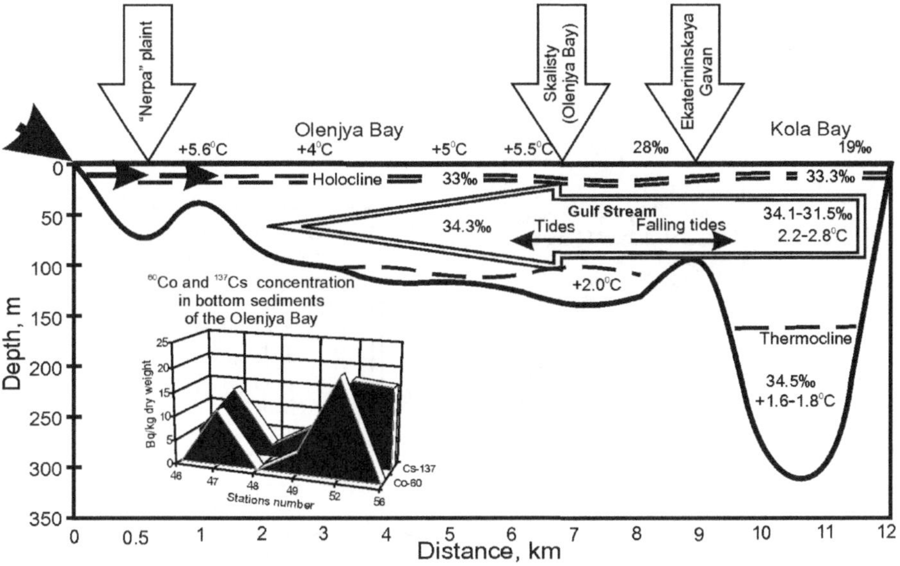

Fig. 5.5. Temperature regime of water masses and ^{60}Co and ^{137}Cs concentration in bottom sediments of the Olenjya Bay and Ekaterininskaya Bay (Kola Bay), Northwest Russia

The distribution of artificial radionuclides in these bays has been examined in a number of studies (Matishov et al. 1996f, 1997b,f; 1999g; Matishov D. et al. 1996; Rissanen et al. 1998). In 1995-1999 bottom sediments in nearly all areas of the Kola Bay contained ^{60}Co on the level 0.5-1.0 Bq / kg and ^{137}Cs - from 1 to 24 Bq / kg. The highest concentration of ^{137}Cs was in the sediments of the Vaenga and Varlamovo Bays. The density of anthropogenic radionuclides in sediments increases toward the nuclear bases (Fig. 5.6, 5.7).

Bottom sediments in Sajda Bay, approximately 2-3 km from the location of submarine nuclear reactor compartments contained ^{137}Cs and ^{60}Co from 3 to 34 Bq / kg and from 0.7 to 12 Bq / kg respectively. $^{239,240}Pu$ is also present in the marine sediments at levels of 1.6 Bq / kg. In Olenjya Bay, 800 m from the ship repairing facility "Nerpa", ^{137}Cs, ^{60}Co and $^{239,240}Pu$ is respectively 2-14, 1-12 and 1.6 Bq / kg. Bottom sediments of the Pala and Ekaterininskaya Gavan Bays near the town of Polarny contains relatively high ^{60}Co, (to 24 Bq / kg), $^{239,240}Pu$ (2.2 Bq / kg) and ^{137}Cs (to 24 Bq / kg) (Fig. 5.5, 5.8).

Higher levels of ^{60}Co were determined in Pala Bay (to 74 Bq / kg) and Ekaterininskaya Gavan Bay (to 28 Bq / kg) during a survey of Kola Bay in 1998. Cobalt was also found in Severomorsk (to 5 Bq / kg). Almost all bottom sediment samples taken from the bay contained 60Co (Matishov et al. 1996d, 1997b,e,f, 1999g). Concentrations in these sediments varied from 0.2 to 3 Bq / kg dry weight. Small quantities of ^{134}Cs, ^{125}Sb and ^{154}Eu were also present in some bottom sediment samples. The short-lived ^{134}Cs (half-life = 2.07 year) possibly is of local origin.

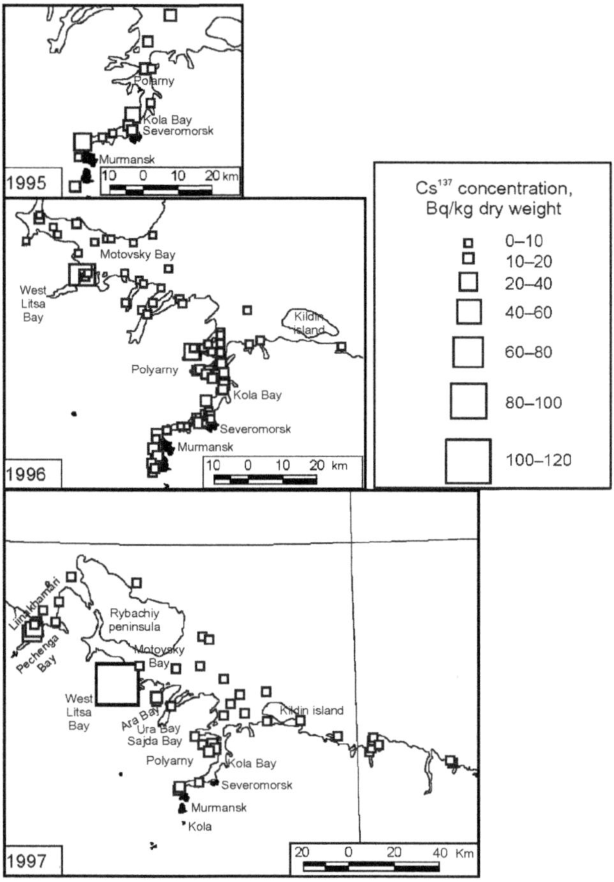

Fig. 5.6. ^{137}Cs distribution in the surface layer of bottom sediments in the bays and inlets of the Kola Bay. Map is consolidation of data from MMBI expeditions carried out between 1995-1997

As a result of a detailed survey (84 stations) carried out adjacent to the nuclear reprocessing facility Atomflot located in the northern part of Murmansk, a spatially heterogeneous radionuclide distribution was revealed (Namyatov 1995). The ^{137}Cs content in the bottom sediments 50-100 m from the facility varied from 2 to 40 Bq / kg and ^{60}Co varied from 2 to 27 Bq / kg. The highest levels of ^{137}Cs, ^{152}Eu and ^{154}Eu (correspondingly 60, 55 and 123 Bq / kg) were found in bottom sediments along the 2 km moorage line of facilities Sevmorput and Atomflot (Fig.

5.9, 5.10). The ^{137}Cs content (to 40 Bq / kg) also increased in aleurite sediments of the fjord near the Abram Cape settlement. $^{239,240}Pu$ was found in bottom sediment samples from eight stations with the content varying from 0.53 to 1.0 Bq / kg (Namyatov 1995).

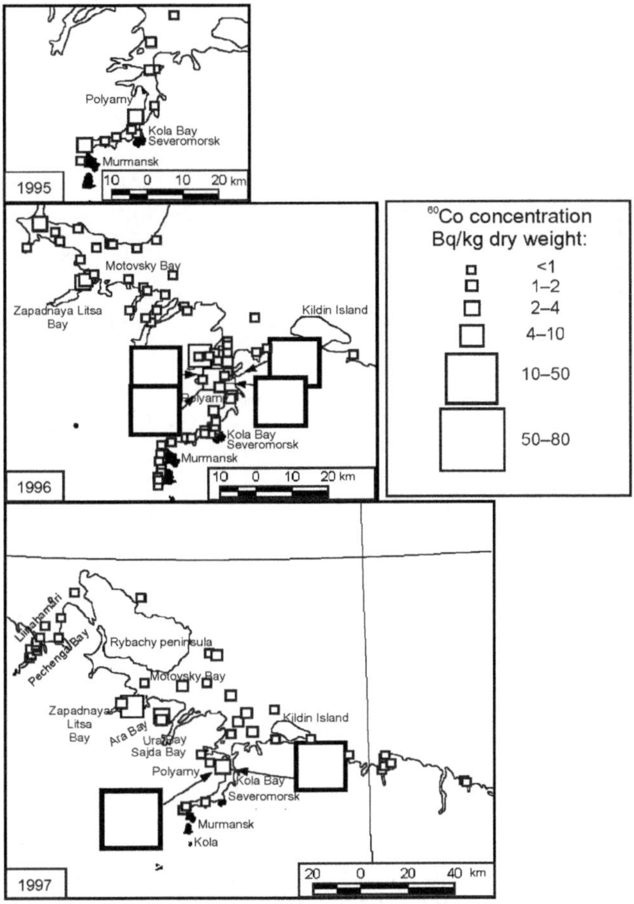

Fig. 5.7. ^{60}Co distribution in surface layer bottom sediments in the bays and inlets of the Kola Bay

Many researchers (Namyatov 1995; Rissanen et al. 1998; Nikitin et al. 1999) note the presence of some peak ^{137}Cs concentrations in bottom sediments (60-630 Bq / kg). These maximum concentrations could be caused by the presence of radioactive particles. The appearance of corrosion products - ^{60}Co (half-life =5.3 years) in all sediment samples between Atomflot and the Barents Sea, and the presence of other anthropogenic radionuclides in sediments near Atomflot and at the exit of the Kola Bay are evidence of the existence of secondary radioactivity from the nuclear installations.

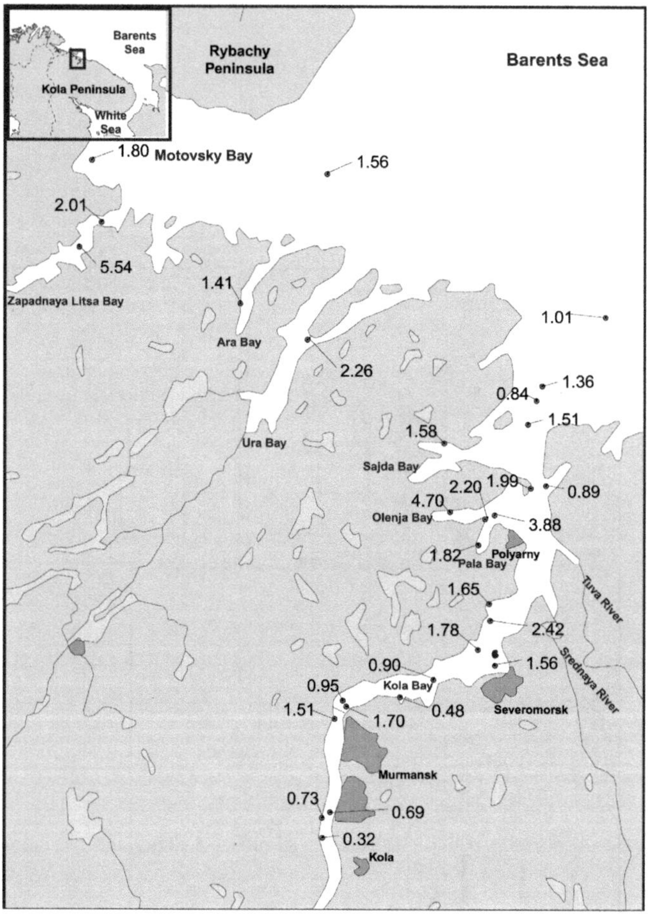

Fig. 5.8. ^{239,240}Pu content in the surface layer of bottom sediments from Kola and Motovsky Bays (by MMBI data), Bq / kg dry weight

It is clear from the available data that the content of anthropogenic radionuclides in bottom sediments of the Kola Bay depressions is several times more than on underwater ridges and banks. In particular, the ^{137}Cs level in aleurites and silts varied at many stations from 9 to 20 Bq / kg. ^{137}Cs and ^{60}Co are traced to depths of 15-20 cm in bottom sediments providing evidence of a downward migration of radionuclides. ^{137}Cs concentrations in surface bottom sediments (0-4 cm) vary with location from 9 to 20 Bq / kg. Evidently, the process of accumulation and redistribution of radionuclides in the fjord takes place by tidal current processes and the general water circulation dynamics which vary seasonally and over longer time scales. It is quite probable that some artificial radionuclides are transported from the fjord to the open sea.

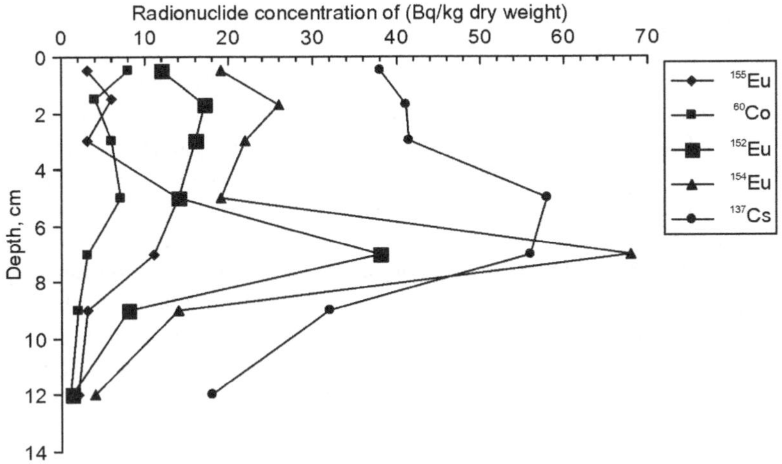

Fig. 5.9. Distribution of gamma-emitting radionuclides with sediment depth near the Atomflot nuclear reprocessing facility in the Kola Bay (Nikitin et al. 1999)

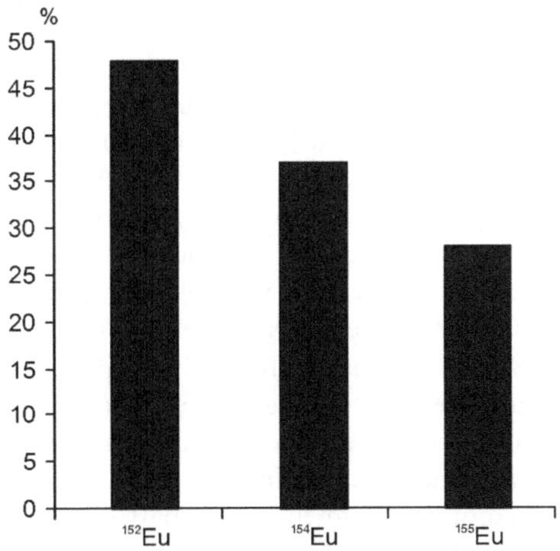

Fig. 5.10. Percentage ratio of some radionuclides connected with a contaminated particle discovered in the 6-8 cm layer of a sediment core collected near the Atomflot reprocessing facility (Nikitin et al. 1999)

The majority of anthropogenic radionuclides is detected in the 4-8 cm layer of bottom sediments from the fjord. These depths are inhabited by benthic organisms. Bottom sediments which serve as habitats for macrophytes and zoobenthos are also important for the cycling of radionuclides in coastal bay ecosystems. In bottom sediment cores taken in depressions (depths 250-340 m), radionuclides are detected in the 10-20 cm depth interval. In sediment cores taken from the mouth of the Kola Bay ^{137}Cs (4.5-7.6 Bq / kg) and ^{60}Co (0.1-1.1 Bq / kg) are presen from the surface to a depth of 10 cm.

The algae *Lithotamnion corimbossum*, collected near Polarny, contained 1.3 Bq / kg of ^{60}Co in 1997. Starfishes from Olenjya Bay contained 0.03 Bq / kg of ^{137}Cs. Very low ^{137}Cs concentrations (less than 0.2 Bq / kg) were detected in the mollusk *Mutilus edulis* and Gastropoda *Littorina saxatilis*. Atlantic cod, haddock, and other bottom fish caught in Motovsky Bay, in the Kildin Island area and at the mouth of Kola Bay contained very little ^{137}Cs - 0.3-1.0 Bq / kg.

The distribution and levels of anthropogenic radionuclides in bottom sediments of Motovsky Bay is related to ocal sources and water circulation in the bayt. The hydrological features of the bay are determined by the hydrodynamic processes in the coastal zone of the Barents Sea. Warm Atlantic waters penetrate freely into this relatively wide bay (10-12 km) with depths to 250-300 m. In spring during the beginning of June in particular a thin (5-10 m) freshened (33.4-33.8 ‰) water layer with temperature 3-4 °C forms at the surface. Below this layer, to depths of 170-190 m, a higher salinity (33.8-34.0 ‰) and colder (2.0-2.8 °C) water mass is found. The coldest water (1.6-1.8 °C) with salinity 34.1 ‰ occupies the bay depressions. The surface layer in the coastal zone of the bay is freshened to 30.7-32.3 ‰. This phenomenon is connected with the runoff of melted and river waters into the bay.

The bays Titovka, Zapadnaya Litsa, Ara and Ura, divide the southern coast of Motovsky Bay and will be briefly characterized here. The lengths and depths of these bays vary from 10 to 20 km and 100-150 to 250-300 m respectively. Runoff currents from small rivers flow into the bays. The freshening of surface waters 29.4-33 ‰ occurs during the spring snow melt period (Fig. 5.11).

The tidal currents set into motion the Atlantic waters with salinity from 33.6 to 34.1 ‰. The tides also create noticeable water exchange between the narrow bays and Barents sea coastal zone. The bottom depressions of the bays are filled by relatively cold (0.5-1.9 °C) sea water with salinities up to 34.4 ‰.

In Motovsky Bay, a nuclear submarine base is situated in Zapadnaya Litsa, and a storage site for reprocessed nuclear fuel and other radioactive wastes is located in Andreev Bay. Measurements taken in1996-1997 indicate that the fine-sandy and aleurite bottom sediments from the outer part of the Zapadnaya Litsa Bay taken by us in 1996-1997 contain ^{137}Cs (from 0.7 to 50 Bq / kg), ^{60}Co (from 0.5 to 4.0 Bq / kg) and $^{239,240}Pu$ (about 2 Bq / kg) (Fig. 98). The ^{137}Cs and ^{134}Cs level in surface seawater from the same area is low, about 4.9 and 0.2 mBq / l correspondingly. Bottom sediments in the outer parts of Ara and Ura Bays contained from 0.5 to 1 Bq / kg of ^{60}Co and from 0 to 11 Bq / kg of ^{137}Cs (Fig. 5.12).

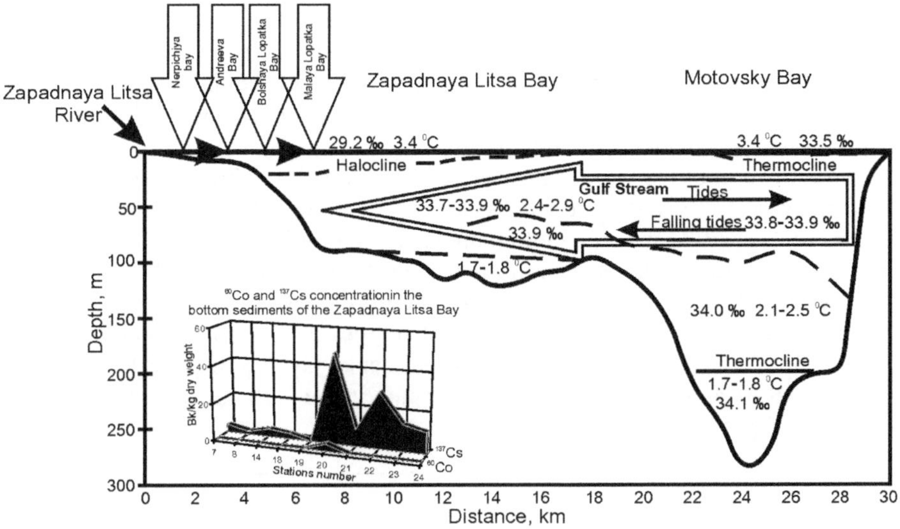

Fig. 5.11. Temperature distribution in seawater and radionuclide concentrations in the bottom sediments of Motovsky Bay

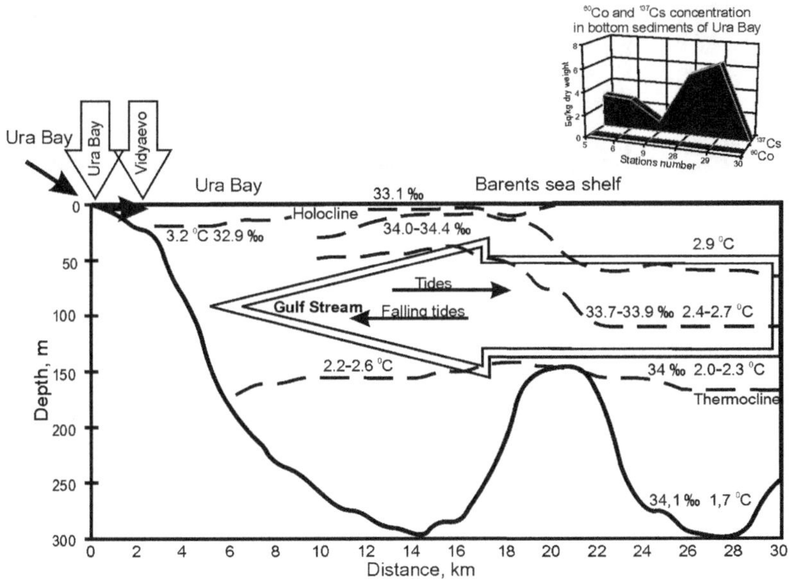

Fig. 5.12. Temperature regime of water masses and radionuclide concentrations in the bottom sediments of Ura Bay

In 1997 ^{137}Cs levels in Zapadnaya Litsa Bay, including Andreev Bay, reached 75-115 Bq / kg and ^{60}Co ranged from 6-33 Bq / kg (Matishov et al., 1997b,f, 1998b,c). The surface 20-cm layer of bottom sediments contains high concentrations of ^{137}Cs (20 to 115 Bq / kg). In Ara Bay, concentrations of ^{137}Cs and ^{60}Co were 5 and 13 Bq / kg correspondingly.

Investigation of the vertical distribution of plutonium in bottom sediments from Zapadnaya Litsa Bay showed two peaks in $^{239,240}Pu$ concentration in the sediment layers (0-1 and 3-4 cm). The plutonium concentration in the deeper layer (3-4 cm) is almost two times higher than in the 0-1 cm layer (8.71 and 55.4 Bq / kg correspondingly). The $^{239,240}Pu$ concentration in the 9-10 cm sediment layer is near the levels found in similar areas without local sources of $^{239,240}Pu$. It should be noted that in the Andreev bay, where bottom sediments have the highest level of ^{137}Cs contamination and the plutonium concentration in the 9-10 cm layer exceeded (3.2 Bq / kg) (Fig. 5.13).

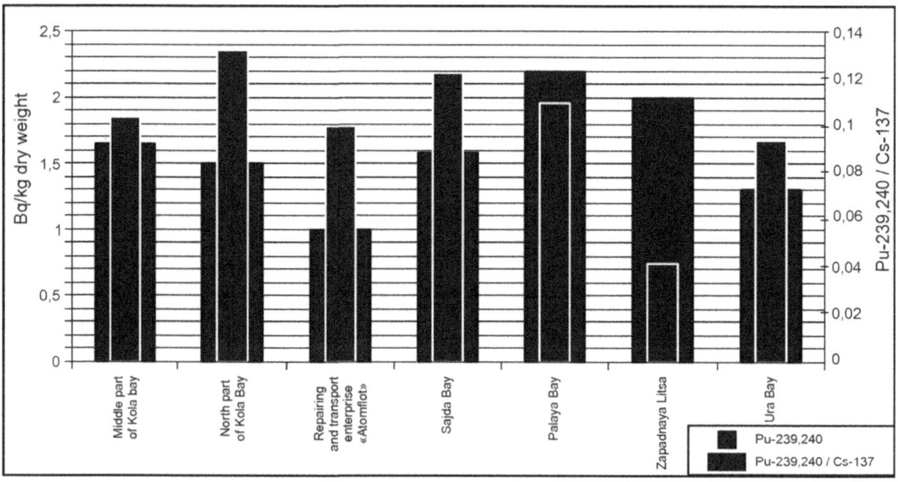

Fig. 5.13. $^{239,240}Pu$ content and $^{239,240}Pu$:137Cs ratio in bottom sediments of some areas of the Kola and Motovsky Bays

Radioactive contamination in Motovsky Bay is noticeably lower than in the adjoining side bays. Bottom sediments in the coastal zone of the bay contain about 0.7-1.0 Bq / kg of ^{60}Co and from 5 to 14 Bq / kg of ^{137}Cs. The rate of contamination of the Titovka Bay is similar. The aleurite sediments in the open part of the bay have accumulated even less ^{137}Cs (1-10 Bq / kg) and ^{60}Co (0.5-1.0 Bq / kg) (Matishov et al. 1997f, 1998c, 1999d).

$^{239,240}Pu$ is characteristic for the sediments of the Kola and Motovsky bays almost everywhere ranging from 0.15-0.40 to 2-9 Bq / kg. The content of this nuclide in clayey sediments near Severomorsk reaches 5.7 Bq / kg. The ^{238}Pu concentration in surface sediments varied from 0.02-0.12 Bq / kg (Rissanen et al.,

1998). The $^{239,240}Pu$ concentration in the surface sediment layer of Ara, Ura and Pala Bays (1.4-2.2 Bq / kg) is close to global fallout levels. High plutonium concentrations in comparison to the global fallout level of approximately 4-6 Bq / kg, was discovered in surface sediments from Zapadnaya Litsa and Olenjya Bays. plutonium concentration was discovered in the surface samples of bottom sediments from Zapadnaya Litsa and Olenjya Bays. A high ^{238}Pu: $^{239,240}Pu$ ratio observed near the Atomflot facility indicates a small quantity of plutonium has been emitted from the enterprise.

The $^{239,240}Pu$:^{137}Cs ratio in a number of cases was essentially equal to the value expected from atmospheric fallout (0.019-0.043), in one case it was 0.067 which is similar to maximum values found in the open parts of the Kara Sea. In all other cases the range was 0.080-0.012. Overall, these data suggest that sources of $^{239,240}Pu$ related to nuclear fleet activities are present in some of the investigated areas. The $^{239,240}Pu$:^{137}Cs ratio in all samples from side bays and some open areas of Kola and Motovsky Bays exceed by several times the ratio for vatmospheric fallout.

The situation for cobalt contamination in bottom sediments from different areas of Kola and Motovsky Bays is somewhat different. Only samples taken near radioactive objects were analyzed. In the side bays of Kola Bay, ^{60}Co and ^{137}Cs concentrations increase exponentially with distance toward the objects. If ^{137}Cs is less than 17 Bq / kg dry weight, the ^{60}Co: ^{137}Cs ratio is typically less than 1. Cobalt contamination is not characteristic for the side bays of Motovsky Bay. The maximum content of cobalt in the surface layer (4.5 Bq / kg dry weight), was observed in the outer part of the Zapadnaya Litsa Bay.

It can be noted that in spring the distinctly pronounced halocline is formed in the pelagic zone of bays (to 15-20 m depth). The velocity of the runoff current of freshened waters (melted and river), gliding over the surface of the saline oceanic waters, can reach 1-4 m/s. Such periglacial conditions persist for not more than 2-3 months. Probably, anthropogenic radionuclides supplied by rivers or reaching the bay basins during planned discharges are transported from the coastal zone into the Barents Sea via the halocline.

5.2.3 The White Sea

Conditions in the White Sea bays, including Kandalaksha, Onega, and Dvina, are related to their hydrodynamic regime, sediment dynamics, biological phenomena and related natural processes. The influx of terrigenous substances to the sea takes place mainly through alluvial releases and coast abrasion processes associated with subpolar climates. The process of sedimentation of fine mineral particles from low density suspensions dominates in the closed sea basins (Pavlidis et al. 1995).

Coarse aleurites and sands found to depths of 30-50 m in Onega Bay and around the Solovetski Islands accumulate 5-13 Bq / kg of ^{137}Cs. ^{137}Cs concentrations are noticeably higher in the sediments of the central part of Onega bay (to 23 Bq / kg). For comparison the ^{137}Cs content of sands from Gorlo, White Sea is comparatively small – up to 4.3 Bq / kg.

Variable levels of ^{137}Cs are observed in Dvina Bay an alluvial sandy and clayey bank with depths between 15-25 m (Fig. 5.14). The ^{137}Cs content in sediments from this area varies from 1.6 to 67 Bq / kg depending on conditions in the bay and on the type of bottom sediments. In surface sediment samples and short (to 6 cm) sediment cores (to 6 cm), radiocesium is detected (Matishov D. et al. 1995a).

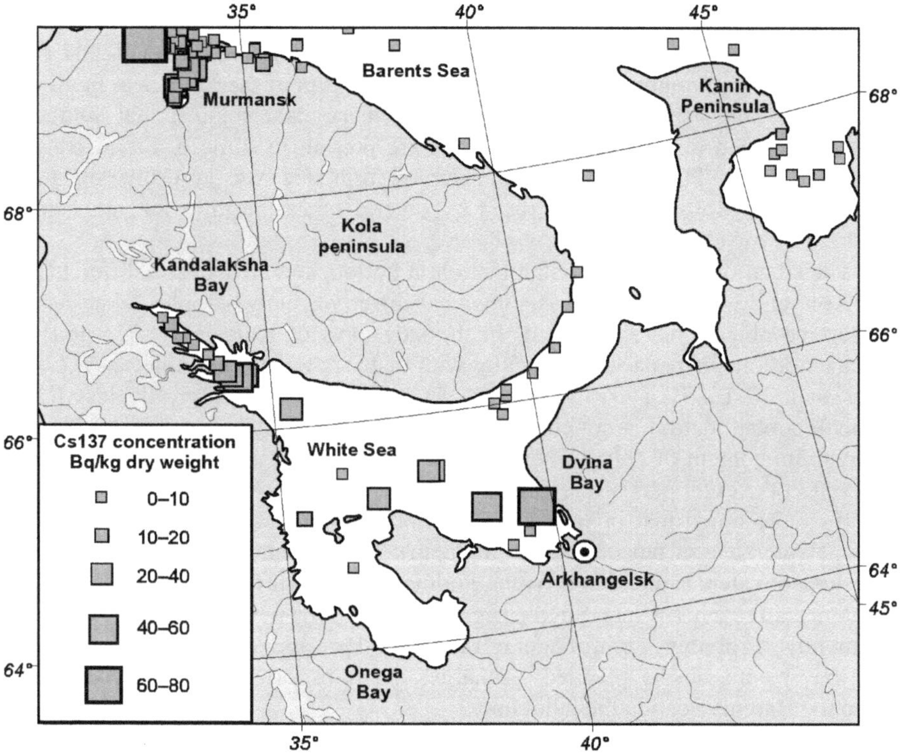

Fig. 5.14. ^{137}Cs distribution in the surface layer of bottom sediments from the White Sea

An anomalously high level of ^{137}Cs is observed in a layer of aleurite sediments in Dvina Bay to the east of Severnaya Dvina mouth. ^{137}Cs is present throughout the entire length of the sediment core from 40 to 67 Bq / kg. Maximum values are found between 6-10 cm. However, the surface layer (0-4 cm) also contains the short-lived ^{134}Cs at levels of 0.9-1.5 Bq / kg. There are three sediment layers with higher concentrations of radiocesium the sediment core. Possibly, the formation of these may be the result of avalanche sedimentation in the Severnaya Dvina delta.

Sediment accumulation, and the associated accumulation of anthropogenic radionuclides beyond the Dvina Bay in water depths of 100-150 m is caused by other hydrodynamic conditions. The ^{137}Cs level in two sediment cores taken on the Severnaya Dvina traverse decreases uniformly from 47 Bq/kg to zero. No traces of radiocesium are found below 10-12 cm depth.

In the depression of Kandalaksha Bay (depths to 300 m), fine-aleurite and silty sediments are encountered. It is interesting that in fluidized surface silts (0-2 cm) ^{137}Cs concentrations up to 45 Bq / kg are observed and also ^{60}Co is present at about 1.7 Bq / kg. Radiocesium concentrations decrease uniformly with sediment depth ut with appreciable concentrations even at a depth of 10 cm (16 Bq / kg). The natural radionuclide ^{40}K in the same sediment core ranges from 600 to 780 Bq / kg. Thus, the general image of the Kandalaksha Bay is that, as water depth increases from 200 to 300 m the slopes of these depression, there is particle sorting of terrigenous material, resulting in the accumulation of the smallest silt particles and a noticeable increase in ^{137}Cs concentration from 6 to 45 Bq / kg ((Matishov D. et al. 1995a).

Some general conclusions may be drawn from the nature of radioactive contamination observed in the bottom sediments of the shelf and the White Sea basin as a whole. It is quite evident that the migration and distribution of clayey mineral particles and radionuclides are focussed in the depressions of the sea bottom. This system is evident in the relief of the depression in Kandalaksha Bay, the central hollow of the White Sea and in the Dvina Bay. River runoff undoubtedly plays an important role in the removal of ^{137}Cs derived from atmospheric fallout, including radioactive Chernobyl dust, from the catchment area to the closed sea basin.

Within the central hollow of the White Sea, the dominant influence of releases from the Severnaya Dvina River can be traced. The characteristic band of clayey minerals with high content of smectite in the surface layer of bottom sediments (0-10 cm) extends from the Dvina Bay through the central hollow of the sea to the Kandalaksha Bay. The radioactivity situation in the underwater channel of the Severnaya Dvina is similar to the Yenisey and Ob Bays.

The relatively high ^{137}Cs concentration and the presence of ^{134}Cs and ^{60}Co in bottom sediments of the underwater depressions located between the Severnaya Dvina River and Kandalaksha Bay indicate the existence of local sources of artificial radionuclides leaks along the White Sea coasts (Rissanen et al. 1997b, 1999b). Emissions from nuclear submarines, built on the Severodvinsk shipyard and tested in the deep-water zone of the Central hollow and Kandalaksha Bay, are possible sources. In addition, there are many obsolete submarines located in Severodvinsk Bay. The increased content of ^{137}Cs in lichens from the White Sea coast between the settlements of Kuzomen and Tetrino indicate the existence of atmospheric sources of artificial radionuclides.

5.2.4 Pechora and Cheshskaya Bays

The scale of radiocesium contamination in sediments of the Pechora and Cheshskaya Bays, located adjacent to one another, is different from the regions previously discussed (Fig. 5.15). In the beginning of the 1990s ^{137}Cs and ^{134}Cs concentrations in Pechora Bay bottom sediments varied from 1 to 11 and from 0.4 to

1.13 Bq / kg respectively and in the Cheshskaya Bay from 0.5 to 5.1 and about 0.6 Bq / kg. The differences in radiocesium levels evidently are caused by the difference in the size of the catchment areas and the volumes of river runoff. The Pechora runoff reaches 130 km^3 / year, while only small rivers flow into the Cheshskaya Bay ((Pesha, Oma, Vizhas).

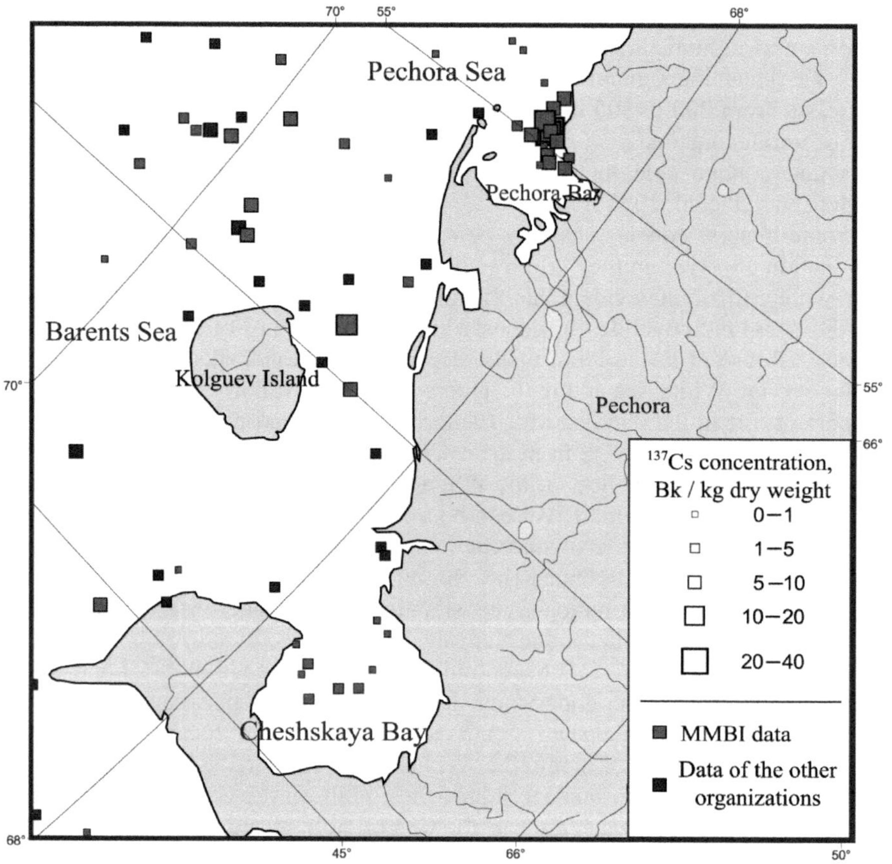

Fig. 5.15. ^{137}Cs concentrations in the surface layer of bottom sediments from the Cheshskaya and Pechora Bays (1990s)

In both bays is that the sands contain no γ-emitting particles at all. In the Cheshskaya Bay only the coarse aleurite sediments found at depths of 30-50 m contained an insignificant quantity of ^{137}Cs (to 5 Bq / kg).

^{137}Cs content is somewhat higher (5-11 Bq / kg) in the clayey sediments along the margin of the Pechora Bay (Fig. 5.16, 5.17) (Ivanov 1999). Precisely these sediments are enriched by organic substances supplied mainly by river runoff and coastal abrasion. The deposition of river suspended sediments carrying radionuclides takes place in the basin within a salinity zone of 2-12 ‰ with out-radionuclides takes place in the basin within a salinity zone of 2-12 ‰ with outflow to the sea in the upper layer of these estuarine waters.

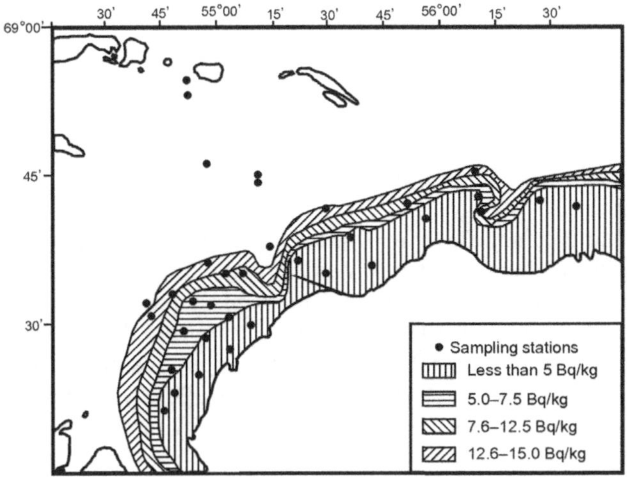

Fig. 5.16. ^{137}Cs concentrations in Pechora Bay bottom sediments (Ivanov 1999), Bq/kg

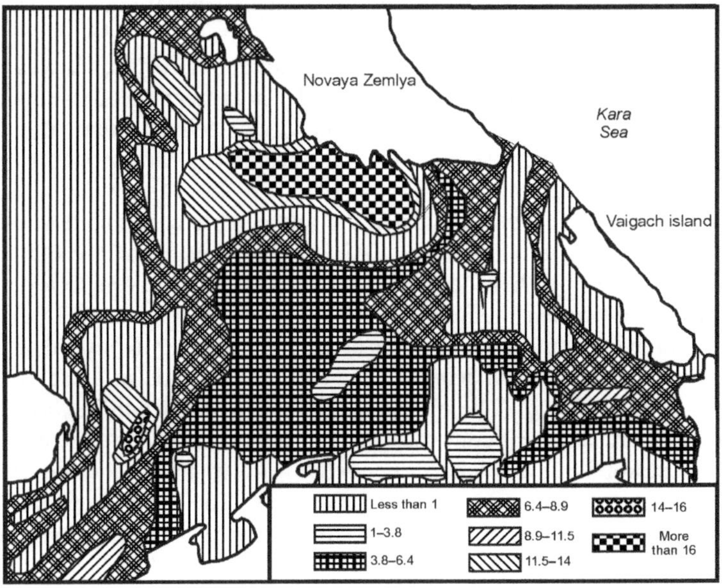

Fig. 5.17. ^{137}Cs distribution (Bq/kg) in the Pechora Sea surficial bottom sediments (Ivanov 1999)

The spreading of reduced salinity waters (to 25 ‰) in the Pechora Sea is directed from the Pechora Bay to the northeast. In the middle of summer 1992 the low salinity waters of low salinity were are observed in the coastal zone between Pechora Bay and Vajgach Island. Probably, the migration of radionuclides also follows this direction. In the shallow southern part of the sea within the 15-20 m depth zone the low salinity waters reach the seabed. Significant vertical and horizontal density gradients are observed here.

Overall, the radioactivity of sediments deposited in the Cheshskaya and Pechora Bays is low. The origin of ^{134}Cs and ^{137}Cs is global fallout within the river basins.

The Ob and Yenisey Bays are typical estuaries. These bays are characterized by large widths, elongation, shallowness, strong wind and tidal mixing. The vertical distribution of salinity is uniform. The Yenisey and Ob Rivers brought to the Kara Sea supply more than 1050 km^3 / year of runoff to the Kara Sea. Transported along with this huge runoff volume are large quantities of radionuclides. Radionuclide releases were carried during 1950-1990s by the nuclear plants Krasnoyarsk, Tomsk and Chelyabinsk (Mayak). In 1971, anomalously high levels of ^{137}Cs were detected in bottom sediments of the Yenisey Bay. They reached 15 kBq / m^2 which is 4-8 times higher than the contamination levels caused by global fallout. The presence of ^{54}Mn and ^{65}Zn was also discovered in these bottom sediments (Vakulovsky et al. 1994a). The spectrum of radionuclides and their concentrations in bottom sediments of the Yenisey indicated a sharp decrease towards the mouth. The level of radioactive contamination of the bottom decreased by an order of magnitude over a distance of 800 km and about 70 % of it was caused by ^{65}Zn and ^{137}Cs. In 1973 bay sediments in the area near the Ust-Port contained more than 70 Bq / kg of ^{137}Cs (Vakulovsky et al., 1994a). Areas of silty bottom deposits are comparatively scarce, but these areas are characterized by increased radionuclide accumulation.

MMBI data (1993, 1996), indicated the appreciable accumulation of ^{137}Cs (4-29) Bq / kg) in silts throughout the entire 200 km length of the Ob Bay in the depth zone of 5-30 m. These levels are maintained in silt layers to a depth of 6-7 cm. ^{60}Co is also present in small amounts in these sediments (0.3-0.6 Bq / kg). Especially high concentrations of ^{137}Cs and ^{60}Co are found in bottom sediments of the Yenisey Bay and the underwater channel that extends into the coastal shelf (Matishov et al. 1998b,c). The characteristic levels of ^{137}Cs throughout the entire length of a 20 cm sediment core were 67-81 Bq / kg and of ^{60}Co were 2-6 Bq / kg. Such concentrations of artificial radionuclides are found on the Novaya Zemlya shelf only in areas near to old nuclear test sites. The technologically enhanced levels of radionuclides were brought to the Kara Sea regularly from the Ob and Yenisey bays in large quantities during several decades (Champ et al. 1994; Kuznetsov 1994; Rissanen et al. 1995; Stepanets et al. 1999).

The surface bottom sediments in the Yenisey Bay are saturated by ^{137}Cs to 40-74 Bq / kg (Fig. 5.18). Two peaks of radiocesium concentration are distinctly pronounced in the sediment layers (layers 4-7 and 13-15 cm). The maximum values for the Ob and Yenisey sediments are approximtely 40 Bq / kg and 100 Bq / kg respectively. The levels for the same layer in the Kara Sea are 11-13 Bq / kg (Stepanets et al., 1999). Judging by the depth of radiocesium penetration in the sediment column, the sedimentation rate in these bays is about 0.5 cm / year.

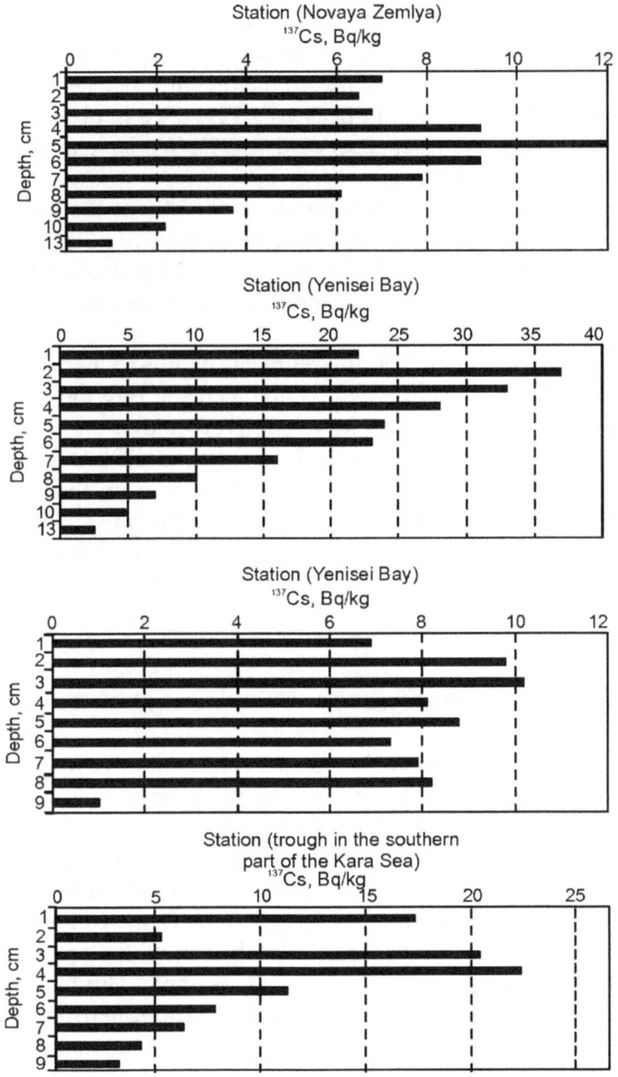

Fig. 5.18. ^{137}Cs levels in sediment cores from the Kara Sea (Krosshavn 1995)

The radiochemical analysis of the sediment samples taken layer by layer from a 20 cm sediment core from the Yenisey Bay in 1993 showed the greatest accumulation of ^{90}Sr in the 2-5 cm layer (10.64 Bq / kg). With increasing depth in the col-

column of sediments, the level of this radionuclide decreases to 0.26 Bq / kg (Matishov et al. 1994c, 1997c). The maximum ^{90}Sr concentration in surface sediments was in 1993, 3-5 Bq / kg (Kuznetsov 1994; Stepanets et al. 1999). These levels are characteristic only for the bay. Strontium levels decrease sharply towards the Kara Sea.

It should be noted that much more ^{90}Sr (2-8 Bq / kg) is accumulated in bottom sediments of the Ob Bay. In some places the quantity reaches 20 Bq / kg (Novitsky et al., 1999). A concentration peak to 16 Bq / kg is observed in the 4-7 cm layer of bottom sediments (Stepanets et al. 1999). We have also noted $^{239,240}Pu$ at levels of 2 Bq / kg in the Ob Bay.

There are relatively high concentrations of $^{239,240}Pu$ in the botton sediments of the Yenisey Bay. The content of this nuclide varies from 0.6 to 6.0 Bq / kg. Silt containing sediment cores taken by MMBI contained plutonium throughout an entire 20 cm core. The highest concentration was found in the 11-14 cm layer (Matishov et al. 1994c).

Thus, during the 1990s the bottom sediments of West Siberian bays contained a significant quantity of ^{137}Cs and ^{90}Sr. In the Yenisey Bay the levels of ^{137}Cs and ^{90}Sr were 50-100 and 0.3-5.0 Bq / kg respectively and in the Ob Bay - 10-40 and 2-16 Bq / kg. 60Co also was present in the Yenisey and Ob Bays at concentrations of 2-6 and 0.3-0.6 Bq / kg respectively. Sediments from the Yenisey Bay contained $^{239,240}Pu$ concentrations of 0.6-6.0 Bq / kg. One must take into account that the depth of penetration of these radionuclides into the clayey sediments reaches 15-20 cm extending across an area of the seabed of about 200-300 km^2 in each of the bays.

5.2.5 Bays of Novaya Zemlya

Many fjords are found on the west and east coasts of Novaya Zemlya. The upper parts of the fjords on the Northern Island of Novaya Zemlay are occupied by glaciers with 5-20 m high precipices. Continuous ice cover develops only on the Northern Island. Short rivers located here are replenished here by glacier melt waters. The river network is more developed on the Southern Island where glaciers are not as prevalent. The length of the largest river, Bezymyannaya, is 77 km.

The hydrological structure of the Nordensheld Bay, the upper part of which is occupied by a discharge glacier, serves as an example. During summer the water mass structure of the bay is characterized by complicated interactions between Atlantic waters (temperature 3-4 °C, salinity 34.0-34.8 ‰) and glacier melt waters (temperature 0.0-0.7 °C) (Fig. 5.19, 5.20). The warm Atlantic waters enter the bay as a powerful stream (70-100 m by vertical) almost reaching the edge of continental ice. Radionuclides of European origin are brought in small quantities to the west coast of Novaya Zemlya bays.

Away from the edge of the Nordensheld glacier is a pronounced frontal zone between warm Atlantic and cold local waters (Matishov et al. 1992a). The upper layer of Atlantic waters (25-40 m) is very freshened (to 29.8-33.0 ‰) because of the ice melting processes. Locally formed cold waters along the glacier front mix with open sea water, sink below them, and spread over the bay bottom towards the sea. A pronounced thermocline is evident in the depth interval 30-60 m near the glacier and 60-90 m further offshore . The vertical temperature gradient in this thermocline is 1.5-1.6 °C / m.

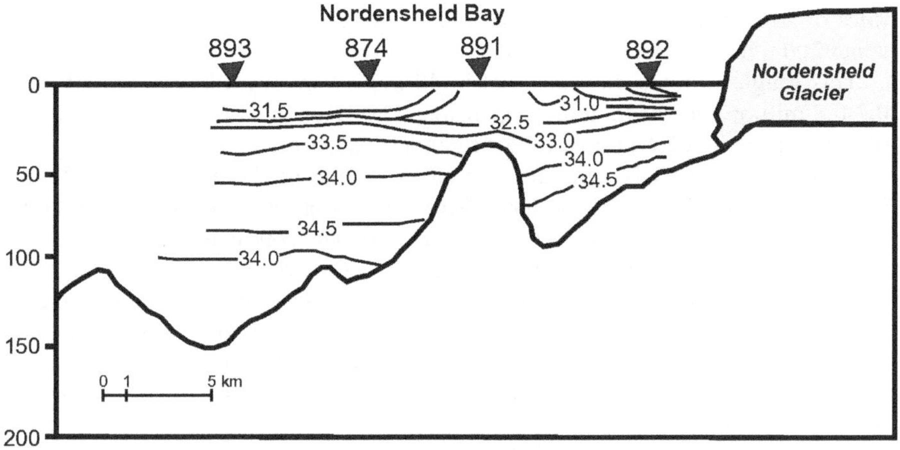

Fig. 5.19. Salinity of water masses in the Nordensheld Bay in September 1991. Salinity near the glacier is almost 5 ‰

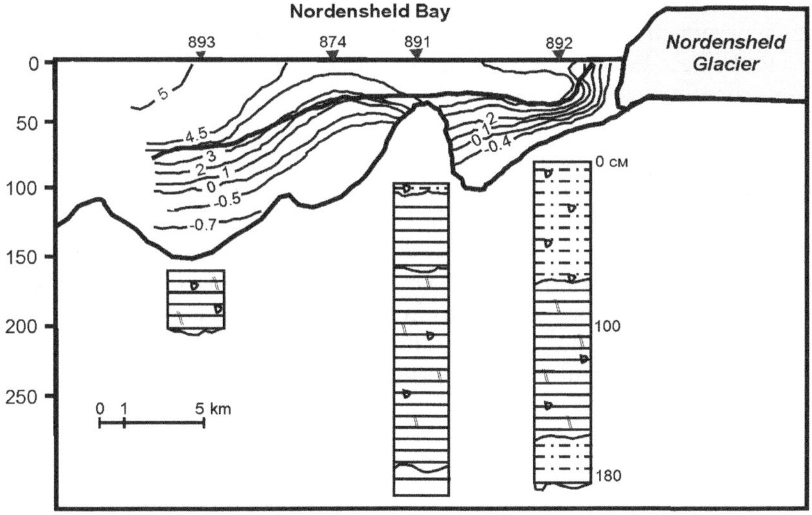

Fig. 5.20. Water temperature distribution and bottom sediment lithology in the Nordensheld Bay during September 1991

Nordensheld glacier is the largest on the Barents coast of Novaya Zemlya. In the upper part of Nordensheld Bay where the Nordensheld glacier discharges, we have noted very high concentration of suspended sediments (about 200 mg / l) (Matishov et al. 1994c). The fjord waters have zero transparency as a result of the high concentration of suspended sediments. Seasonal streams and rapid rivers from the glaciers as well as under-glacier streams of melt waters carry huge quantities of suspended sediments. About 0.2 g is contained in one liter of water and about 200 tons of sediment in 1 km^3.

This rough estimate indicates that in 40 km^3 of muddy waters contains about 0.7-0.8 million tons of mineral suspension. The sedimentation rate in the fjord is of highly because of the powerful terrrigenous discharge.

A similar thermohaline structure is characteristic of the east Novaya Zemlya bays and also for the coastal waters where the Nansen, Rozhdestvensky and other glaciers (15-20 m high) reach the Kara Sea coast. The continental glaciation of the Novaya Zemlya archipelago, the active feeding of the rivers from the glaciers and snow- field basins and the presence of icebergs in fjords are indicative of the accumulation of terrigenous and radioactive substances. The radioactive contamination of bottom sediments of a number of bays is linked to the direct influence of nuclear explosions, radioactive contamination from the atmosphere, or are the burial places of containers with nuclear wastes.

Chernaya Bay is the location of underwater, atmospheric and underground nuclear tests in 1955 and 1961. This bay stretches 40 km to the south, in the direction of the Pechora Sea. The primary mechanism of radionuclide migration in Chernaya Bay is by sedimentation processes. The radioactive products of the nuclear explosions quickly descended to the bottom of this small water body and were incorporated into the abiotic ecosystem components. The influx of radionuclides to the near-bottom water and especially to the bottom sediments led to their accumulation in benthic organisms (Matishov et al. 1993b, 1994b,c, 1995a). Tests in Chernaya Bay supplied only a small part (less than 1 %) of the total global fallout (Smith et al. 1995a, b). However, they have caused already and still may cause significant regional contamination of the marine environment.

Two surveys carried out in 1992 in Chernaya Bay showed that the transuranic elements - ^{239,240}Pu, ^{241}Am and others - are the main contaminating elements in the bottom sediments (Matishov D., 1993a; Matishov et al., 1993b, 1994a,b,c; Smith et al., 1995a). ^{239,240}Pu concentrations in the bottom sediments varies from 400 to 11148 Bq / kg. The highest levels of this nuclide are in the aleurite sediments of the bay depression (depth 80-100 m) (Fig. 5.21-5.23). ^{241}Am reaches about 3000 Bq / kg.

The levels of ^{239,240}Pu activity in surface sediments of the Chernaya Bay (0-5 cm) are close to the levels of plutonium in sediments at Enivetok Island in the Pacific Ocean where U.S. nuclear tests were conducted. It should be noted that ^{60}Co in Chernaya Bay bottom sediments is also quite high: from 20 to 618 Bq / kg (Matishiov et al. 1994a,c; Smith et al. 1995a). ^{137}Cs concentrations in sandy and clayey sediments ranges from 100 to 1450 Bq / kg.

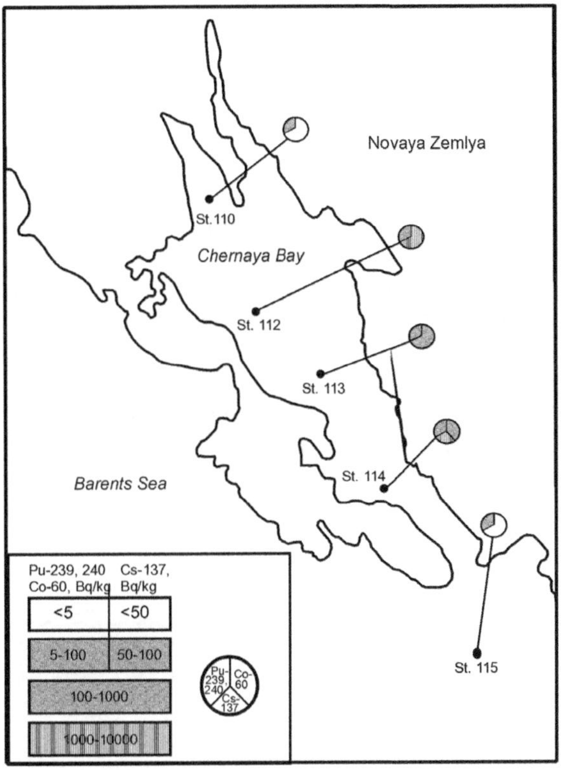

Fig. 5.21. $^{239,240}Pu$, ^{137}Cs and ^{60}Co content in the surface layer of bottom sediment in Chernaya Bay (Smith et al. 1995)

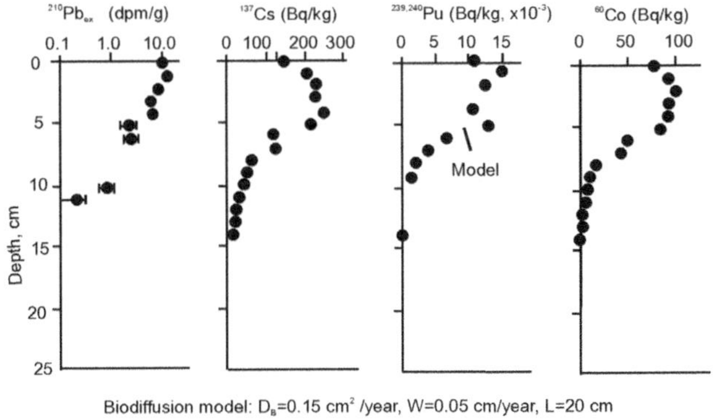

Fig. 5.22. Vertical distribution of radionuclides in bottom sediment cores from Chernaya Bay (Smith et al. 1995)

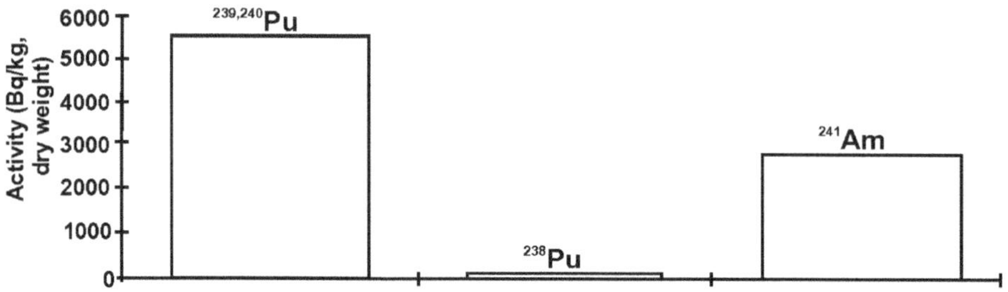

Fig. 5.23. Activity of gamma-emitters in Chernaya Bay bottom sediments collected near the underwater nuclear test site. Novaya Zemlya (July 1992)

As a result of nuclear explosions, the bottom sediments in the bay are contaminated to a high degree by radioactive plutonium, cesium and other radioactive isotopes. However, the mobility of radionuclides in bottom sediments is low, and their influence on the Pechora Sea is not significant at present (Fig. 5.24).

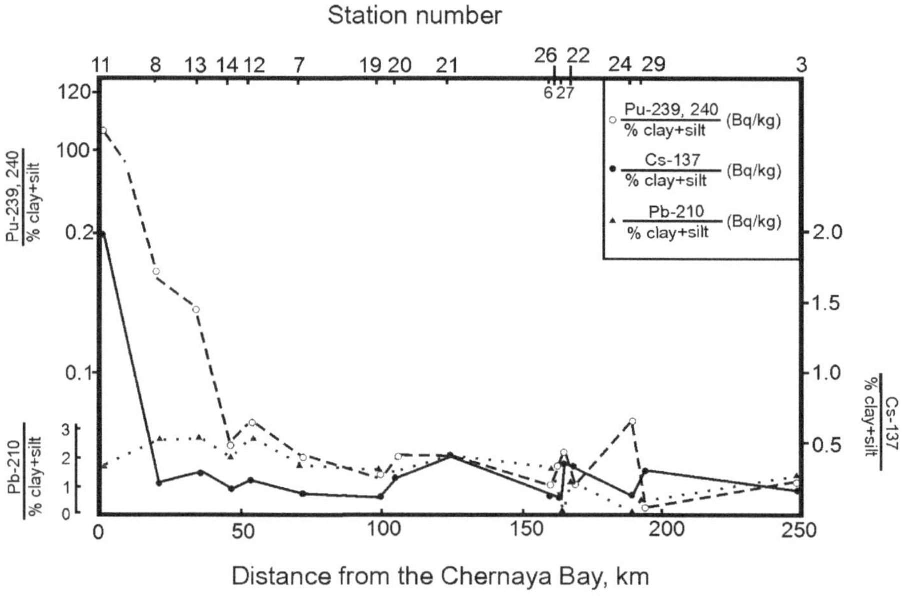

Fig. 5.24. Ratios ^{239,240}Pu : (clay+silt), ^{137}Cs : (clay+silt), ^{210}Pb : (clay+silt) depending on the distance from Chernaya Bay (Smith e al. 1995)

The Nordensheld, Inostrantsev and Glazov Bays, situated on the northwest of the archipelago, are characterized by insignificant (from 1 to 13 Bq / kg) ^{137}Cs accumulation in bottom sediments composed mainly of iceberg silt. The distribution of cesium in a 1.5 meter sediment core taken in the Nordensheld Bay is not uniform (Fig. 5.25). As a result of glacial melting the process of sediment accumulation is very active. The concentration of suspended sediments throughout the bays reaches 200 mg / l. The maximum ^{137}Cs concentration (11 Bq / kg) is at a water depth of 1 m (Fig. 5.25) and decreases with depth below 1m (Matishov et al. 1994c).

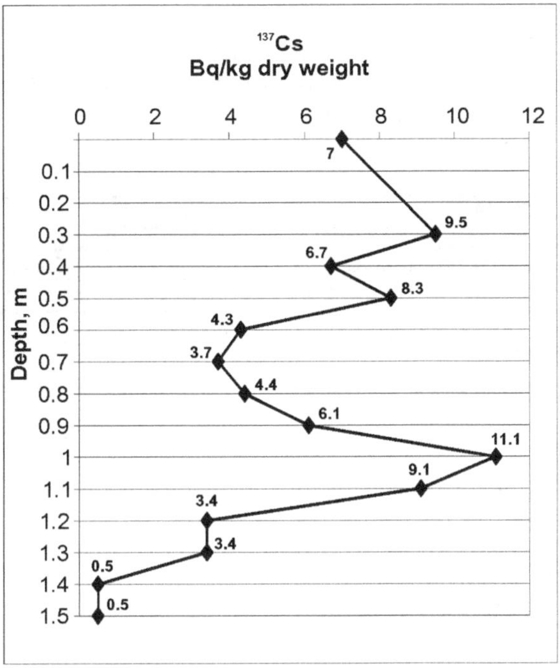

Fig. 5.25. Vertical distribution of ^{137}Cs in a 1.5 meter sediment core in Nordensheld Bay (Novaya Zemlya), Bq/kg dry weight

The distribution of radiocesium in surface sediments of the Glazov and Inostrantsev Bays are similar to Nordensheld Bay. Low levels of ^{137}Cs (1-12 Bq / kg) are observed in bottom sediments to 15 cm. Clayey sediments below this depth contain no artificial radionuclides.

Probably, the level and distribution of ^{137}Cs in the glacial-marine sediments of the Nordensheld, Inostrantsev and Glazov Bays are influenced by several factors simultaneously. There is a very high contemporary rate of sediment accumulation in the fjords. Maximum sediment accumulation takes place during the period of summer melting of glaciers. Such conditions do not contribute to the high nuclide concentration in sediments. The influx of radionuclides to the bays was from Novaya Zemlya sources (atmosphere, glaciers), from global fallout and also partly with warm Atlantic waters carrying radionuclides from the West European sources.

The Abrosimov, Stepovoy and Tsivolka Bays on the eastern coast of the archipelago was a location for the underwater storage of barges, ships and many containers with radioactive wastes. The increased concentration of ^{137}Cs, ^{90}Sr, ^{60}Co and $^{239,240}Pu$ is identified in the aleurite sediments taken near the buried objects (Fig. 5.26-5.29, Table 5.1) (Strand end Nielsen 1995; Strand 1998). Contamination levels are most pronounced immediately near the containers. The buried ships have relatively low activities of radionuclides. On the coastal shelf to the west of the archipelago, in particular, near Inostrantsev, Nansen and Rozhdestvensky glaciers, the quantity of radiocesium in sandy and aleurite sediments changes from 3 to 10 Bq / kg.

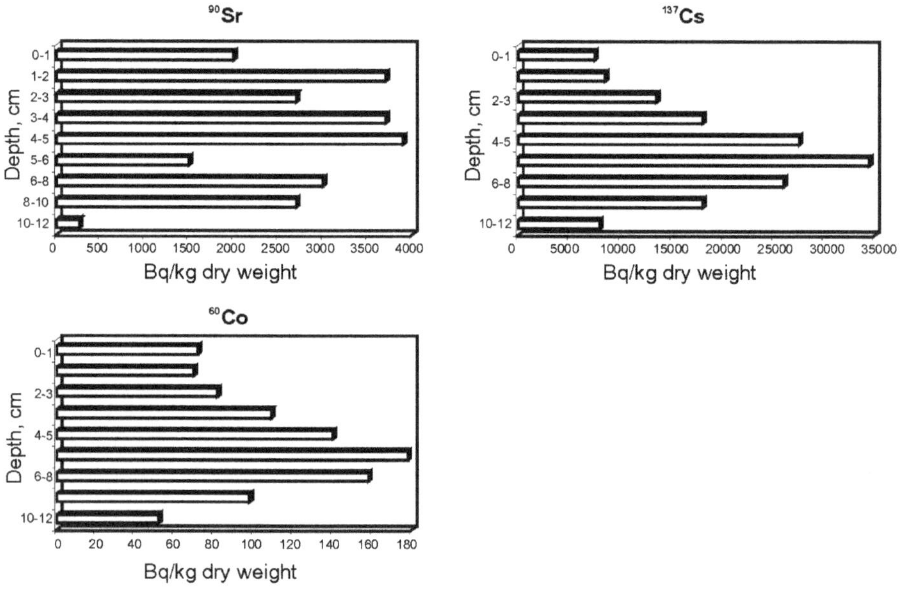

Fig. 5.26. Distribution of ^{137}Cs, ^{90}Sr and ^{60}Co with depth in Abrosimov Bay bottom sediments (Strand et al. 1994)

Abrosimov and Stepovoy are narrow (1-3 km) and short (8-14 km) bays with depths to 20-50 m. The rivers of the same names, which flow into these bays are active during summer only,. In the surface bottom sediments of these bays (beyond the zone of sunken objects the ^{137}Cs content varies from 4 to 40 Bq / kg and ^{90}Sr - from 0.4 to 2 Bq / kg. Radionuclide levels in sediments from the outer parts of the bays are similar to those on the Kara Sea shelf (Strand et al. 1992; Strand end Nielsen, 1995; Salbu et al. 1997).

The contamination of sediments in Abrosimov Bay is the most pronounced near buried containers (Strand, Nielsen, 1995). The maximum activity of ^{137}Cs in bottom sediments (5-10 cm layer) reached 8445, ^{90}Sr - 3800, ^{60}Co - 172 and $^{239,240}Pu$ - 18 Bq / kg. The level of radionuclides in sediments decreases noticeably the further one gets from the objects. The ^{137}Cs, ^{90}Sr, ^{60}Co and $^{239,240}Pu$ activity is correspondingly 23, 4, 0.4 and 0.7 Bq / kg. Such mixed distributions of radionuclides in sediments are connected mainly with the presence of «hot» radioactive particles.

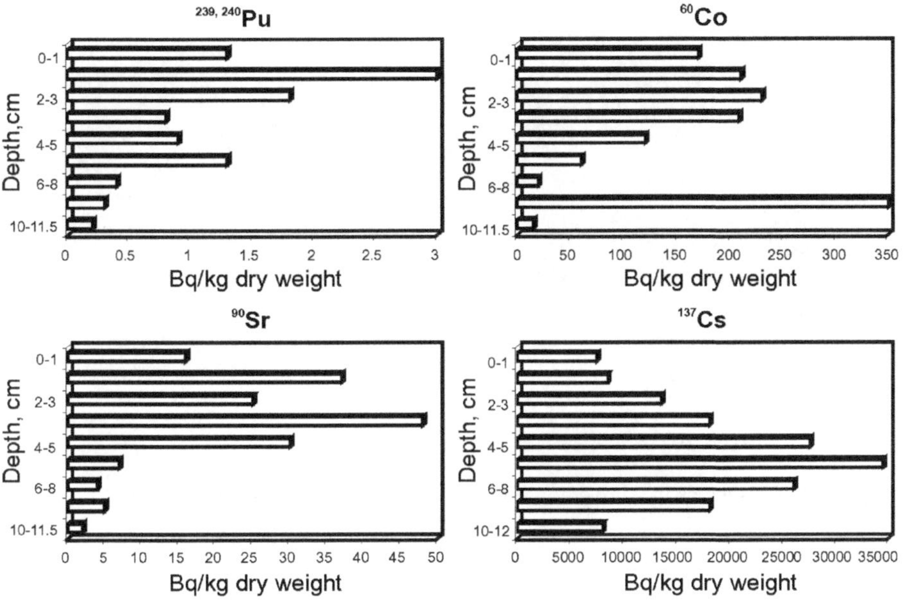

Fig. 5.27. Distribution of $^{239,240}Pu$, ^{137}Cs, ^{90}Sr and ^{60}Co with depth in Stepovoy Bay bottom sediments (Strand et al. 1994)

In the inner part of Stepovoy Bay the bottom sediments around reactors and containers are distinguished by high levels of ^{137}Cs (1.6-109 Bq / kg), ^{60}Co (0.3-3.2 Bq / kg), ^{90}Sr (47-310 Bq / kg) and $^{239,240}Pu$ (6-15 Bq / kg). The silty sediment deposits in bay depressions normally contain large quantities of ^{137}Cs (to 90 Bq / kg).

Artificial radionuclides are traced only in the 5-cm layer of sediments. The concentration of radionuclides decreases many times with increasing distance from containers to the outer part of the bay. For example, the sediments near the exit of the bay contain very few $^{239,240}Pu$ (0.3-0.6 Bq / kg).

Tsivolka Bay is a typical fjord with depths to 100-150 m. A glacier is situated in the narrow (3 km) corner of the bay. On the bottom of the bay, the radioactive wastes from the icebreaker Lenin are buried. In spite of this, the radionuclide level in the water and sediments is low: the ^{137}Cs quantity in sediments is 4-30 Bq / kg (Strand end Nielsen 1995). However, traces of ^{60}Co in sediments (4 Bq / kg) testify to the emission of radiation.

As it was noted, already, the sea water in Abrosimov, Stepovoy and Tsivolka Bays had the same radionuclide level as the open part of the Kara Sea. The ^{137}Cs

and ^{90}Sr content varied from 2 to 11 Bq / m^3 (Strand end Nielsen 1995), and only the bottom waters in the area of sunken containers had increased concentrations of ^{137}Cs (to 220 Bq / m^3) and ^{90}Sr (26 Bq / m^3).

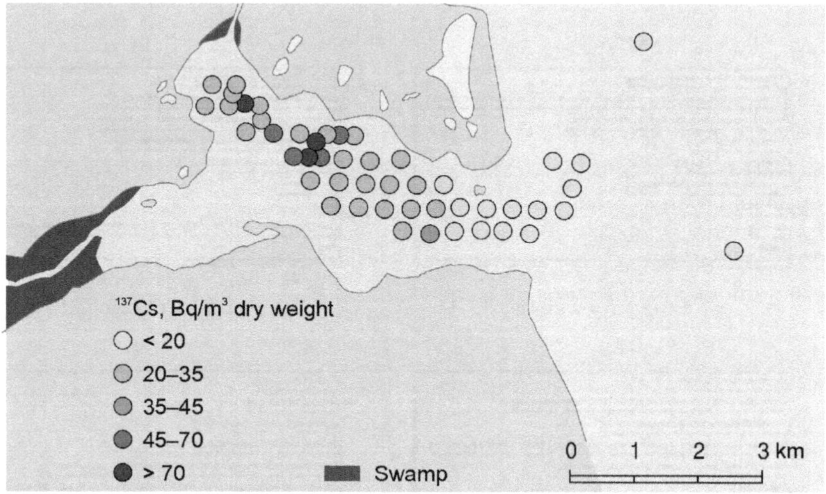

Fig. 5.28. ^{137}Cs levels in Abrosimov Bay bottom sediments (Strand et al. 1998)

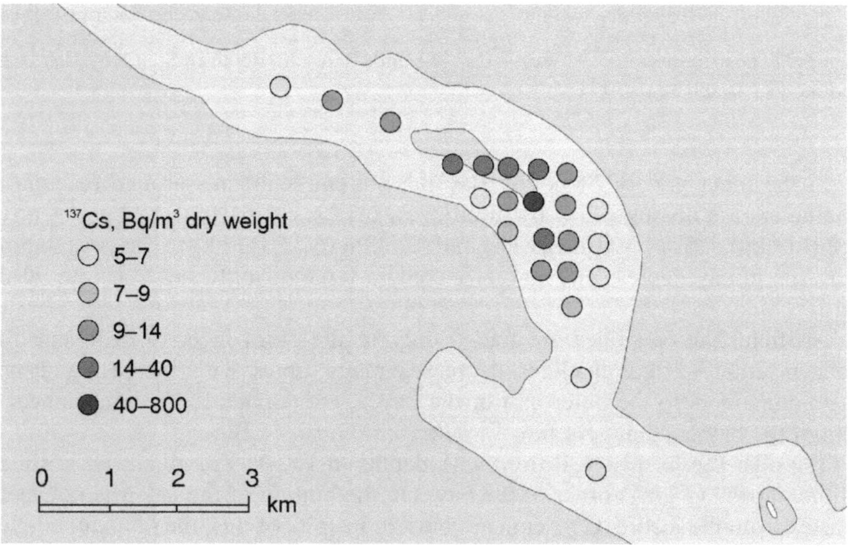

Fig. 5.29. ^{137}Cs levels in Stepovoy Bay bottom sediments (Strand et al. 1998)

Table 5.1. Concentration of radionuclides in sea water (Bq/m^3) and sediments (Bq / kg dry weight) in the bays and coastal shelf of Novaya Zemlya (Kara Sea, locations of buried nuclear waste; Strand 1998)

Area, buried object	^{137}Cs	^{90}Sr	^{60}Co	$^{239,240}Pu$
	Bottom sediments			
Abrosimov Bay:				
containers	22–8300	1–3250	0.4–56	1–18
containers (1st cm of a column of sediment)*	29–31000	1–8900	0.4–180	1–14
ship 1	46–196	0.3–3	1–2.5	1–2.6
ship 2	38–120	0.3–2	0.5–1.5	1.0–1.1
ship 3	39–54	1.31–51	–	0.7–1.0
reactor (1st, 2nd compartments)	33–8445	8–3250	1–61	1–5
reactor (3nd compartments)	43–61	0.4–22	1–11	1.0–1.3
Stepovoy Bay:				
containers	13–109000	0.6–310	2–3150	<0.1–18
containers (1st cm of a column of sediment)*	4–7350	1.6–47	0.3–345	0.2–3
Atomic submarine	4–1670	0.4–8	<0.1–6	<0.1–0.6
	Water			
Abrosimov Bay, 1994:				
surface water	4–7	2–4	–	–
bottom water	4–9	2–4	–	–
Stepovoy bay, 1993, 1994:				
surface water	4–9	2–7	–	–
bottom water	6–32	3–26	–	–
Tsivolka Bay, 1993:				
surface water	4–6	4–6	–	–
bottom water	5–14	3–4	–	–
Novaya Zemlya Trough, 1993:				
surface water	4–6	2–3	–	–
bottom water	7–14	2–3	–	–
Kara Sea, 1992:				
surface water	3–8	3–11	–	–
bottom water	4–18	3–6	–	–
	Bottom sediments, 0-2 cm			
Abrosimov Bay, 1994	1–60	–	0.5–3	–
Stepovoy bay, 1993, 1994	8–200	–	<1–20	–
Tsivolka Bay, 1993	4–30	–	<1	–
Novaya Zemlya Trough, 1993	7–30	–	<1	–
Kara Sea, 1992	2–33	–	<1	–

* to 10 cm See note above

5.2.7 Sea of Azov

The Azov Sea is an inland basin with an arid climate. The hydrochemical regime and sedimentation processes are controlled mainly by river runoff. Owing to the small size and shallowness, the basin contains low salinity water (1-13 ‰). This water is transformed river runoff originating mainly from the Don River. The phenomena of mechanical, chemical and biological sedimentation in the sea are very dynamic (Khrustalev, 1989) and result in the differentiation of sedimenting radioactive particles on the shelf.

Salinity in the central part of the estuarine basin varies from 8.06 ‰ in the surface layer near the Kuban River mouth to 11.7 ‰ in the bottom layer near the Kerch Strait. In September the salinity in Taganrog Bay varies from 0.3-0.5 ‰ near the Don River mouth to 10 ‰ near the exit to the Azov Sea (Belosarayskaya Spit). The hydrological and hydrochemical regimes of the basin are formed by advection of saline Black Sea waters and by river runoff (Fig. 5.30) (Matishov et al. 1998a, 1999a,b).

The rivers of the eastern part of the Azov Sea - Kagalnik, Mokraya Chumburka, Yaseni, Chelbas, Kirpily and others supply only small quantities of river-derived materials to the Sea of Azov. The river arteries of the Northern Azov region - Bolshoy and Maly Utluk, Molochnaya, Korsak, Losovatka, Obitochnaya, Berda, Kiltichiya, Kalmius, Sukhoy Elanchik, Mius, Sambek - are shallow, often only dry narrow rivers or brooks. These rivers have slow currents and discharge small quantities of suspended material. The influx of suspended particles is impeded by numerous dams which regulate the river runoff during flood periods.

The largest river of the Northern Azov region, the Don River, transports much larger quantities of suspended particles. The suspended sediment runoff of the Don is equal to 6.5 million tons annually, 92-95 % of it is discharged during the summer flood period. The water turbidity in this period is 345 g / m^3, whereas during the rest of the year it is 47 g / m^3. The annual discharge of suspended material to the sea from the Kuban River is approximately two times lower than for the Don River.

The physiogeographic conditions of sedimentation are determined by the underwater relief of the Azov Sea. The steep coastal slopes of this sea grade into an even and flat sea bottom. Water depths increase slowly and smoothly with distance from the coast. This uniformity is disturbed only by elongated sandbanks reaching of elongated spits extending kilometers into the open sea. Maximum water depths are found in the centre of the sea. The central region of the sea bottom is a large plain with depths 10-13 m. Its length and width are more than 120 km. A deeper area (12-13 m) is located in the southern part of the shelf adjoining the Kerch and Taman Peninsulas. The bottom relief of the Taganrog Bay contains a hollow with very slope borders and depths of 4-10 m. The distinguishing features of the coastal shelf are the Don (or Rostov) Sea Channel, underwater channels of the Taganrog, Temryuk, Berdyansk Ports and also of the Azovstal Port and Mariupol coal harbour. The lengths of these channels vary from 5 to 30 km and the depths - from 5 to 8 m.

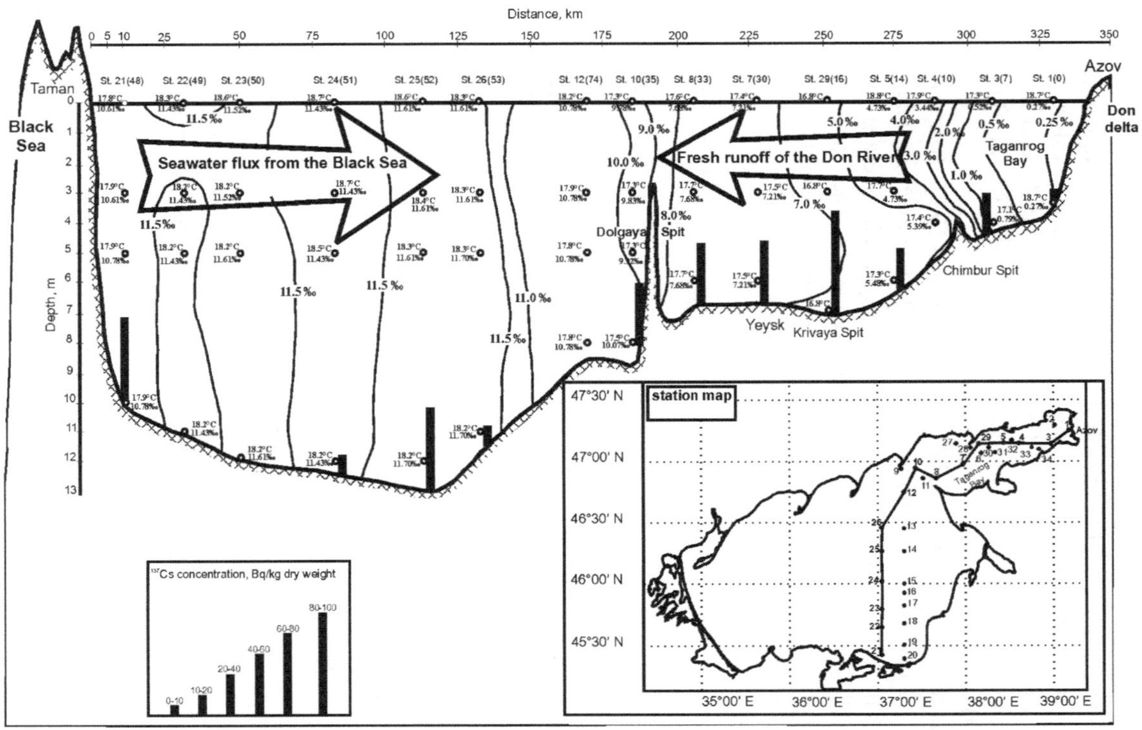

Fig. 5.30. Thermohaline structure and ^{137}Cs concentrations in bottom sediments in the Azov Sea 1997

Until quite recently information on anthropogenic radionuclides in the Azov Sea was limited. It was known that the ^{90}Sr content in sediments varied from 0.6 to 3.6 Bq / kg in the lower part of the Don and in the Tsimlyanskoye Reservoir in 1961-1962. The ^{90}Sr concentration in the Azov Sea water in 1985 and in January - April 1986 was on the order of 11 Bq / m^3. However, already in May 1986 the ^{90}Sr content in the northeast of the sea and in the lower part of the Don increased to 50-90 Bq / m^3.

There are also fragmentary data from the Institute of Biology of Southern Seas (INBUM) on radionuclides in the Tagangog Bay and southwestern part of the sea. In 1986 coastal sands contained ^{137}Cs and ^{90}Sr from 2 to 60 and from 1 to 3 Bq / kg correspondingly. Clayey sediments also contained ^{137}Cs (26 Bq / kg) and ^{90}Sr (3 Bq / kg).

The range of ^{137}Cs content in bottom sediments in 1987-1988 was from 16.3 to 264 Bq / kg dry weight (Ryabinin et al. 1992). The average value in 1987 was (100.9±66.8) and in 1988 - 74.0 Bq / kg dry weight. The average value of ^{134}Cs in 1987 was 22.1 and in 1988 - 14.4 Bq / kg dry weight. The ^{90}Sr content in bottom sediments in 1986 was small: (1.37±0.86) Bq / kg dry weight with a maximum value of 2.89 and a minimum of 0.65 Bq / kg dry weight (Fig. 5.31, 5.32).

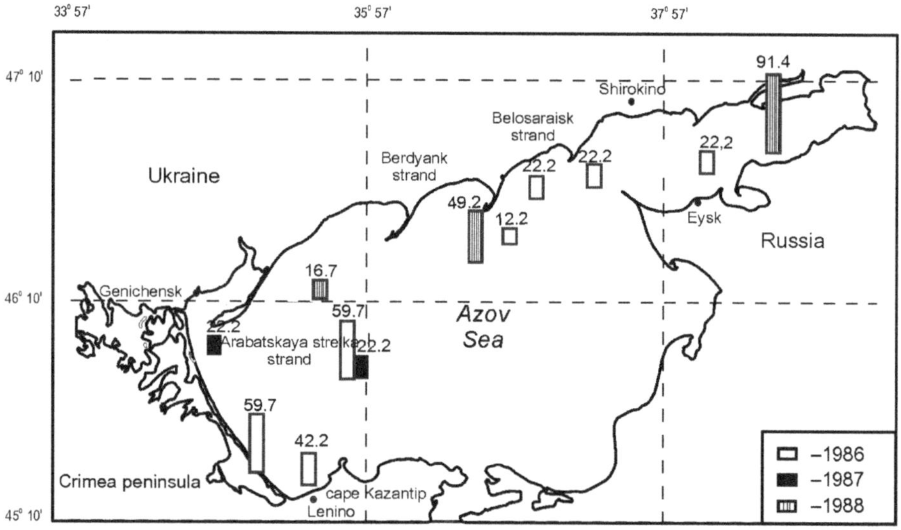

Fig. 5.31. ^{137}Cs concentrations (Bq / kg dry weight) in Azov Sea bottom sediments

As the ^{137}Cs and ^{90}Sr content in seawater decreased markedly from 1986 to 1989, the content of these nuclides in the Azov Sea algae didn't change accordingly. The range of ^{137}Cs content in different algae species was from 0.8 to 12.5 Bq / kg wet weight. The ^{90}Sr content was more uniform and changed from 2 to 1.4 Bq / kg wet weight.

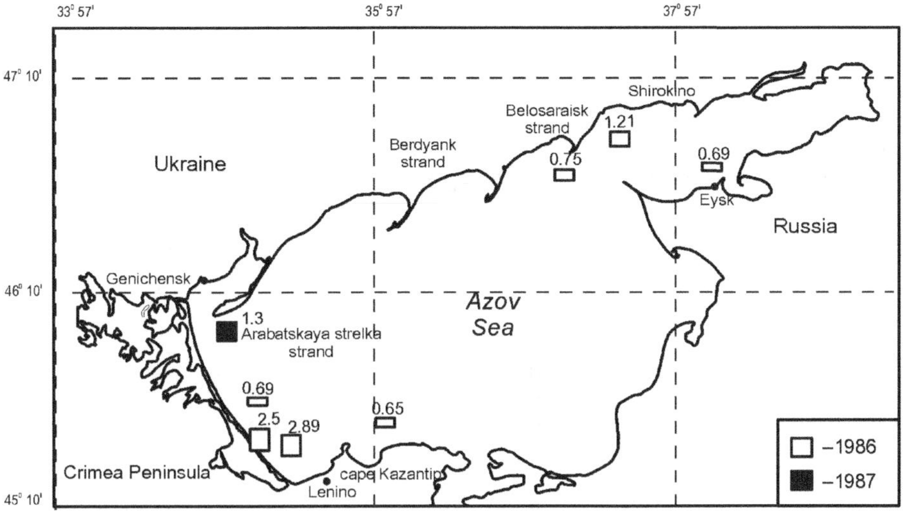

Fig. 5.32. ^{90}Sr concentrations (Bq / kg dry weight) in Azov Sea bottom sediments

Anthropogenic radionuclides were investigated in the sediments of the Black Sea northern shelf. The maximum ^{137}Cs concentration in the surface layer (to 7 cm) of shelf bottom sediments was 42 Bq / kg (Domanov et al. 1996). The origin of these radionuclides is connected with the Chernobyl fallouts.

Additional information about the radionuclide content in the Sea of Azov was obtained in 1997-2000 (Tables 5.2, 5.3) (Matishov et al. 1998a, 1999a, 2000b,c). These studies showed that the ^{137}Cs content in Azov Sea bottom sediments ranged from 1 to 100 Bq / kg. ^{134}Cs was not detected in sediments. The general distribution of radiocesium on the shelf is related to the geomorphological and lithological characteristics of the sea bottom, water salinity in the basin, and also by the hydrodynamic conditions. Strong wind mixing mixes the entire water column resulting in uniform salinities with depth. At the same time water salinities increase rapidly away from the Don mouth to the Kerch Strait as a result of the strong river discharge(Fig.5.33).

In the eastern part of the Azov Sea shelf (depth 9-13 m), fine- and coarse aleurite silts contain insignificant quantities of ^{137}Cs (11-27 Bq / kg). Higher radiocesium levels (to 80 Bq / kg) are noted in the clayey silts in the central, deepest portion of the Azov Sea. The ^{137}Cs content in fine-aleurite silts from the relatively deep (10-11 m) Temruck Bay is from 30 to 80 Bq / kg. It is quite normal that the ^{137}Cs content in samples with 93-95 % pelite are 75-85 Bq / kg dry weight. The content decreased to 10-20 Bq / kg dry weight in samples consisting of 10-20 % pelite.

The radioactive contamination of bottom sediments from Taganrog Bay was investigated by collecting samples at distances of 5-25 km from the steep shores. The lowest 137Cs concentrations (0.6 to 20 Bq / kg) are noted in the coarse aleurites deposited in shallow areas (2-6 m) where the hydrodynamic regime is very active. The level of radiocesium accumulation increases uniformly (30-60 Bq / kg) in the fine-aleurite silts of the axis hollow on the bottom of the bay.

Table 5.2. ^{137}Cs, ^{90}Sr, $^{239,240}Pu$, ^{40}K (Bq / kg dry weight) and aleurite -pelite silts (volume %) in the surface layer of Azov Sea bottom sediments (MMBI expeditions in 1997, 1998 and 1999)

Layer, cm	^{137}Cs	^{90}Sr	^{90}Sr:^{137}Cs	$^{239,240}Pu$	^{40}K	Pelite
1	2	3	4	5	6	7
			1997			
Stations:						
1	0.64±0.13	0.21±0.06	0.33	–	–	15
2	32.2±0.8	2.5±0.3	0.08	–	–	76
3	35.9±1.6	2.5±0.3	0.07	0.31±0.05	–	34
4	35.8±0.8	2.9±0.3	0.08	–	–	76
5	37.7±0.9	4.4±0.3	0.12	–	–	79
6	4.7±0.4	0.65±0.13	0.14	–	–	15
7	52.3±0.8	0.86±0.15	0.02	–	–	96
8	58.2±2.6	1.6±0.2	0.03	0.51±0.6	–	65
8	60.9±1.2	1.7±0.2	0.03	–	–	–
9	37.7±0.8	5.5±0.5	0.15	–	–	65
10	50.2±1.1	0.79±0.2	0.02	–	–	93
11	48.8±0.9	0.84±0.2	0.02	–	–	88
12	17.2±0.7	2±0.2	0.12	–	–	25
13	27.1±0.7	2.3±0.3	0.08	–	–	–
14	13.5±0.5	3.1±0.3	0.23	–	–	5
15	11.3±0.6	1.7±0.2	0.15	–	–	–
16	19.8±0.7	3.1±0.4	0.16	–	–	10
17	74±1.0	1.3±0.2	0.02	–	–	95
18	78.6±1.4	5.7±0.5	0.07	–	–	–
19	50.6±1.0	5.6±0.5	0.11	–	–	80
20	27.9±0.7	3.3±0.3	0.12	–	–	–
21	78.2±1.2	2.1±0.2	0.03	–	–	93
22	87.5±1.3	0.77±0.2	0.01	–	–	–
23	74.8±1.2	1.3±0.3	0.02	–	–	95
24	18.4±0.6	2.6±0.2	0.14	–	–	15
25	79.7±1.4	0.72±0.2	0.01	–	–	–
26	15.1±0.6	0.74±0.3	0.05	–	–	18
27	60.9±1.1	1.2±0.2	0.02	–	–	–
28	68.3±1.2	1.7±0.2	0.02	–	–	–
29	100±1.1	2.8±0.3	0.03	–	–	–
30	17.3±0.5	0.93±0.15	0.05	–	–	–
31	17.8±0.5	0.86±0.15	0.05	–	–	15
32	33.9±0.8	2.9±0.3	0.09	–	–	–
33	11.3±0.4	0.93±0.15	0.08	–	–	–
34	51.8±0.9	1.9±0.2	0.04	–	–	–

Table 5.2. (Cont.)

1	2	3	4	5	6	7
			1998			
Stations:						
1	17.8±1.4	–	–	–	–	–
2	53.2±1.5	–	–	–	–	–
3	49.5±3.3	–	–	–	–	–
4	11.9±1.3	–	–	–	–	–
5	30.7±2.0	–	–	–	–	–
6	42.4±2.7	–	–	–	–	–
7	6.0±0.6	–	–	–	–	–
8	47.4±3.0	–	–	–	–	–
9	57.9±3.6	–	–	–	–	–
10	<0.5	–	–	–	–	–
11	15.9±1.1	–	–	–	–	–
12	4.8±0.57	–	–	–	–	–
13	3.6±0.9	–	–	–	–	–
15	14.6±1.1	–	–	–	–	–
16	6.5±0.85	–	–	–	–	–
17	13.7±1.0	–	–	–	–	–
18	3.6±0.51	–	–	–	–	–
19	<0.6	–	–	–	–	–
20	3.3±0.69	–	–	–	–	–
21	2.1±0.5	–	–	–	–	–
22	9.7±0.86	–	–	–	–	–
23	13.8±1.3	–	–	–	–	–
24	32.5±3.1	–	–	–	–	–
			1999			
Stations						
18–1						
0–1	1.2±0.2	–	–	–	120±5	–
1–2	1.25±0.20	–	–	–	110±5	–
2–3	1.0±0.1	–	–	–	90±4	–
3–4	1.0±0.1	–	–	–	102±4	–
4–5	<0.9	–	–	–	76±4	–
5–6	<0.6	–	–	–	72±6	–
6–8	<0.8	–	–	–	84±9	–
8–10	3.8±0.6	–	–	–	136±9	–
10–12	3.1±0.6	–	–	–	144±9	–
12–14	3.5±0.5	–	–	–	119±7	–
14–16	6.1±0.6	–	–	–	246±10	–
16–18	7.5±0.6	–	–	–	270±10	–
Station 34 Port Katon						
0–1	<1.1	–	–	–	534±16	–
1–2	<1.4	–	–	–	505±19	–
2–3	<1.1	–	–	–	557±15	–
3–4	<1.6	–	–	–	549±21	–

Table 5.2. (Cont.)

1	2	3	4	5	6	7
4–5	<1.5	–	–	–	457±19	–
5–6	<1.0	–	–	–	501±14	–
6–8	<1.1	–	–	–	492±15	–
8–10	5.1±0.7	–	–	–	458±13	–
10–12	3.7±0.6	–	–	–	461±11	–
12–14	6.4±0.6	–	–	–	452±12	–
14–16	8.5±0.7	–	–	–	476±12	–
Stations:						
2	10.1±0.7	–	–	–	274±11	–
3	9.3±0.6	–	–	–	218±9	–
4	10.6±0.6	–	–	–	244±9	–
5	10.7±0.7	–	–	–	351±11	–
6	5.1±0.5	–	–	–	163±8	–
7	13.3±1.3	–	–	–	320±19	–
10	13.6±0.6	–	–	–	300±8	–
12	23.0±1.0	–	–	–	382±14	–
14	5.1±0.4	–	–	–	192±7	–
15	11.0±0.6	–	–	–	350±10	–
17	7.7±0.4	–	–	–	208±7	–

The maximum radiocesium content (100 Bq / kg) was found in the hollow at depths of 8-9 m (to the east of the Krivaya Spit). The process of accumulation of radioactive substances in Taganrog Bay occurs as a result of intensive mixing of fresh river water with Black Sea water. Water salinity changes from 0.15-0.5 ‰ in the upper part of the bay to 9-11 ‰ at the entrance to the open sea.

One can notice the presence of small quantities of ^{90}Sr and ^{239,240}Pu in the Azov Sea bottom sediments (Fig. 5.34, 5.35). The ^{90}Sr concentration varies from 0.3 to 9 Bq / kg and is related to the grain size composition of bottom sediments. The largest accumulation (4-5 Bq / kg) of ^{90}Sr is noted in the deepest (7-13 m) open part of the Azov Sea. Precisely in this part of the basin the fine aleurites and clayey silts accumulate after repeated sedimentation and mixing (Khrustalev 1989; Matishov et al. 1998a).

The ^{239,240}Pu content in the Azov Sea bottom sediments varies from 0.2-0.5 to 1.5 Bq / kg (Fig. 5.35). Plutonium concentrations level are highest in a trough located along the axis of the bay. Clayey silts in the trough contain a maximum plutonium content of about 1.5 Bq / kg and are found to the south from Begletskaya Spit at a depth of 6 m. Overall the plutonium activity in sediments from the bay is low. The ^{239,240}Pu concentration in the silts of the Don delta (Svinoye Channel) is about 0.35 Bq / kg.

It should be noted that sediments from several 10-cm cores were taken for the analysis of radionuclide distributions with sediment depth. From these cores it was determined that artificial radionuclide concentrations decrease uniformly with depth For example, the ^{137}Cs level in one sediment core decreased from 36 to 10 Bq / kg. The ^{239,240}Pu content in the aleurite sediments decreased from 0.6 to 0.4

Bq / kg and of ^{90}Sr from 2.8 to 2.2 Bq / kg with sediment depth. However, in areas of the seabed where the sedimentation rate is high, ^{90}Sr concentration increases with sediment depth from 1.9 to 9 Bq / kg. As a rule, the radionuclide content increases from fine sands to clayey silts (pelite fraction) supplied by the rivers.

Table 5.3. ^{137}Cs in the surface layer of bottom sediments in the Azov and Black Seas (MMBI expedition, June 2000), Bq/kg dry weight

Station	Area	Coordinates		Water depth, m	Layer, cm	^{137}Cs
		Lat. (N)	Long. (E)			
1	2	3	4	5	6	7
1	Taganrog Bay, near Veselo-Voznesenka	47°08.269'	38°18.427'	0.92	0–2	1.4±0.1
2	Don mouth	47 08.250	39 14.450	1.50		7.5±0.6
3	Martyshka Island	47 11.386	39 12.998	2.50		1.6±0.1
4	Kamyshevatskaya Village, Azov Sea	46 23.393	37 56.713	0.94		0.7±0.2
5	Dolgaya Spit, Taganrog Bay	46 41.264	37 43.652	0.88		<0.3
6	Shabelskoe settlement, Taganrog Bay	46 51.852	38 28.494	0.81		–
7	Kuterma Channel	47 11.792	39 14.249	2.20		5.3±0.5
8		47 12.274	39 17.307	3.00		4.5±0.4
9		47 11.138	39 19.872	2.00		14.2±1.0
10	Kalancha Channel	47 08.388	39 27.457	2.50		19.8±0.6
11		47 09.656	39 24.205	6.00		11.2±0.4
12	Kuterma Channel at the exit to the Taganrog Bay	47 11.149	39 13.227	2.50		4.1±0.3
13	Malye Dvoryane - Taganrog section (sand)	47 10.626	39 12.186	2.50		<0.1
14		47 10.525	39 09.475	2.50		1.6±0.3
15	Chumburskaya Spit, Taganrog Bay	46 57.976	38 54.012	0.30		1.5±0.3
16	Markitanskoe Lake, Taman Bay	45 14.400	36 46.080	1.00		13.1±0.5
17	Tuzla system of lagoons, Tuzla Cape	45°11.400'	36°36.300'	1.00		14.1±0.5
18	Bugaz Spit, «Salt Lake», Burgaz Lagoon	45 06.600	36 47.100	1.00		29.6±0.7
19	Neftyanoe Lake, Burgaz Lagoon	45 07.020	36 49.140	0.30		28.0±1.0
20	Kerch Strait, Azov Sea, CTD-St.48	45 24.320	36 45.012	2.66		1.2±0.1
21	Chushka Spit, near the camping, Azov Sea, CTD-St.64	45°23.440'	36°44.407'	0.85	Column of sediment	–

Table 5.3. (Cont.)

1	2	3	4	5	6	7
22	Chushka Spit, the end, Kerch Strait	45 18.328	36 37.299	2.01	0–2	38.0±1.0
23	Chushka Spit, near the camping, Dinskoy Bay	45 23.370	36 44.523	1.00		3.4±0.5
					2–4	5.4±0.7
					4–6	5.8±0.7
					6–8	8.5±0.7
					8–10	11.5±0.8
					10–12	26.3±1.1
					12–14	36.6±1.3
					14–16	25.6±1.1

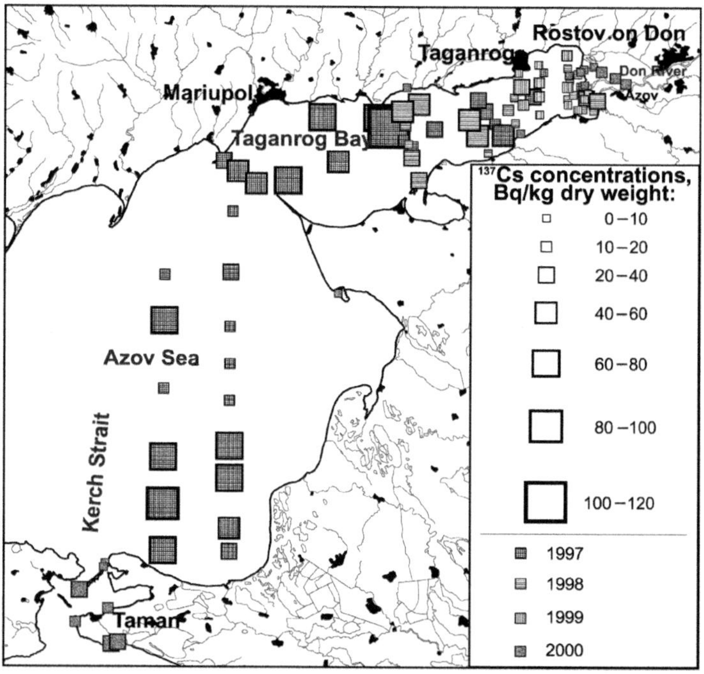

Fig. 5.33. ^{137}Cs content in the surface layer of bottom sediments of the Azov Sea, Don mouth, lakes and lagoons of the Taman Peninsula (MMBI radioecological investigations 1997-2000)

The analysis of the depth distributions of ^{137}Cs, ^{90}Sr and ^{239,240}Pu in sediment cores revealed two peaks in ^{137}Cs concentration above a background of general ly decreasing concentration with depth: the first peak is observed in the 4-6 cm layer (127 Bq / kg dry weight) and the second is observed in the 8-10 cm layer (88.4 Bq / kg dry weight. The first peak is likely caused by the Chernobyl disaster and reflects the period of maximum atmospheric fallout. The second peak most probably is connected with the influence of earlier atmospheric nuclear tests (Matishov et al. 1999a,b, 2000c).

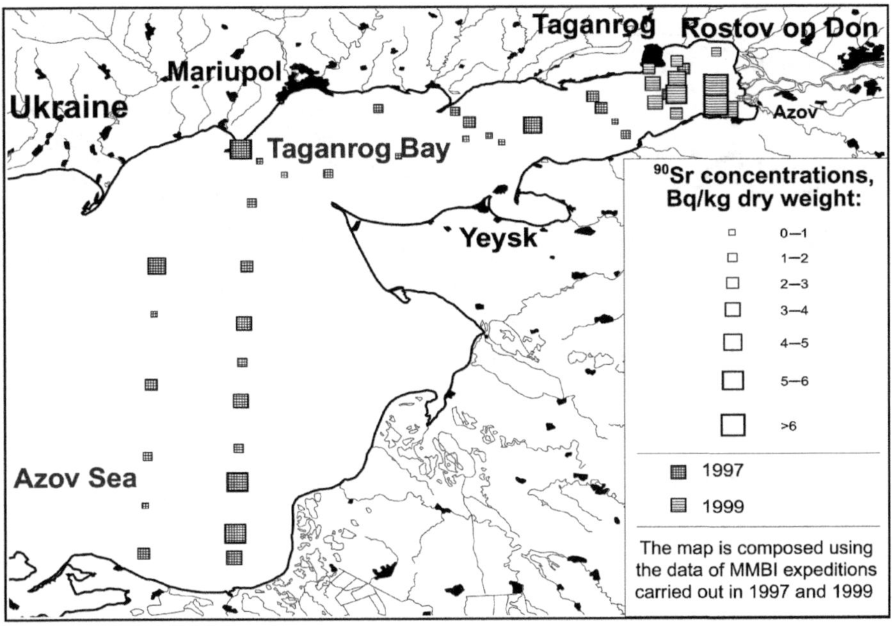

Fig. 5.34. ^{90}Sr content in the surface layer of the Azov Sea bottom sediments

Thus, the modern bottom sediments of the Azov sea contain significant quantities of anthropogenic radionuclides. The relatively high concentrations of ^{137}Cs in the sea bottom sediments are especially noticeable. The ^{137}Cs concentration 100 Bq / kg was fixed in the fine-aleurite silts of the Taganrog Bay. However, there are no direct sources of artificial radionuclides in the Azov region. It can be noted for comparison that in the coastal zone of the Kola Peninsula the highest levels of ^{137}Cs (120 Bq / kg) were determined in the bottom sediments of the Andreev Bay. The ^{239,240}Pu concentration in Azov Sea sediments is less by an order of magnitude than the content in bottom sediments of the Barents and Kara Seas (from 0.1 to 20 Bq / kg).

The origin of radionuclides in the Azov Sea bottom sediments is radioactive dust from the Chernobyl disaster. Radioactive (^{239,240}Pu, ^{90}Sr, ^{137}Cs) aerosol particles (of size 50 μm and less) had formed in the accident area and after the disaster the contamination was spread by the prevailing winds. The aerosols spread all over the Azov region, in particular along the coast of Taganrog Bay. Considering that all of the Azov rivers are controlled by dams impeding the influx of suspended particles, one can conclude that the major part of the Chernobyl fallout was received by the sea from the air during the disaster in 1986. In addition, ^{137}Cs,

^{90}Sr, $^{239,240}Pu$ evidently reached the Azov Sea through the process of re-sedimentation and migration within the river systems of the Azov basin. As a result, the role of global radioactive fallout from atmospheric nuclear tests during the 1950-1960s is less evident today for the Azov Sea.

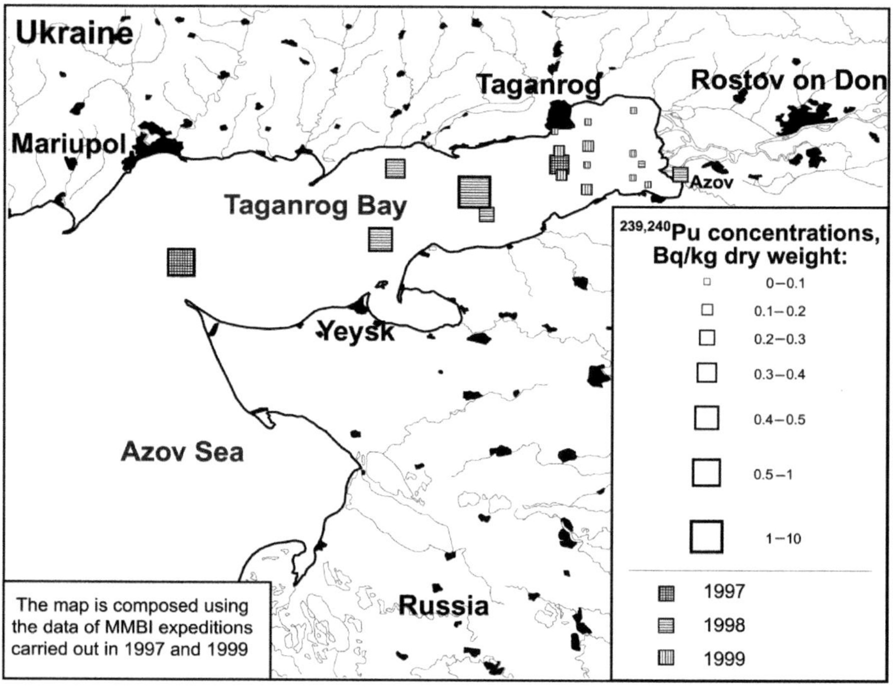

Fig. 5.35. $^{239,240}Pu$ content in the surface layer of the Azov Sea bottom sediments

5.3 Contaminant classification of coastal sea areas

Bottom sediments of bays, fjords and estuaries in the northern seas are formed depending on their geomorphological and oceanographic conditions and also on the capacity of the land provinces to supply the bays with sediments containing radionuclides (Matishov, 1984, 1987; Matishov et al. 1994c). In the course of the dispersion of sediment particles, suspended in the surface and deep water layers, a size separation occurs.

As it was noted already, the spectra of artificial radionuclides and ranges of their concentration in sediments vary among the different systems investigated. ^{137}Cs is most widespread in bays. ^{90}Sr and $^{239,240}Pu$ are less so. ^{60}Co and ^{134}Cs are found in sediments near local sources. The general behavior of all anthropogenic radionuclides is that their level increases from coarse sands to clayey deposits.

Radionuclides in sediment cores are observed to depths on the order of 10-20 cm. Bays with a high rate of glacial-marine sedimentation are an exception. Radionuclides are traced in such bays down to depths of about 1.5 m.

A general analysis of γ-spectrometric and radiochemical data for the 1990s provides an image of the radioactive contamination in more than 40 bays (Table 5.4). Based on these data and information on potential sources of pollutants and physiogeographical characteristics, we have developed a classification system for bays in relation to the concentration and composition of artificial radionuclides in sediments (Fig. 5.36).

In this classification, we divide the bay into seven categories based on the character of radioactive pollution. The **first type** includes such bays as Yarnyshnaya, Teriberka, Titovka, Ura, Motovsky (Kola Peninsula), Nordensheld, Glasov, Inostrantsev (Novaya Zemlya Archipelago). These bottom sediments have very low ^{137}Cs content (1-15 Bq / kg) and ^{60}Co and $^{239,240}Pu$ are absent. Low radiocesium levels are the result of influxes from small river catchment areas.

The **second type** includes the Dvina, Onega and Pechora Bays and is characterised by higher ^{137}Cs concentrations (20-70 Bq / kg). ^{60}Co is not present in the sediments. The increase of radiocesium in sediments of these bays reflects the effect of strong river runoff. The Pechora and other rivers have vast drainage basins which received large quantities of global fallout.

We suppose that the global atmospheric fallout of radionuclides were the main source of ^{137}Cs influx to the White Sea. The migration of radionuclides from land to sea is carried out by the Severnaya Dvina River system. However, judging by the presence of ^{134}Cs and ^{60}Co in bottom sediments, there are obviously local sources of artificial radionuclides which also plays a significant role in some cases (for example, in the Kandalaksha Bay). Technologically enhanced radionuclides from Sellafield evidently are less important in the restricted White Sea basin (Matishov et al. 1996f).

The **third type**, to which the Taganrog, Finnish and Bothnia Bays belong, is distinguished by relatively high ^{137}Cs levels in sediments (from 80 to 400 and locally 3600 Bq / kg). Traces of $^{239,240}Pu$ (0.2-3.0 Bq / kg), ^{90}Sr (0.3-9.0 Bq / kg) are present in sediments, but ^{60}Co is absent. This radionuclide compostion obviously is connected with the effect of drainage by rivers draining large territories and where Chernobyl fallout was deposited (Ilus et al. 1988a, b, 1993; Ilus end Ikaheimonen 1989, 1991; Matishov et al. 2000c). For example, the active water erosion of the soil cover and the discharge of suspended sediments from the drainage basin caused the migration of both terrigenous and radioactive substances to the Azov shelf (Matishov et al. 1998a, 1999a). Additional radiochemical and radioecological investigations of this sea will be needed in order to reconstruct the discharge, biogeochemical cycles and accumulation of artificial radionuclides. The information is also needed in order to trace the mechanism of radionuclide incorporation by plankton, benthos, fish and other biota.

Table 5.4. Concentration of artificial radionuclides in bottom sediments of bays in the northern and southern seas

Area	Salinity, ‰	^{137}Cs	^{60}Co	$^{239,240}Pu$	^{90}Sr
Cheshskaya Bay		0.5–6.0	Absent	No mesure-ments	–
Yarnishnaya Bay		1–10	The same	–	–
Titovka Bay	29–33	1–10	"	–	–
Hornsunn Bay (Spitsbergen)		11	"	–	–
Nordensheld Bay	29–34.8	1–13	"	–	–
Glazov Bay	29–34.8	1–13	"	–	–
Inostrantsev Bay	29–34.8	1–13	"	–	–
Teriberka Bay		3–13	"	–	–
Ura Bay	32–34	1–13	"	1.4–2.2	–
Motovsky Bay	33–34.5	5–14	"	1.4–2.2	–
Olenjya Bay	33–34.5	2–14	1–2	2–6	–
Is-Fjord		16			
Ara Bay	32–34	2–20	1–13	1.4–2.2	–
Kola Bay	20–35	1–24	0.5–5	0.15–5	–
Pechora Bay	2–12	2–24	Absent	–	–
Pala Bay	33–34.5	10–25	20–74	2.2	–
Ekaterininskaya Gavan bay	33–34.5	10–25	28	2.2	–
Pechenga Bay		10–26	Absent	–	–
Onega Bay		5–26	The same	–	–
Tsivolka Bay	33.5–34.8	4–30	0.9–4.0	0.03–0.5	0.4–6.0
Sajda Bay	26–34.4	3–34	0.7–12	1.6	–
Stepovoy Bay	20–34.8	4–40	0.3–6.2	6–15	0.3–47
Ob Bay	1–28	4–40	0.3–0.6	2	2–16
Kandalaksha Bay	24–30	20–45	1.7	4.0	0.5–3.8
Zapadnaya Litsa Bay	29–34	3–50	0.5–4.0	2–9	–
«Atomflot» enterprise	10–25	2–60	2–27	0.5–1.0	–
Dvina Bay	21–27	2–67	Absent	–	–
Azov Sea	8–12	10–80	The same	0.2–1.0	0.3–9.0
Yenisey River	1–28	60–100	2–6	0.6–6.0	3–11
Taganrog Bay	0.5–10	10–100	Absent	1.5	0.3–9.0
Andreev Bay (Zapadnaya Litsa)	28–33	70–120	6–33	3.2	–
Abrosimov Bay	32–34.2	4–120	0.4–61	0.7–18.0	4–22
Chernaya Bay		100–250 (Up to 1450)	20–200 (Up to 618)	400–2500 (Up to 11 ths.)	

The **fourth type** includes the Yenisey and Ob Bays. The Siberian rivers flowing into the bays drain huge catchment basins. These rivers receive radioactive discharges from Mayak and other chemical plants. The concentration of all radionuclides is relatively high: ^{137}Cs - 40-100, ^{60}Co - 0.5-6.0, ^{90}Sr - 3-16 and $^{239,240}Pu$ - 1-6 Bq / kg. The role of global radioactive fallout is probably of minor importance.

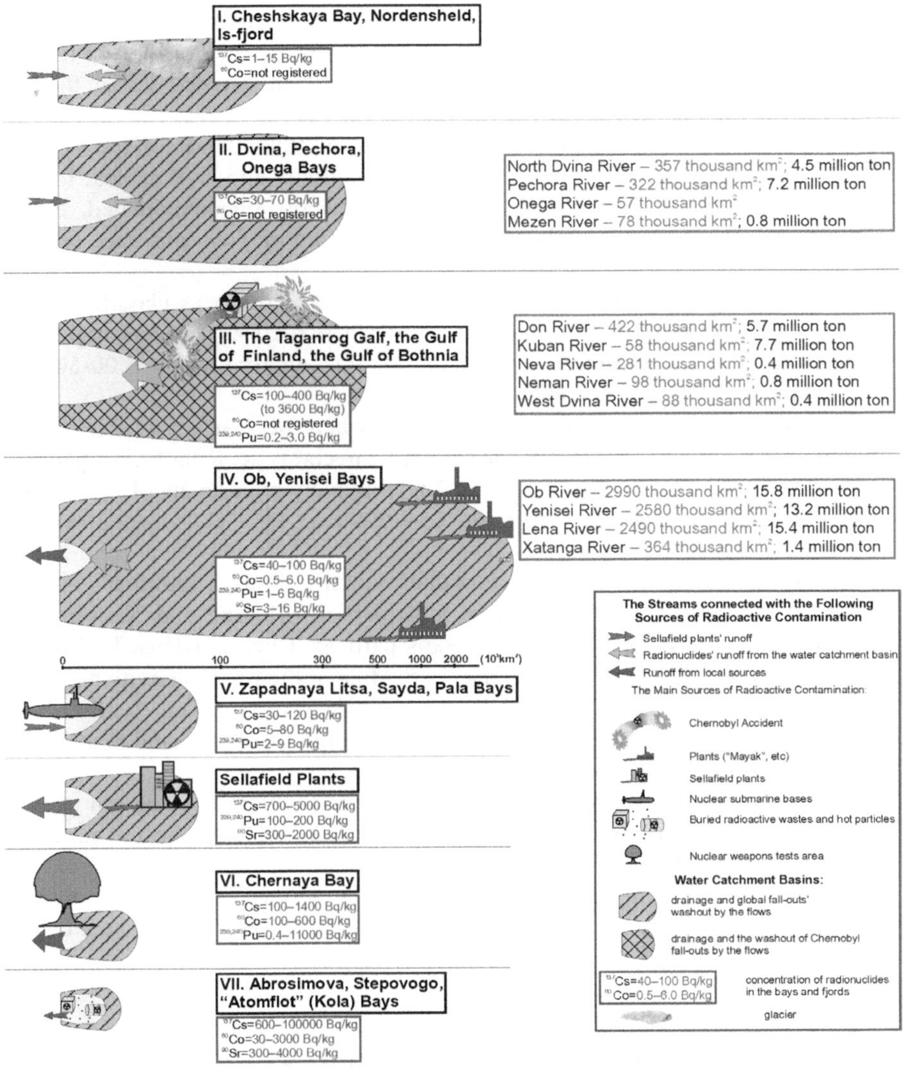

Fig 5.36. Classification of the bays of northern seas in relation to the degree of radiation contamination (1990s)

The **fifth type** is rather specific because it unites some bays in the Kola and Motovsky Bays area (Zapadnaya Litsa, Sajda, Pala) where naval bases are situated. The bottom sediments in the deep parts of these bays have accumulated ^{239,240}Pu from 2 to 9 Bq / kg, ^{137}Cs from 30 to 120 Bq / kg, ^{60}Co from 4-10 to 80 Bq / kg. The east bays of Novaya Zemlya (Abrosimov, Stepovoy and Tsivolka also can be classified as this type, in areas zones of impacted by dumped nuclear materials.

Overall the relatively high radioactive background caused by emissions from local sources occurs only in the vicinity of nuclear bases. The transport of radioactive materials from these areas into the Barents Sea basin is noted only in the Kola Peninsula coastal zone.

We also note that the measured concentrations of radionuclides in the Kola Bay are low, considering the existence of numerous military and civil nuclear installations. The bays without military bases are practically uncontaminated. Bottom sediments in these bays contain from 1 to 12 Bq / kg of ^{137}Cs and trace levels of ^{60}Co.

The Chernaya Bay where nuclear tests were carried out is classified as the **sixth type**. At present the contamination of Chernaya Bay remains high: ^{137}Cs from 50 to 1440, ^{60}Co from 92 to 618 and $^{239,240}Pu$ – approximately 1000-5000 Bq / kg (Matishov et al. 1997f, 1998b).

The Abrosimov, Stepovoy, Tsivolka Bays, where nuclear waste materials are buried in the bottom sediments is the **seventh, special type**. Kola Bay, adjacent to Atomflot, is also included in this type. The extremely high levels of ^{137}Cs - from 0.6 to 100 000 Bq / kg, ^{90}Sr - from 0.3 to 4 000 Bq / kg, ^{60}Co - from 30 to 3000 Bq / kg are observed here.

The various images of radionuclide distribution in bottom sediments is partly related to the presence of radioactive (hot) particles (Strand end Nielsen 1995; Nikitin et al. 1999). The term 'hot' means particles (fuel, condensed, absorbed) with specific activity more than 0.1 Bq / μg. Most of the f radioactivity associated with hot particles is fixed onto polydispersed particles of sizes from tenths to hundreds of micrometers. These are the fractions ranging from fine sand to small pellite, thus, 0.1 to 0.0001 mm.

At present the discharges or leaks of hot particles directly to the sea occur in very limited locations of the mentioned bays. They indicate the occurrence of radionuclide emissions.

The radioecological situation in the bays of northern seas shows that the concentration of artificial radionuclides in these bottom sediments is much higher than on the arctic shelf.

5.4. Artificial radioactivity in macrophytes

Phytobenthos which intake nutrients from seawater during growth, are a the main component of the coastal ecosystem. All groups of macrophytes are represented in the coastal zone, but the green, blue-green, red and brown algae and the sea weeds (zostera, thalassia) prevail. The community of microorganisms also contributes to the biological cycle in the photic zone.

The brown algae (Fucus, Laminaria, Alaria) which cover the rocks in the surf zone are the most prevelant among coastal macrophytes. They descend into water approximately to a depth of 15 m. These macrophytes can have biomass to 6-20 kg / m^2 and an annual production from 400 to 1500 g C / m^2 (Zenkevich 1963;

Vozhinskaya et al. 1994). Algae serves as (1) a feeding and spawning area for a number of fish and invertebrate species, and (2) a mechanical filter of material discharged to the sea with melt- and river- waters. Mineral, organic and radioactive suspensions, are also associated with marine algae.

Besides the flux of radioactive contamination from land, the photic ecosystem receives radionuclides from the coastal currents. During the 1990s the coastal water masses of the Arctic and Subarctic Seas contained relatevely small quanties of artificial radionuclides. In particular, ^{137}Cs, ^{90}Sr and ^{239,240}Pu concentration in water varied from 2-11 Bq / m^3, 1-10 Bq / m^3 and 2-15 mBq / m^3 respectively. Only near the Sellafield facility, the Atomflot enterprise in the Kola Bay and in the Novaya Zemlya bays near nuclear burial sites did ^{137}Cs and ^{239,240}Pu levels in water increase to 15-30 Bq / m^3 and 20-100 or more mBq / m^3 correspondingly. During 1960-1980s the content of these radionuclides in the coastal waters was many times more.

Routine monitoring of radioactive contamination in macrophytes of the northern seas was carried out during 1980-1990s (Ilus et al. 1988a,b; Gurevich 1992; Matishov D. 1993a; Matishov et al. 1994c, 1996f; Foyn end Swaren 1995; Strand end Nielsen 1995; Rissanen et al. 1997b). MMBI analzyed more than 10 algae species with life cycles varying from 2 to 6 years. The accumulation of radionuclides in macrophytes varied due to a number of factors. The level of radionuclide accumulation by algae correlates with the concentration of these substances in the water and sediment, the salinity of seawater, and the time duration of atmospheric exposure during falling tides. The age of plants is important as well as the specific area of their surfaces through which radionuclides are sorbed. Contamination of macrophytes occurs also in the process of permanent sediment mixing by tide and storm currents.

^{137}Cs is the most toxic element among the artificial radionuclides found in macrophytes. Data on the accumulation of ^{137}Cs, ^{90}Sr, ^{60}Co and ^{239,240}Pu by Baltic fucuses (*Fucus vesiculosus*) were obtained in 1983-1990. Before Chernobyl the ^{137}Cs concentration in water and littoral fucoids along the Finnish and Bothnia Bays was correspondingly 9-50 and 6-30 Bq / kg (Ilus et al. 1988a; Ilus end Ikaheimonen, 1989). A month after the accident the radiocesium level in water and fucus increased sharply to 1400 Bq / m^3 and 1300-4900 Bq / kg. In autumn 1986 the level of this nuclide decreased to 300-500 Bq / m^3 in water and to 190-210 Bq / kg in algae. This trend continued through 1987 (Fig. 5.37; Table 5.5). At the same time the ^{137}Cs concentration in the White Sea laminarias *Laminaria Saccarina* reached 100 Bq / kg (Gurevich 1992).

Observations on 14 experimental sites along the coast of Norway in the North, Norwegian and Barents Seas during the period 1980-1983 showed that the radioactive contamination of *fucus vesiculosus* displayed a trend of decreasing concentrations (Fig. 5.38). ^{137}Cs levels in fucus from the Barents Sea (Vardo area) reached a maximum value of 10 Bq / kg in the beginning of the 1980s and also in 1987, after the Chernobyl accident (to 7 Bq / kg) (Sickel et al., 1995). In 1993 the ^{137}Cs concentration in these algae varied from 0.8 Bq / kg in Finmark to 9.4 Bq / kg in Oslo Fjord.

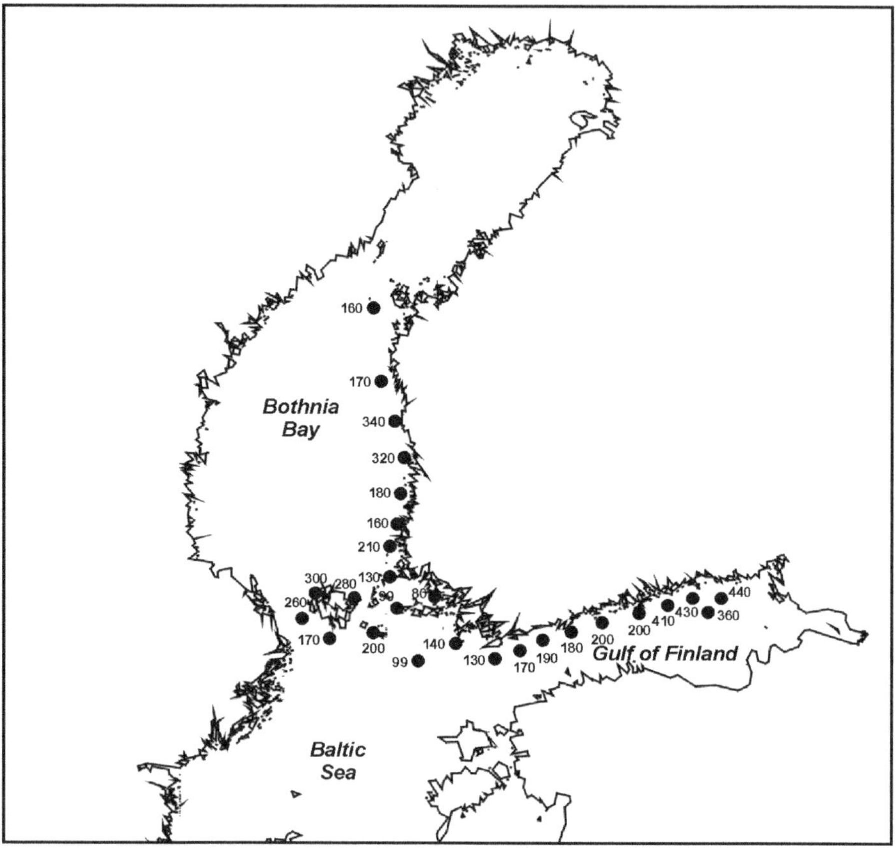

Fig. 5.37. ^{137}Cs content in *Fucus vesiculosus* along the Finnish coast in 1987, Bq/kg dry weight (Ilus et al. 1988)

During the 1990s the ^{137}Cs content in the White, Pechora, Barents and Kara Seas macrophytes varied from 0.4-1.0 to 3-5 Bq/kg, depending on the algae species and the location along the coast (Matishov et al. 1994b,c, 1997f; Rissanen et al. 1996, 1997a,b). In Abrosimov Bay of Novaya Zemlya the ^{137}Cs concentration in *Fucus vesiculosus* was 2-3 Bq / kg and in *Laminaria digitata* - about 1.5 Bq / kg (Strand et al., 1994; Strand, 1998). The radiocesium levels in the algae *Laminaria saccarina, Fucus vesiculosus* collected in the Barents Sea along the Kola Peninsula coast were 0.45-2.3 Bq / kg wet weight (Table 5.6). ^{137}Cs was present in similar quantities (0.5-2.5 Bq / kg) in macrophytes collected in the southern bays of Novaya Zemlya and the Vajgach Island area.

The content of radionuclides in the biota of the Kola Bay is not well-studied (Matishov et al. 1996f; Rissanen et al. 1996). The macrophytes *Ascophyllum nodosum*, *Fucus vesiculosus*, *Laminaria Sp*. in the littoral zone of the west coast of the bay contained a small quantity of ^{137}Cs (0.3-2.0 Bq / kg). In Olenjya Bay, near a ship repairing facility for nuclear submarines, the laminaria algae accumulated about 0.33 Bq / kg of ^{137}Cs in leafs and about 1.9 Bq / kg - in stalks. We can note for comparison that Fucus vesiculosus from Porchnikha Bay contains almost no radiocesium (0.06 Bq / kg).

Table 5.5. ^{89}Sr, ^{90}Sr and g-emitting nuclides in Fucus vesiculosus from the Baltic Sea (1986) (Ilus et al. 1988a), Bq/kg dry weight

Nuclide	Olkiluoto A			Loviisa A	
	28.04	21.05	14.08	29.05	29.08
^{7}Be	74	0	43	0	180
^{40}K	590	690	690	1100	1100
^{54}Mn	18	37	57	31	19
^{58}Co	0	0	14	0	10
^{60}Co	70	110	94	44	35
^{65}Zn	3.3	23	26	0	7.1
^{89}Sr	–	–	25	710	53
^{90}Sr	–	–	14	83	28
^{95}Zr	690	110	4.2	210	17
^{95}Nb	920	220	7.6	400	36
^{95m}Nb	290	0	0	0	0
^{103}Ru	570	1900	52	5900	260
^{106}Ru	110	590	61	2100	410
^{110m}Ag	2.4	170	62	230	130
^{124}Sb	0	0	0	0	16
^{125}Sb	0	24	4	97	24
^{129m}Te	22	2800	64	5600	230
^{131}I	410	29000	6.8	13000	0
^{132}Te	500	300	0	0	0
^{134}Cs	8.3	710	140	2700	550
^{136}Cs	0	150	0	210	0
^{137}Cs	26	1300	280	4900	1100
^{140}Ba	1000	4700	0	1900	0
^{140}La	810	3700	17	1800	7.1
^{141}Ce	700	94	1.7	250	8.1
^{144}Ce	360	130	7.1	330	35
^{147}Nd	280	0	0	0	0
^{148m}Pm	5.5	0	0	0	0
^{237}U	38	0	0	0	0
^{239}Np	3900	0	0	0	0

The ^{137}Cs content in algae can increase by 20-40 times in the proximity of sources of artificial radionuclide discharges (e.g. adjacent to Atomflot and near the 'Lepse' nuclear submarine). The ^{137}Cs content in the macrophytes from the area near Atomflot reached 20-46, ^{134}Cs - 1.2, ^{60}Co - 1.6 and ^{152}Eu - 4.6 Bq / kg wet weight (Fig. 5.40; Table 5.7). In addition, other radionuclides discharged by

Atomflot were observed in the trace quantities (Matishov et al., 1996d). At the same time high ^{137}Cs levels (to 260 Bq / kg) are noted in the thalluses of laminaria *Laminaria saccarina.* Such radioactivity obviously is caused by hot particles from Atomflot discharges (Nikitin et al. 1999).

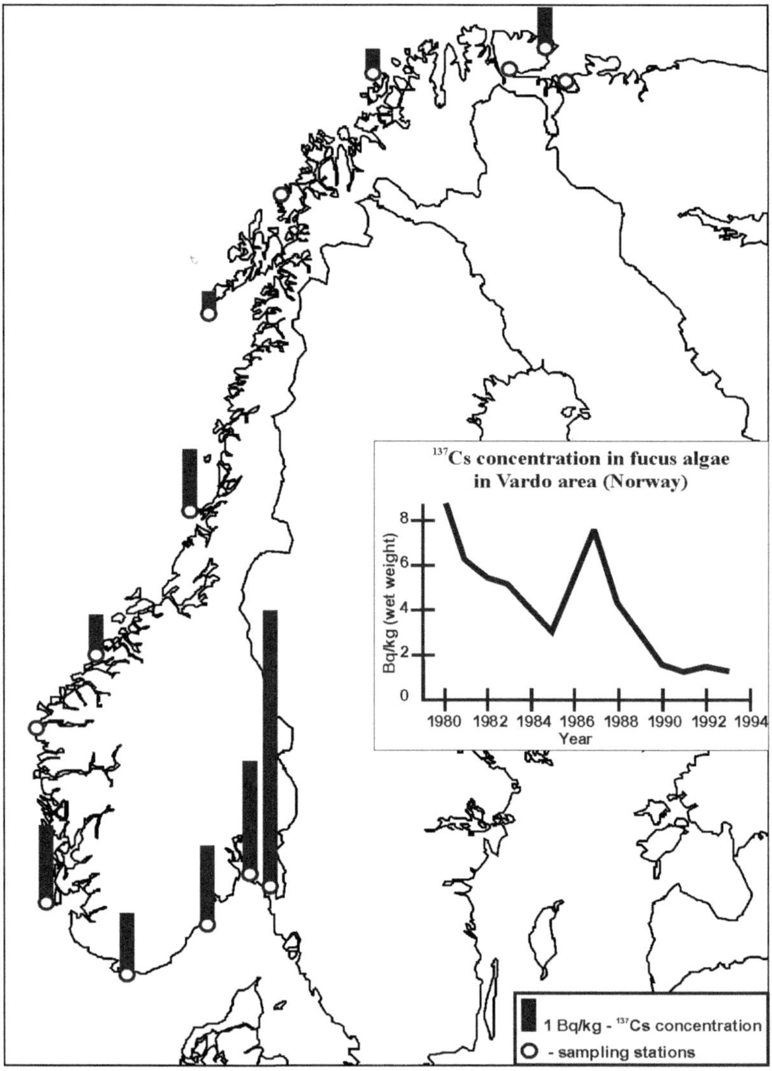

Fig. 5.38. ^{137}Cs concentration in marine algae and seawater from the Norwegian coast

Table 5.6. ^{137}Cs concentration in algae from the White and Barents Seas (1993) (Rissanen et al. 1997b), Bq/kg dry weight

Species	White Sea		Barents Sea			
	Kuzreka	Kandalaksha Bay	Vajgatch Island, Varneck Island, 69°43′N, 60°01′E	Dalnie Zelentsy settlement area, 69°08′N, 36°06′E	Teriberka settlement area: Meteorological station	Teriberka settlement area: Strait
Fucus sp.	–	–	1.7	–	–	–
F. vesiculosus	2.6	2.3	–	0.8	–	1.8
F. distichus	–	–	–	0.5	–	–
F. inflatus	–	–	–	–	0.66	2.3
Ascophyllum nodosum	–	2.2	–	0.45	–	–
Laminaria sp.	–	–	1.4	–	–	–

^{60}Co is found in macrophytes from areas in close proximity to radionuclide sources. Immediately after the Chernobyl accident in June-August 1986 fucus collected from the Finnish and Bothnia Bays littoral zone contained high levels of ^{60}Co (35-110 Bq / kg) (Ilus et al. 1988a). A year later, the concentration of this nuclide in algae decreased many times ((to 0.4-3.0 Bq / kg). In 1993 fucus contained little ^{60}Co (0.3-0.8 Bq / kg) except in a few locations along the Norwegian coast (Oslo Fjord, Tromso area). Still less ^{60}Co was found in the red alga *Ptilota plumosa* to the east from the Kolguev Island (Rissanen et al. 1995). It is difficult to explain how this short-lived nuclide was present in 1994 in the Pechora Sea region.

^{60}Co was present in the littoral vegetation of the Kola and Motovsky Bays (Matishov et al. 1996f, 1997b,c; Rissanen et al. 1996). This element accompanies the discharges from nuclear bases. In Olenjya Bay, ^{60}Co was fixed in the stalks of laminaria which contained 1.6 Bq / kg. This nuclide is especially toxic in macrophytes living in the coastal zone adjacent to Atomflot (Namyatov 1998; Nikitin et al. 1999). In this area 60Co reached 5-13 Bq / kg in laminaria and 0.2-1.1 Bq / kg in fucus.

^{90}Sr in macrophytes of northern seas is also not well investigated. The analysis of several samples of fucus from the Finnish and Bothnia Bays showed that in the end of May 1986 fucus accumulated up to 83 Bq / kg of ^{90}Sr (Ilus et al. 1988a). In August 1986 the level decreased to 14-28 Bq / kg and in 1987 - to 13-22 Bq / kg. The accumulation of ^{90}Sr in vegetation was caused by enrichment of the Baltic waters by Chernobyl discharges. Strontium, an analogue of calcium, participates actively in the metabolism of macrophytes and phytoplankton cells. In 1993 a small quantity of ^{90}Sr (0.8-2.0 Bq / kg) was detected in fucus from Northern Norway in the Barents Sea (Sickel et al. 1995).

$^{239,240}Pu$ is present today in brown and red algae in a very small quantities in all northern seas (from 0.02 to 0.3 Bq / kg). In the Finnish and Bothnia Bays fucus has accumulated 0.036-0.17 Bq / kg of $^{239,240}Pu$ (Ilus et al. 1988a). *Fusus vesiculo-*

sus located along the Norwegian coasts also contained $^{239,240}Pu$ in very low concentrations (0.08-0.15 Bq / kg) (Sickel et al. 1995). The $^{239,240}Pu$ content in the algae Laminaria saccarina and Fucus vesiculosus of the Barents and White Seas (Pechora Sea, Kanin Cape, Kolguev Island, Novaya Zemlya Archipelago, Dalnezelenetskaya bay and also areas of the Chupa Bay and Solovetskye Islands) was 0.02-0.28 Bq / kg dry weight (Table 39). The ^{238}Pu content was less than 0.01 Bq / kg dry weight (Rissanen et al. 1997a). All concentrations of ^{238}Pu were below the detection level of 0.01 Bq / kg dry weight.

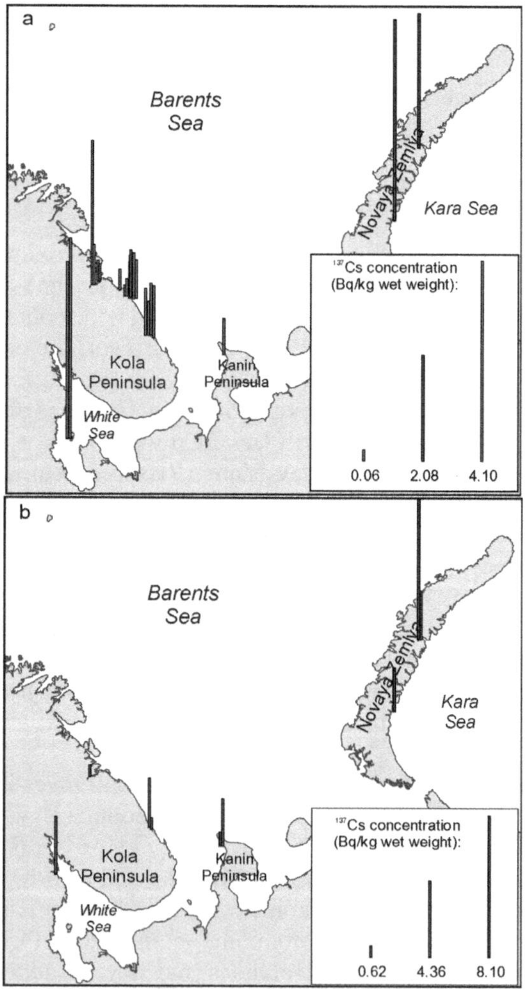

Fig. 5.39. ^{137}Cs content in (**a**) fucus and (**b**) laminaria (1992-1999)

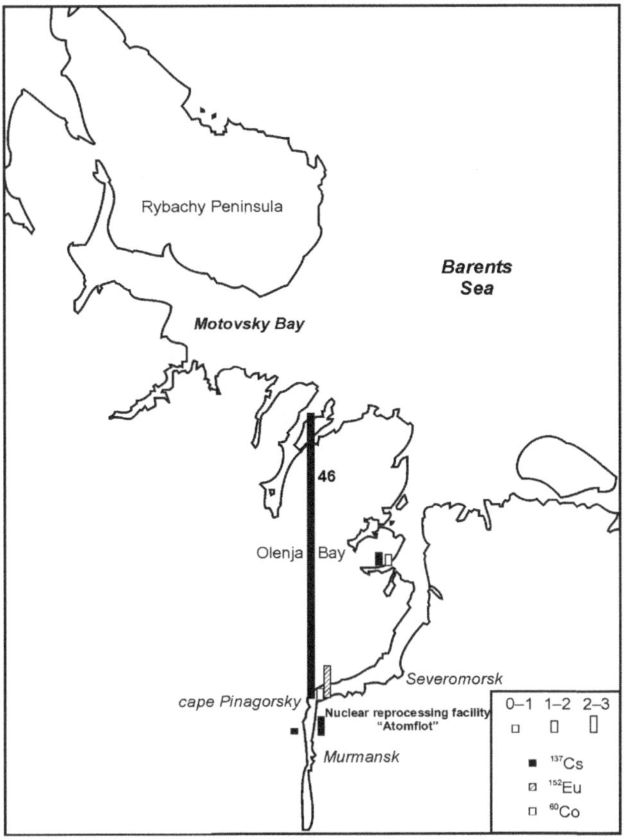

Fig. 5.40. Artificial radionuclides in marine algae from the Kola Bay (1996) (MMBI unpublished)

For comparison, in 1986-1988, algae from the west part of the Azov Sea (zostera *Zostera marina)* contained ^{137}Cs and ^{90}Sr from 1.5-20 and 0.4-3.6 Bq / kg. As the ^{137}Cs and ^{90}Sr content in An observed decrease in water concnetrations from 70.4 to 7.5 and from 57.1 to 34.5 Bq / kg ^{137}Cs and ^{90}Sr respectively was not mirrored by the Azov Sea algae. The largest ^{137}Cs content was registered in 1986 in the green alga (Chlorophyta). The content of this nuclide in zostera Zostera marina was 1.48-7.62 Bq / kg wet weight, and in entheromorpha (*Entheromorpha* Sp.) - 0.8-12.5 Bq / kg wet weight. The ^{90}Sr content was more uniform: in zostera 0.4-1.4, in enteromorpha - 0.2-0.7 Bq / kg wet weight (Fig. 5.41).

The species composition of macrophytes plays a very important role in the absorption of radionuclides because it determines the value amount of their primary sorption area both from the marine environment and from aerosol fallout during falling tides. *Laminaria algae* are the largest among the various species found in the aforementioned seas. Their length can reach 3-5 m and the width of individual lamina more than 1 m. Laminaria beds form the greatest total biomass in the littoral zone.

Table 5.7. Concentration of radionuclides in algae near «Atomflot», Bq/kg dry weight

Species	^{152}Eu	^{154}Eu	^{155}Eu	^{137}Cs	^{60}Co	^{99}Tc	^{40}K
			By: Nikitin et al., 1999				
Laminaria (*L. saccharina*)	<0.6–10.6	<1.5–19.2	<0.2–4.0	3.0–260	<0.5–5.1	2.8–3.6	1850–3950
Fucoids	<0.3	<0.7	<0.4	2.8	<0.2	32	830
			MMBI data, 1998				
Fucus	3.1±0.7	7.1±0.8	–	19.9±0.4	1.1±0.3	–	590±10
Laminaria	74.8±2.0	112±3	14.1±0.9	228±2	12.5±0.6	–	2690±20
Fucus	–	–	–	4.2±0.9	–	–	2000±30
	–	–	–	4.5±0.8	–	–	697±16
	–	–	–	<0.2	–	–	979±15
	–	–	–	0.8±0.4	–	–	550±12
Fucus, «Lepse» area	–	–	–	30.3±1.5	–	–	3028±39
	–	–	–	3.9±0.4	–	–	1433±23

Table 5.8. ^{238}Pu and $^{239,240}Pu$ concentrations in red and brown algae of the Barents, Pechora and Kara Seas (Rissanen et al. 1997a), Bq/kg dry weight

Species	Area	Date	^{238}Pu	$^{239,240}Pu$
Ptilota plumosa	Barents Sea (n=5), Kolguev Island	18.07.1994	<0.01	0.28
Laminaria saccharina	Barents Sea (n=2), Seven Islands archipelago	12.07.1993	<0.01	0.057
	Barents Sea (n=4), Kanin Peninsula	21.07.1994	<0.01	0.029
	White Sea (n=8), Chupa Bay	23.07.1994	<0.01	0.039
Fucus vesiculosus	Barents Sea (n=1), Dalnezelenetskaya Bay	11.07.1993	<0.01	0.076
	Barents Sea (n=2), Seven Islands archipelago	12.07.1993	<0.01	0.11
	White Sea (n=9), Solovetsky Islands	26.07.1994	<0.01	0.035
	Barents Sea (n=3), Nokuev Island	01.08.1994	<0.01	0.23
Fucus sp.	Pechora Sea (n=7), Vajgatch Island	20.07.1993	<0.01	0.041
	Pechora Sea (n=6), near Novaya Zemlya	23.07.1993	<0.01	0.12
Laminaria sp.	Pechora Sea (n=6), near Novaya Zemlya	23.07.1993	<0.01	0.042
Chorda filum	Barents Sea (n=3), Nokuev Island	01.08.1994	<0.01	0.020

By MMBI data, sea algae radionuclide sorption decreases in the following order: *Laminaria saccarina* → *Fucus vesiculosus* → *Fucus inflatus* → *Ascophyllum nodosum* → *Palmaria palmata* → *Fucus* Sp. → *Chorda filum*. The most intensive accumulation of $^{239,240}Pu$, probably, takes place in the red alga *Ptilota plumosa.*

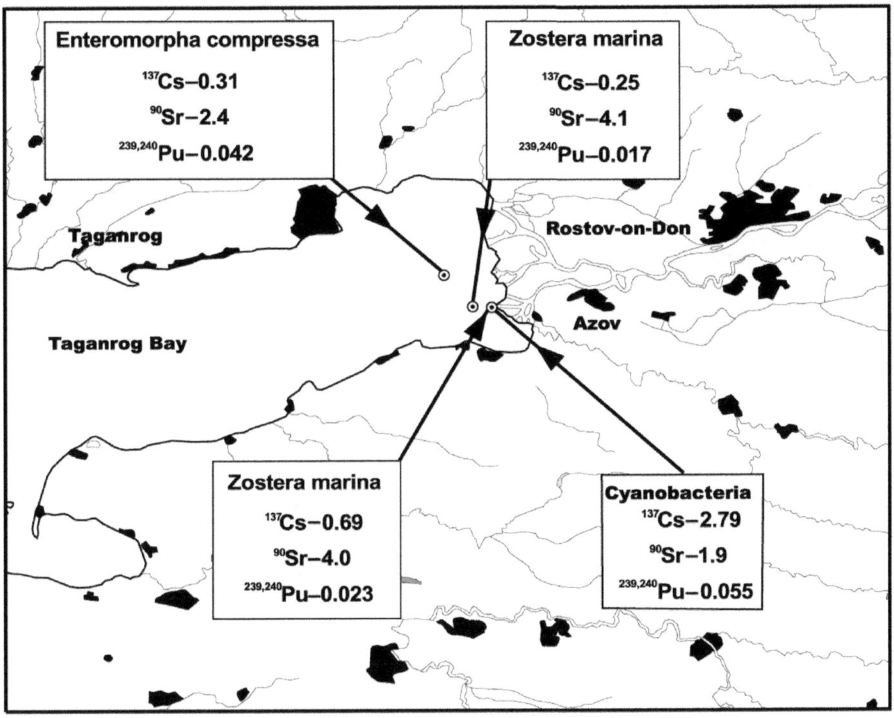

Fig. 5.41. Artificial radioniuclides in sea weeds and algae of the Azov Sea (MMBI expedition, July 1999), Bq / kg wet weight

It should be noted, that a constant decrease of ^{137}Cs and ^{90}Sr levels in macrophytes has occurred since the middle of 1the 1960s i.e. several years after the ban of atmospheric nuclear tests. Only Sellafield and Chernobyl discharges have broken the general downward dynamics of radionuclide accumulation levels.

Thus, during the 1990s macrophytes of the polar seas were distinguished by very low concentrations of all artificial radionuclides. ^{137}Cs, ^{90}Sr and $^{239,240}Pu$ levels varied from 0.5 to 4.9, 0.4 to 2-3 and 0.02 to 0.2-0.3 Bq / kg respectively. As the life span of algae on the shelf is several years, there is in fact no long term accumulation of radionuclides. The background contamination levels of the Arctic phytobenthos does not differ from the concentrations of ^{137}Cs (0.4 Bq / kg) and $^{239,240}Pu$ (0.02 Bq / kg) in macrophytes of the Antarctic (Fig. 5.42; Table 5.9) (Roos et al. 1993).

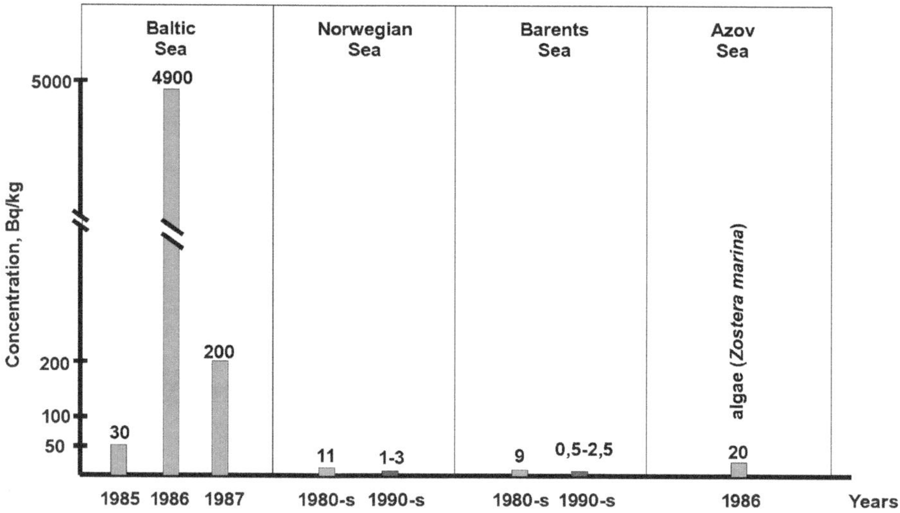

Fig. 5.42. Maximum ^{137}Cs levels in Fucus vesiculosus

Table 5.9. Concentration of radionuclides in algae of the Antarctic Peninsula in 1988-1989, Bq/kg dry weight

Nuclide	Concentration
^{137}Cs	0.36±0.07
$^{239,240}Pu$	0.021±0.006
^{238}Pu	0.004±0.001
^{241}Am	0.014±0.005
^{210}Po	17±5
^{210}Pb	3.4±0.9

The existence of local emissions from nuclear bases and the underwater burial of nuclearwaste in bays causes the accumulation of ^{60}Co (1-13 Bq / kg) and ^{137}Cs (40-260 Bq / kg) in macrophytes. However, these levels are many orders of magnitude less than those formed under the direct influence of discharges from Chernobyl after the accident. For example, in 1986 the ^{137}Cs accumulation by algae in rivers near Chernobyl reached 6-40 thousand Bq / kg (Merezhko et al., 1996; Pavlyutin, Babitskaya, 1996). It is quite evident that *Fucus vesiculosus* and *Laminaria Saccarina* are reliable indicators of radioactive contamination of the sea. Their life span is of order 2-4 years. Judging by the small quantity of ^{137}Cs and ^{90}Sr in modern macrophytes from the coastal waters during the 1990s were practically uncontaminated. However, the fact should be considered that lichens, growing on land near macrophytes, contain, for example, as much as 100-900 Bq / kg of ^{137}Cs. The 1-2 order difference obviously indicates the different mechanisms of the radionuclide accumulation by land and water vegetation.

5.5 Filtration of anthropogenic radionuclides in coastal ecosystems

The coastal ecosystem is distinguished by its ability to efficiently filter anthropogenic radionuclides from seawater. Based on the radioecological results presented above, the maximum level of radioactive contamination for the northern seas is fixed in bottom sediments, macrophytes and zoobenthos of the coastal zone. This is the result of the accumulative properties of clayey sediments, and brown and red algae which are characteristic for the large bays. Many ecosystems of the northern seas are contaminated by non-radioactive materials. These include toxic substances which often exercise cumulative influence on representatives of the flora and fauna (Matishov end Denisov 2000).

The filtering abilities of marginal zones are a function of a number of factors including the saturation of river and melted waters by ^{137}Cs, ^{90}Sr, ^{239,240}Pu, the rate of sedimentation and biological productivity, and the velocity of runoff currents in the surface layer of coastal basins. The stratification of estuarine waters, where seawater is diluted by freshwater from land, also is an important factor. The different sorption potentials of clay minerals depends on the content of illite, smectite, chlorite and kaolinite in the sediment fraction less than 0.001 mm.

^{137}Cs enters the environment from nuclear tests in the atmosphere or through the use and/or accidents of nuclear reactors. The content of radiocesium in bottom sediments of the various bays discussed previously varies from 1 to 200 Bq / kg. For comparison, ^{137}Cs accumulation in the surface layer of bottom sediments on shelves of the Barents, Pechora, Kara and Laptev Seas are much lower, on the order of 0-15 Bq / kg.

^{60}Co with a half-life of 5.3 years is typically not found in atmospheric fallout, or in bottom sediments of the open part of northern shelf seas. ^{60}Co isotopes are formed in nuclear reactor operations as a result of the irradiation of natural ^{59}Co by neutrons. It is quite clear, that ^{60}Co migration in the environment of a number of bays is connected with leakage from deposits of nuclear wastes or with discharges from nuclear submarine facilities. We have studied ^{60}Co in sediments of almost 20 bays. The concentration ^{60}Co in these bays varies from 1 to 200 Bq / kg (Fig. 5.43).

^{239,240}Pu has also been investigated in bottom sediments of a number of the bays. The range of levels of this radionuclide varies from 0.2 to 18 Bq / kg. ^{239,240}Pu accumulations in marine sediments from 1-3 Bq / kg are connected with global fallout. Higher concentrations of this nuclide in sediments are caused by local sources.

Tests of nuclear weapons, discharges from nuclear fuel production and processing facilities, accidents at nuclear burial sites are the main local sources of ^{239,240}Pu. The life-time of ^{239,240}Pu is many thousand years. Plutonium enters the marine food web to a lesser extent. It is a continuous source of contamination in the near proximity of discharge locations because of its long half-life and strong adsorption to marine sediments.

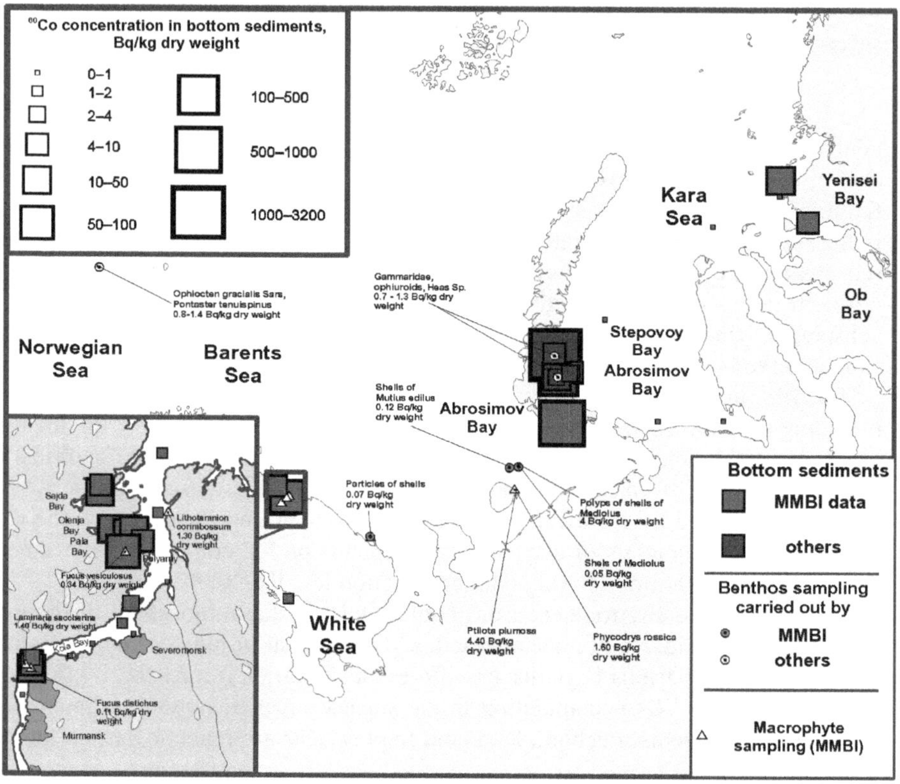

Fig. 5.43. ^{60}Co content in the surface layer of bottom sediments, benthos, macrophytes of the Norwegian, Barents, White and Kara Seas

^{90}Sr was analyzed in bottom sediments of a limited number of bays. Its concentration during the 1990s changed from 0.3-4 to 16-22 Bq / kg (Fig. 5.44). The origin of ^{90}Sr in the environment is nuclear weapon tests, discharges from nuclear industries and leakage from nuclear burial sites.

Approximately 80-90 % of $^{239,240}Pu$ and 60Co that is released into the sea is buried in the bottom sediments of bays near the discharge sources (Nies et al. 1998). This is illustrated by the above-mentioned classification system for artificial radionuclide contamination in bottom sediments. ^{60}Co, ^{137}Cs, $^{239,240}Pu$, radioactive (hot) particles are found in sediments adjacent to nuclear bases, in areas of sunken nuclear submarines, and at underwater burial sites of nuclear wastes. The radionuclide composition of the sediments indicates the existence of a local emission source (Matishov et al. 1996f, 1998b, 1999e).

The fine-dispersed fractions of particles containing may be widely dispersed throughout the sea environment. In particular, cobalt is detected in sediments 40-70 km away from known sources in the Kola Bay. Discharges of anthropogenic radionuclides during the period of 1960-1990s resulted in significant increases in radionuclides during the period of 1960-1990s resulted in significant increases in ^{137}Cs to 40-100 Bq / kg in bottom sediments of the Kola and Motovsky Bays (Fig. 5.45). The transport of radioactive substances beyond the bays to the Barents and Kara Sea basins is noted only in the Kola Peninsula and Novaya Zemlya coastal zones. The content of artificial radionuclides in shelf sediments is 5-7 times less than in the bays.

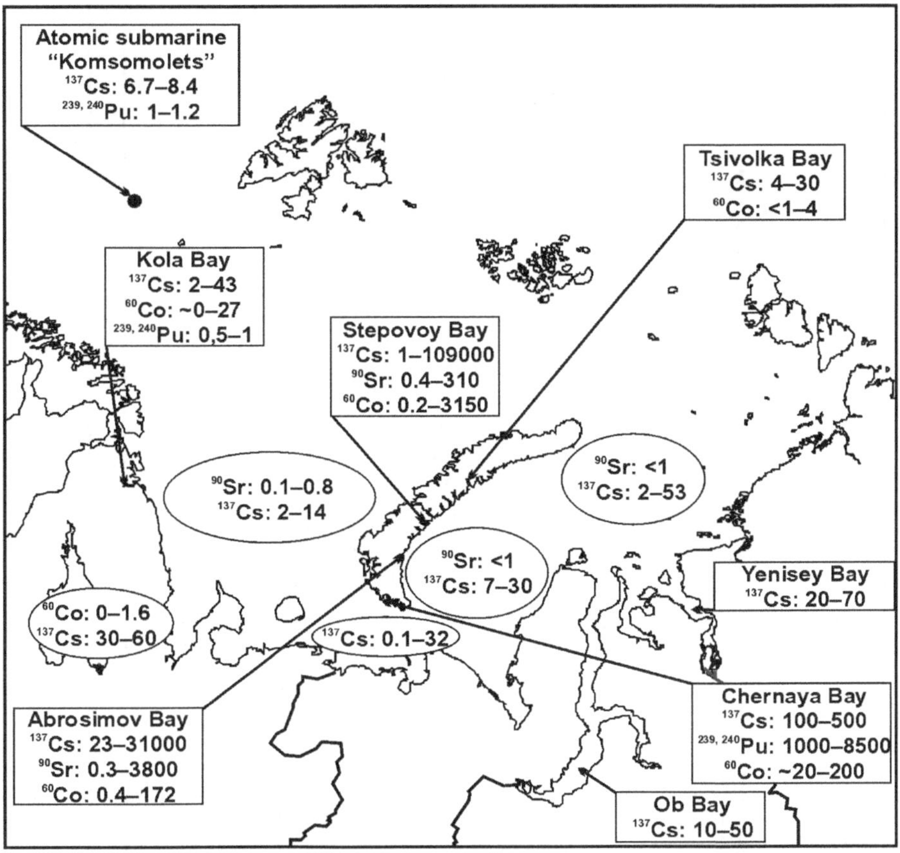

Fig. 5.44. ^{90}Sr, ^{137}Cs and $^{239,240}Pu$ concentrations (Bq /kg dry weight) in bottom sediments from the bays and open sea area of Northern Europe (Baklanov et al. 1996)

V.P. Kilezhenko and V.N. Podymakhin (1964) have established that the rate of renewal of water masses in the different bays of the Murman coast varies from 5 to 30 days, depending on topography and other related factors. The result is that every contaminant that enters the surface waters of bays essentially is immediately in contact with the open sea environment where it may mix with coastal water masses and be transported to the open sea. In cases of radioactive contamination

222

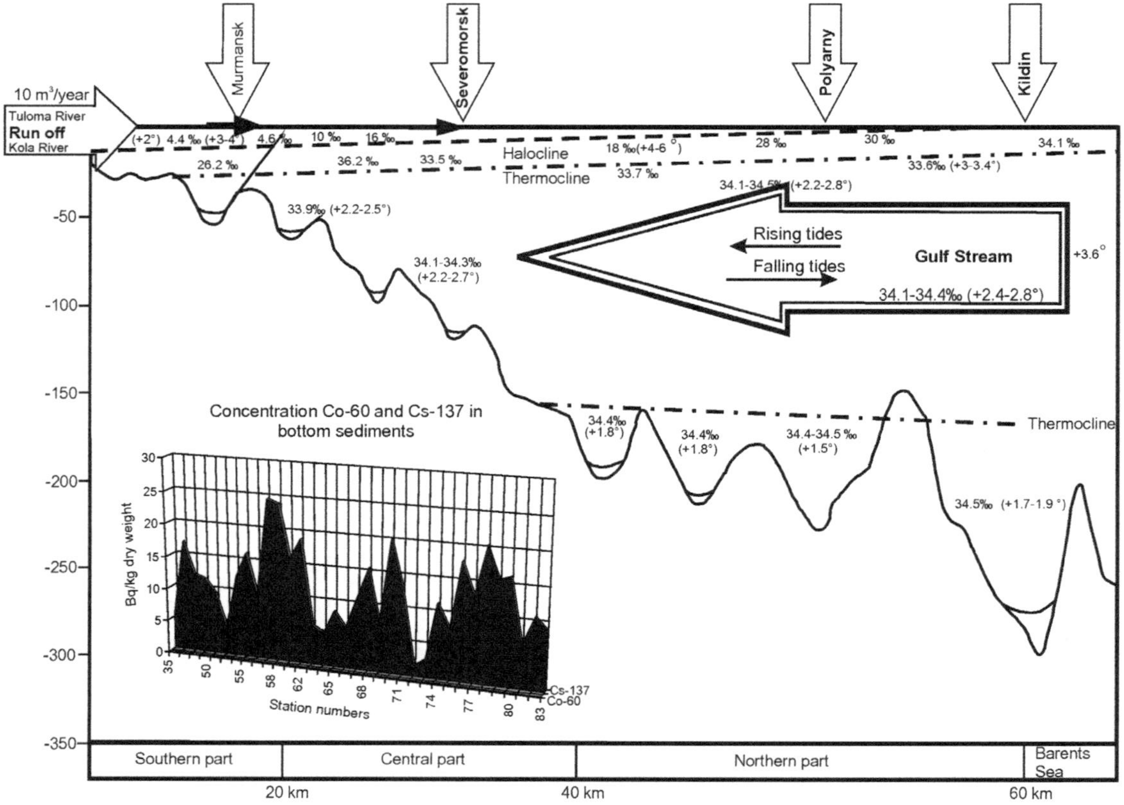

Fig. 5.45. Oceanographic conditions in the transfer of artificial radionuclides from the Kola Bay to the coastal shelf

in bays having no direct outlet to the open sea, the main factor controlling the distribution of radionuclides in these areas will biology and not hydrology.

Macrophytes play a key role in the biological cycle of coastal ecosystems. They filter radionuclides migrating from land and along the coasts. During the well- studied post-Chernobyl period the highest ^{137}Cs assimilation was observed in brown and red algae. During the 1990s the ^{137}Cs concentration in algae varied, depending on the species, from 0.4-1.0 to 3-5 Bq / kg and in some individual places (Oslo Fjord) - to 7 Bq / kg. Immediately after Chernobyl (June 1986) fucus in some bays of the Baltic Sea contained to 1-5 thousand Bq / kg dry weight of ^{137}Cs.

Thus, the intensive filtration of artificial radionuclides in the coastal ecosystem is expressed by their increased concentration in algae relative to the radioactive contamination of bottom sediments and zoobenthos. It is normal that releases of radioactive substances from local and regional sources can be traced distances from several to 10-70 km away from the source. However, the concentrations of radionuclides in sediments, macrophytes, and mollusks are one-two orders of magnitude higher near places of discharge than in the shelf troughs. Practically all short-lived radionuclides (^{60}Co, ^{134}Cs and others), passing through the filter of the coastal ecosystem, are sedimented or sorbed. In cases of large accidents at nuclear bases or the loss of integrity of containers of nuclear wastes red and brown algae would serve as active biofilters of artificial radionuclides. The scenario for radionuclide accumulation in macrophytes will be similar to the situation which took place in the Finnish and Bothnia Bays after the Chernobyl accident. Several weeks after the accident fucus in some places of the Baltic coast accumulated as much as 5,000 Bq / kg dry weight of ^{137}Cs and to 29,000 Bq / kg of ^{131}I (Ilus et al. 1988a,b).

6 Marine vertebrates

6.1 Marine fish

Pelagic and bottom (demersal) fish fauna are the most important links within the marine trophic ecosystem in Arctic and Sub-Arctic latitudes. Fish, on the one hand, concentrate energy from lower trophic levels and, on the other hand, serve as a food source for sea mammals and some birds. The fish population of the Barents Sea includes more than 150 species; about 20 of them are of marketable value. During the last decades the annual catches of Barents Sea fish have varied from 2 to 4 million tons (Matishov end Denisov 2000). During the period 1950-1970s the catches of Barents Sea cod reached 1.0-1.4 million tons annually.

Fish is a valuable raw material for the food industry. As a result, the radioactivity of fish, especially in the Barents-Kara region, is monitored by agencies responsible for the control of food quality and safety. The monitoring of marketable fish for radioactive contamination began during the period of atmospheric nuclear testing on Novaya Zemlya (Nansen Centennial ... 1992; Saetre et al. 1992; Foyn end Nikitin 1993; Matishov et al. 1994b,c, 1995a; Foyn end Swaren 1995). The highest β-activity of Barents Sea fish (more than 60-70 Bq / kg wet weight) was noted in the beginning of the 1960s (Fig. 6.1). Until 1986 radionuclide levels in fish were less than 10 Bq / kg wet weight.

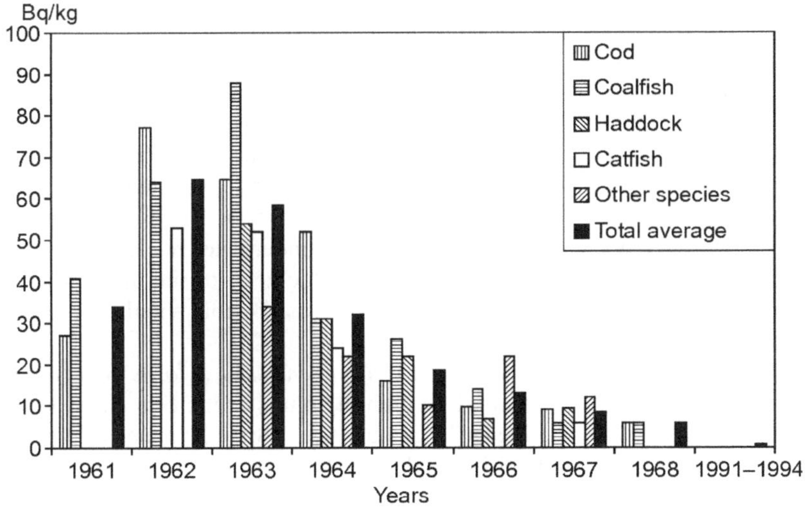

Fig. 6.1. Total average gamma-emitting radionuclide concentrations minus ^{40}K measured in different Barents Sea fish species (Foyn and Swaren 1995). ^{137}Cs data from 1991-1994 are from Sickel et al. 1995

During the 1990s, a number of investigations of fish fauna were carried out in connection with the nuclear submarine «Komsomolets» accident and releases and discharges of technologically produced radionuclides in the sea bays and on the shelf. As a result, it became obvious that the marketable fish species contained insignificant concentrations of artificial radionuclides. However, the question of the role of radioactive contamination on the environment and biota in relation to the decrease in fish catches in the Barents and Norwegian Seas is still open.

Macrophytes, phytoplankton, zooplankton, and benthic organism of the northern seas do not contain dangerous concentrations of anthropogenic radionuclides. However, their participation in the transfer to fish through the food web is quite probable. At the same time, the presence of suspended and dissolved ^{137}Cs, ^{90}Sr and $^{239,240}Pu$ in the sea environment also contributes to the incorporation of long-lived radionuclides in fish fauna. The level of radionuclide accumulation in fish organs and tissues can be influenced by their ecology- habitat, spawning and interspecies relationships.

Boreal fish species of moderate latitudes and exclusively Arctic species are characteristic for the West Arctic Seas. The spawning grounds of herring, cod, haddock and other boreal species are situated along the south-west Norwegian coast, in the Lofoten Islands area. During periods of Arctic warming the spawning of herring and cod is noted in the open Barents Sea, and in Motovsky Bay, as well as near the East Murman coasts (Zenkevich, 1963). Herring reach sex maturity at 4-5 years, while cod, haddock, and sea perch reach maturity in 7-10 years. The life span of sea perch can be 20-50 years. Thermophilic species attain sexual maturity and then migrate back to spawning locations.

Larvae hatched in distant western spawning grounds are passively carried away to the north. After 5-7 months of drift in the streams of Nordcape current, the larvae are introduced into the Barents Sea ecosystem. These surface Norwegian waters were saturated by ^{137}Cs during the 1970-1980s due to discharges to the sea of radioactive wastes from Sellafield. Herring fry penetrate with the streams of warm Atlantic water into the Kola Bay, the coastal waters of Novaya Zemlya and the Novaya Zemlya bays. During atmospheric and underwater nuclear weapons tests, the coastal waters of Novaya Zemlya were in the zone of the direct impact from radioactive fallout. The fishes have formed the marketable shoals during both the seasonal and spawning migrations.

The Arctic-boreal species Capelin (*Mallotus villosus villosus)* is central to the pelagic ecosystem and lives entirely within the limits of the Barents Sea. In spring capelin approach the coastline in large quantities. Spawning occurs on the shallows, and then the spawning fish mainly die. The total stock of capelin during the 1950-1970s changed from 4 to 7 million tons. In the beginning of the 1980s the total and spawning stocks began to decrease sharply, and in 1986 the fishery was closed.

Polar cod *Boreogadus saida* is the most abundant fish in the polar seas. It extends to the 88° N. In the autumn and winter period polar cod are concentrated in the eastern part of the Barents Sea. In early spring (March) mass spawning occurs near the south-eastern coasts. Many eggs are fastened to the lower surface of ice. After spawning these fish migrate to the Kara Sea.

The small fish navaga (*Eleginus navaga),* which belongs to the herring family, is widespread in the coastal zone of the Pechora and White Seas. Navaga is a bottom fish that feeds on chaetopods and crustaceans.

Trophic relationships among fish species permit similar fauna groups to concentrate radionuclides similarly. Fish species may be divided by feeding method into planktivorous, benthivorous and omnivorous. Herring, capelin, polar cod, and partially sea perch are planktivorous. For the other fish species, plankton is the main food source of their fry. Copepods and euphausiids are the preferable food for the majority of fish species. Benthivorous fish include haddock, plaice, and wolffish. The food of benthivorous species is very diverse. For, example, haddock consumes about 200 different species of benthic organisms but feeds mainly on bivalve molluscs. Polychaetes are the main food source for sea plaice which buries itself into the seafloor temporarily.

Cod (*Gadus morhua morhua)* is a main fish species in the Barents Sea ecosystem. Although cod are omnivorous, pelagic fish are its main food source (Zenkevich 1963). Herring fry, capelin, and polar cod constitute 60% of the total amount of food consumed by Cod. Cod also consumes Gammaruses (decapods). In summer decapods are the primary food source for cod (75-100 %) feeding both in the central and south-western parts of the sea. The benthos plays a relatively modest role as a food source for cod. However, cod food preferences change depending on conditions in the area. In summer cod consumes Gammaruses, and in autumn it often transfers to feeding on bottom crustaceans and molluscs.

The habitats, behavior, feeding strategy and the duration of life of different fish species influences to a great extent the level and rate of accumulation of artificial radionuclides in these organisms. The most general conclusion is that the radioactivity of marketable fish from the northern seas is relatively low. At the end of the 1980s and during the 1990s, the radionuclide levels in fish tissues (Bq / kg) varied as follows, : ^{137}Cs - 0.3-6, ^{90}Sr - 0.002-0.1, ^{134}Cs - 0.08-0.7 (Kanisch end Nagel 1992; Matishov et al. 1994b,c, 1997b,c,f, 1999e; Foyn end Swaren 1995; Matishov D. et al. 1995a,b; Miretskiy et al. 1995; Strand end Nielsen 1995; Sickel et al. 1995; Christiansen end Bjornstad 1996). The radiocesium level in fish organs and tissues is nearly two times higher than in their bones.

Pelagic fish feed on phyto- and zoo- plankton. In the beginning of 1960s and in the middle of 1980s, when radioactivity levels in the Barents, White and Kara Seas reached their maximum values, ^{137}Cs concentrations in plankton would have increased to 20-60 Bq/kg. The radioecological situation in the shelf ecosystem during nuclear tests in the atmosphere and seawater at Novaya Zemlya would be approximately similar to the situation in the Baltic Sea and lakes of Northern Scandinavia during the Chernobyl event. For example, the main Chernobyl fallout in Norwegian lakes at the beginning of May 1986 increased the ^{137}Cs concentration in water to 23 Bq / m^3, in phytoplankton - to 830 Bq / kg, in zooplankton - to 50-180 Bq / kg, and in macrophytes - to 2-25 thousand Bq / kg (Gunnerod et al. 1990).

During the last decade the radioactive contamination of plankton living in northern seas has continued to decrease steadily. In particular, the ^{137}Cs level in plankton varies from 1 to 5 Bq / kg, reaching 10-15 Bq / kg in only a few coastal bays (Ilus et al. 1988a,b, 1993). It can be noted for comparison that at the end of the 1980s the Antarctic krill, an analogue of the Barents Sea euphausiids, contained about 1.7 Bq / kg wet weight of ^{137}Cs (Ross et al. 1993). Such contamination levels are the result of global fallout.

^{137}Cs is accumulated in fish (Fig. 6.2) inhabiting the pelagic zone through feeding. During the 1990s the ^{137}Cs concentration in capelin, polar cod, herring (*Clupea harengus harengus*), sea perch (*Sebastes marinus*) of the Barents-Kara region varied correspondingly, Bq / kg wet weight: 0.4-0.8, 0.5-2.2, 0.5-1.3, 0.4-0.9 (Tables 6.1, 6.2). Polar cod in Abrosimov and Stepovoy Bays, on the east coast of Novaya Zemlya, contained about 1.4 Bq / kg of radiocesium (Strand et al., 1994). The Azov sabrefish (*Pelecus cultratus*) from Taganrog Bay had a comparable level of ^{137}Cs (1.2 Bq / kg) in 1999. Based on these data it is clear that the food consumed by the planktivorous fishes, contains insignificant quantities of anthropogenic radionuclides.

Table 6.1. Artificial radionuclides in marine organisms, Bq/kg wet weight

Species	^{137}Cs	^{40}K
Autumn 1993 and 1994, coastal waters of northern Norway, Barents Sea		
Gadus morhua morhua	0.3–1.1	–
Melanogrammus aeglefinus	<0.05–0.7	–
Pollachius virens	0.3–0.7	–
Clupea harengus harengus	0.2–0.6	–
Salmo salar	0.4–1.2	–
Pandalus borealis	0.1–0.4	–
Sebastes marinus	0.4–0.7	–
Chlamys islandica	–	–
Anarhichas minor	0.5	–
Reinhardtius hippoglossoides	0.7	–
Pleuronectes platessa	0.7	–
Summer 1999, RV «Dalnie Zelentsy» (MMBI), Barents Sea		
Mallotus villosus villosus	<0.5	74±6
Mallotus villosus villosus, Murmansk Lobe, depth 280 m	<0.4	81±5
Mallotus villosus villosus, collections by fishery trawl	<0.4	63±6
Mallotus villosus villosus, depth 315 m	<0.6	49±1
Gadus morhua morhua	<0.5	–
Shark	0.1	36±2

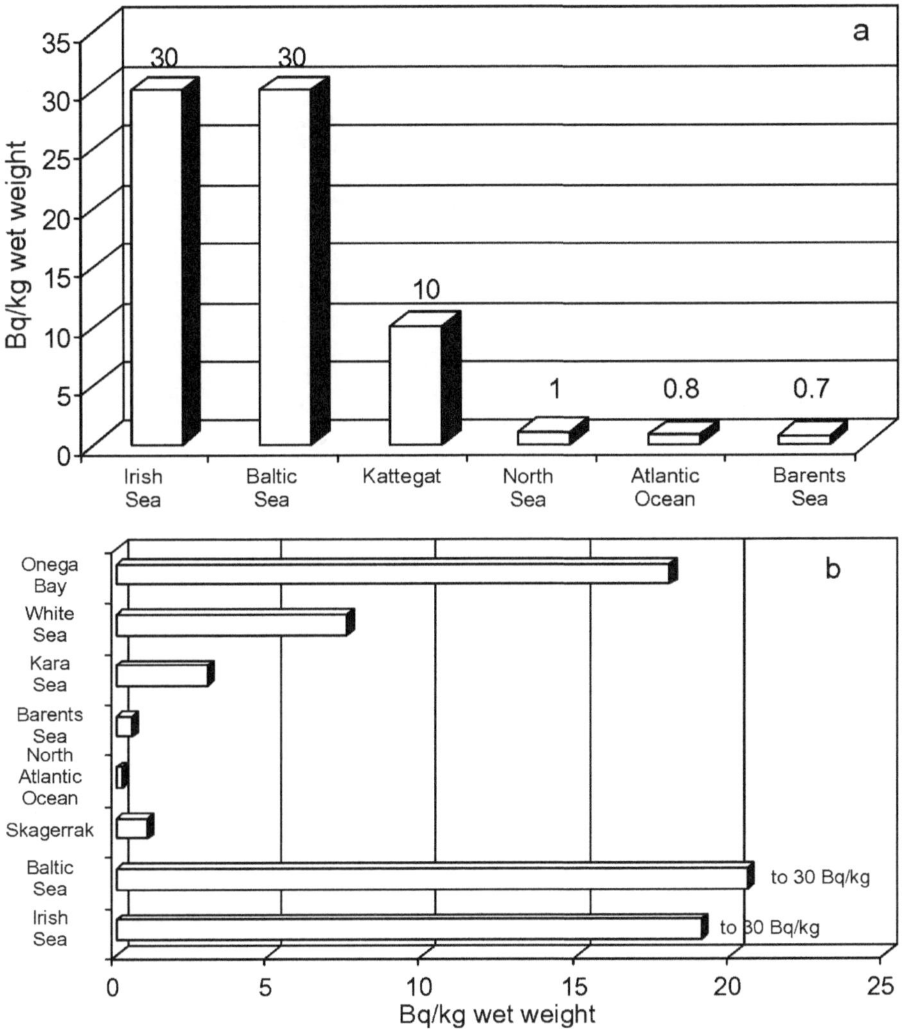

Fig. 6.2. ^{137}Cs concentration in pelagic fish species in the first half of 1990s: **a** - in muscle tissue; **b**- in carcasses

Table 6.2. Ecotoxicological investigations of the Barents Sea commercial species in 1995, 1996 and 1999

Species	Month	Pb	Cd	Cu	As	Zn	Hg	ΣDDT	α-HCH	γ-HCH	Geptachlorine	PCB	Aldrine	^{137}Cs	^{90}Sr
		mg/kg						mkg/kg						Bk/kg	
1	2	3	4	5	6	7	8	9	10	11	12	13	14	15	16
							1995								
Herring	XI	0.1	0.02	1.2	0.1	4.7	0.03	30.0	3.2	0.5	N/d	–	N/d	–	–
		0.15	0.02	0.9	0.2	3.6	0.02	39.0	3.7	0.7	"	–	"	–	–
Cod	XII	0.1	0.01	2.3	0.1	5.7	0.09	4.0	2.0	1.0	"	–	"	2.6	–
Salmon		0.07	0.01	2.4	0.07	6.5	0.01	11.0	3.3	1.5	"	–	"	–	–
							1996								
Cod	I	0.4	0.1	4.0	0.5	14.7	0.005	8.2	0.9	0.2	"	–	"	3.476	–
Haddock		0.3	0.07	2.1	0.8	15.4	0.05	3.5	1.2	0.9	"	–	"	2.1	–
Blue catfish	I–II	0.1	0.02	0.6	0.1	7.5	0.02	6.8	2.4	1.9	"	–	"	2.8	–
Cappelin	I–III	0.1	0.02	1.5	0.2	12.0	0.03	17.1	5.0	1.0	"	–	"	–	–
Haddock	II	0.1	0.005	2.7	1.6	2.6	0.03	6.0	1.5	2.0	"	–	"	2.8	–
	III	0.1	0.005	0.8	0.05	1.0	0.01	3.4	2.3	0.9	"	–	"	2.5	–
Cod	IV	0.1	0.01	1.2	0.2	4.0	0.01	3.0	1.9	0.8	"	–	"	6.0	–
	IV–V	0.3	0.05	2.1	1.0	17.4	0.05	4.5	2.4	1.2	"	–	"	–	–
Blue catfish		0.2	0.01	1.2	0.2	14.5	0.02	16.5	1.7	1.0	"	–	"	2.3	–
Plaice		0.4	0.01	0.8	0.3	10.1	0.02	19.0	3.0	0.5	"	–	"	2.6	–
Cod	V	0.2	0.05	0.9	0.2	4.9	0.05	3.7	2.3	0.09	"	–	"	2.4	–
		0.5	0.05	2.4	1.0	20.0	.005	4.8	1.7	1.8	"	–	"	2.4	–
Haddock		0.2	0.01	0.8	0.05	2.5	0.03	11.9	3.2	3.1	"	–	"	–	–
Cod	VI	0.5	0.005	6.2	0.9	19.8	0.05	4.1	2.3	1.2	"	–	"	2.1	–
	VI–VII	0.1	0.02	0.8	0.2	8.5	0.01	5.7	2.4	1.3	"	–	"	4.5	–
		–	–	–	–	–	–	–	–	–	–	–	–	2.2	–

Table 6.2. (Cont.)

1	2	3	4	5	6	7	8	9	10	11	12	13	14	15	16
Shrimp	VII	0.2	0.01	4.0	0.1	7.5	0.02	7.0	0.8	0.6	–	–	–	–	–
Catfish	VII	0.3	0.02	0.7	0.4	6.5	0.01	10.6	2.8	1.7	N/d	–	N/d	2.1	–
Cod		0.2	0.01	2.1	0.3	4.4	0.01	5.6	3.2	1.7	”	–	”	2.3	–
		0.1	0.02	0.8	0.1	3.7	0.02	9.6	2.8	1.8	”	–	”	–	–
Sea comb-bearer		–	–	–	–	–	–	–	–	–	–	–	–	2.1	2.238
Haddock	VII–VIII	0.2	0.03	0.7	0.1	4.8	0.01	15.4	2.6	3.3	N/d	–	N/d	2.4	–
Cod	VIII	0.1	0.01	0.9	0.2	6.5	0.01	11.2	2.9	1.9	”	–	”	–	–
		0.2	0.05	2.0	0.7	13.0	0.005	9.8	2.6	1.9	”	–	”	2.4	–
		0.3	0.01	0.8	0.1	3.5	0.02	12.2	2.6	2.0	”	–	”	–	–
Shrimp	VIII–IX	0.05	0.03	6.5	0.2	10.5	0.02	–	–	–	–	–	–	–	–
	IX	0.1	0.04	8.5	0.2	12.5	0.03	–	–	–	–	–	–	–	–
Sea perch	IX–X	0.1	0.02	0.9	0.1	2.5	0.01	7.1	1.7	1.1	N/d	–	N/d	6.1	–
Haddock	X	0.1	0.01	0.4	0.1	1.0	0.02	46.8	1.6	5.2	”	–	”	5.5	–
		0.1	0.01	0.8	0.1	1.1	0.02	10.8	2.2	1.4	”	–	”	6.3	–
		0.1	0.01	1.1	0.1	1.0	0.02	11.8	2.2	1.5	”	–	”	6.7	–
		0.1	0.005	0.9	0.1	1.2	0.02	11.5	2.1	1.5	”	–	”	3.3	–
Cod		0.1	0.02	0.9	0.1	2.5	0.01	10.0	2.5	1.5	”	–	”	3.02	–
Arctic cod		0.05	0.01	0.7	0.1	1.0	0.01	11.9	2.3	1.6	”	–	”	16.78	–
Arctic cod		0.05	0.01	0.5	0.1	0.9	0.01	9.8	2.4	1.6	”	–	”	17.83	–
Atlantic herring		0.2	0.01	1.1	0.1	6.2	0.01	19.1	2.8	1.2	”	–	”	4.6	–
Mackerel		0.2	0.01	2.7	0.3	6.5	0.01	12.6	3.3	2.2	”	–	”	3.4	–
Polar cod		0.05	0.01	0.8	0.1	1.2	0.01	10.1	2.4	1.6	”	–	”	7.5	–
Cod	XI	0.2	0.03	0.9	0.1	2.6	0.005	12.3	2.3	1.6	”	–	”	4.4	–
		0.2	0.01	1.2	0.1	1.6	0.005	11.2	2.3	1.6	”	–	”	4.4	–
		0.1	0.02	0.9	0.2	2.5	0.03	–	–	–	–	–	–	–	–
Blue whitings		0.2	0.01	0.9	0.1	3.2	0.005	10.6	2.0	1.5	N/d	–	N/d	2.8	–

Table 6.2. (Cont.)

1	2	3	4	5	6	7	8	9	10	11	12	13	14	15	16
Herring		0.3	0.02	1.8	0.2	5.4	0.02	9.1	2.1	1.4	”	–	”	4.4	–
		0.2	0.01	2.5	0.2	3.3	0.01	10.1	2.8	1.4	”	–	”	3.4	–
		0.1	0.03	1.9	0.2	4.0	0.01	9.3	2.6	1.4	”	–	”	–	–
		0.1	0.03	2.7	0.1	6.8	0.01	9.1	2.1	1.4	”	–	”	4.2	–
Mackerel	XI	0.3	0.02	2.1	0.1	4.5	0.02	9.5	2.4	1.7	”	–	”	13.0	–
Haddock		0.2	0.01	0.9	0.1	3.2	0.005	8.5	2.1	1.6	”	–	”	16.0	–
Cod	XII	0.1	0.03	1.5	0.2	2.1	0.01	15.5	2.2	1.6	”	–	”	3.8	–
Haddock		0.2	0.01	0.9	0.1	2.5	0.1	12.9	1.9	1.4	”	–	”	4.5	–
							1999								
Blue catfish	V	0.1	0.01	0.6	0.5	2.9	0.04	10.0	1.3	1.2	–	–	–	–	–
American plaice		–	–	–	–	–	–	–	–	–	–	11.0	–	2.6	5.1
Mackerel	VII	0.1	0.02	1.1	0.2	3.4	0.03	12.0	1.7	1.8	–	13.1	–	2.5	3.0
Atlantic herring	VIII	0.1	0.02	0.7	0.4	4.6	0.03	11.0	1.8	1.5	–	13.0	–	3.6	3.0
	IX	0.1	0.02	0.7	0.3	3.9	0.03	10.0	1.3	1.5	–	10.6	–	2.6	3.4
American plaice		–	–	–	–	–	–	–	–	–	–	–	–	–	–
Blue catfish		0.1	0.02	0.7	0.4	2.8	0.04	10.0	1.4	1.5	–	11.0	–	4.0	2.5
Atlantic herring	IX–X	0.1	0.02	0.9	0.4	4.9	0.03	10.0	1.4	1.4	–	7.7	–	2.5	3.8
Plaice	X	0.2	0.01	0.8	0.3	3.2	0.01	8.0	0.8	0.8	–	–	–	–	–
American plaice		0.2	0.02	0.9	0.2	3.6	0.01	16.0	0.6	0.4	–	11.0	–	–	–
		–	–	–	–	–	–	–	–	–	–	13.0	–	–	–
		–	–	–	–	–	–	–	–	–	–	–	–	–	–
		–	–	–	–	–	–	–	–	–	–	–	–	3.2	2.6
		–	–	–	–	–	–	–	–	–	–	–	–	–	–
Blue catfish		–	–	–	–	–	–	–	–	–	–	13.0	–	2.4	3.0
		–	–	–	–	–	–	–	–	–	–	–	–	–	–
Spotted catfish		–	–	–	–	–	–	–	–	–	–	–	–	3.0	3.4

Table 6.2. (Cont.)

1	2	3	4	5	6	7	8	9	10	11	12	13	14	15	16
Polar cod		0.1	0.02	0.6	0.3	5.2	0.03	9.0	1.0	1.0	–	12.0	–	3.7	2.3
		0.1	0.02	0.4	0.7	2.1	0.01	7.0	1.3	1.2	–	12.0	–	3.0	3.4
		0.1	0.01	0.5	0.2	5.5	0.02	10.0	1.3	1.4	–	9.7	–	2.5	3.1
Horse mackerel	X	0.1	0.03	1.0	0.6	3.2	0.06	9.0	0.9	0.9	–	28.0	–	2.4	5.6
Capelin		0.1	0.02	0.7	0.3	4.2	0.02	9.0	1.1	1.2	–	9.0	–	2.5	4.4
		0.1	0.02	0.8	0.3	4.3	0.02	9.0	1.5	1.2	–	9.4	–	3.4	3.5
Blue whitings		0.1	0.02	0.8	0.3	3.8	0.03	8.0	1.0	0.9	–	9.3	–	3.1	2.8
		0.1	0.02	0.5	0.3	4.9	0.02	11.0	–	–	–	–	–	4.1	3.6
Cod		0.2	0.01	0.8	0.3	4.8	0.03	14.0	1.6	1.3	–	–	–	–	–
		0.1	0.01	1.2	0.4	2.6	0.02	9.0	1.5	1.4	–	13.0	–	2.5	2.5
		0.2	0.01	0.8	0.3	4.7	0.03	10.0	1.6	1.5	–	11.7	–	2.4	4.1
Haddock		–	–	–	–	–	–	–	–	–	–	14.4	–	3.1	3.9
Halibut		–	–	–	–	–	–	–	–	–	–	–	–	2.5	2.8
		0.1	0.02	1.4	0.4	5.7	0.03	11.0	2.0	1.9	–	12.2	–	5.0	2.9
Atlantic herring		0.1	0.02	0.8	0.3	3.7	0.03	10.0	1.4	1.4	–	12.0	–	2.4	5.3
		0.1	0.02	0.7	0.3	3.9	0.02	10.0	1.4	1.3	–	11.9	–	5.4	3.2
Sea perch		–	–	–	–	–	–	–	–	–	–	–	–	–	–
		0.1	0.02	0.8	0.4	5.7	0.03	9.0	1.0	1.1	–	9.0	–	3.9	5.5
Polar cod	XI	0.1	0.01	0.5	0.2	2.8	0.02	14.0	0.6	0.7	–	9.3	–	4.0	2.6
		0.1	0.02	0.5	0.3	6.2	0.03	7.0	1.0	0.8	–	22.3	–	2.4	2.2
		–	–	–	–	–	–	–	–	–	–	12.0	–	3.4	2.7
		0.1	0.02	0.5	0.3	5.9	0.03	10.0	1.1	1.1	–	11.0	–	2.4	3.3
		0.1	0.01	0.6	0.2	5.6	0.02	6.0	1.1	0.9	–	12.0	–	2.4	2.7
		0.1	0.02	0.8	0.3	3.9	0.03	9.0	1.3	1.1	–	9.1	–	3.9	4.6
Cod		–	–	–	–	–	–	–	–	–	–	16.0	–	3.3	3.9
		–	–	–	–	–	–	–	–	–	–	–	–	2.8	3.5

Table 6.2. (Cont.)

1	2	3	4	5	6	7	8	9	10	11	12	13	14	15	16
		0.2	0.01	0.9	0.4	5.1	0.03	8.0	1.4	1.3	–	12.0	–	4.1	4.7
		0.1	0.01	1.1	0.3	2.7	0.02	9.0	1.5	1.4	–	11.0	–	3.2	5.9
Arctic cod	XI	0.1	0.02	0.5	0.3	5.1	0.02	10.0	1.0	1.1	–	10.0	–	2.9	2.8
		0.1	0.02	0.5	0.2	4.1	0.02	9.0	1.2	1.3	–	10.0	–	4.0	7.4
Haddock		0.2	0.01	0.8	0.4	2.8	0.03	9.0	1.4	1.1	–	11.3	–	2.5	2.9
		0.2	0.01	0.8	0.2	2.8	0.02	10.0	1.6	1.4	–	13.2	–	–	–
		–	–	–	–	–	–	–	–	–	–	23.0	–	4.1	2.8
Sea comb-bearer		0.1	0.01	1.4	0.2	11.3	0.03	–	–	–	–	–	–	5.4	5.8
American plaice		0.1	0.01	0.5	0.3	3.3	0.02	9.0	1.3	1.4	–	14.6	–	2.7	3.9
		0.1	0.01	0.5	0.3	3.1	0.02	10.0	1.2	1.3	–	–	–	–	–
		0.1	0.01	0.6	0.3	3.6	0.02	10.0	1.1	1.1	–	–	–	–	–
Capelin		0.1	0.02	0.7	0.2	4.1	0.03	10.0	1.4	1.5	–	12.3	–	3.7	7.7
		0.1	0.02	0.7	0.3	4.1	0.02	11.0	2.1	1.6	–	14.0	–	3.3	6.3
Sea perch		0.1	0.03	0.8	0.3	3.2	0.01	8.0	1.5	0.9	–	10.4	–	2.8	4.1
		–	–	–	–	–	–	–	–	–	–	10.0	–	5.6	4.9
Spotted catfish		–	–	–	–	–	–	–	–	–	–	18.4	–	–	–
Halibut		0.1	0.02	1.2	0.3	5.7	0.03	11.0	1.9	1.8	–	15.2	–	–	–
Sea perch		–	–	–	–	–	–	–	–	–	–	11.0	–	3.1	5.0
		0.1	0.02	0.7	0.4	5.2	0.03	10.0	1.1	1.2	–	11.7	–	3.1	5.7
Cod	XII	0.1	0.02	0.6	0.4	4.9	0.03	9.0	1.1	0.8	–	12.0	–	6.0	5.5
		0.1	0.02	0.6	0.4	4.8	0.03	11.0	1.3	1.2	–	9.6	–	3.0	2.5
Blue catfish		0.1	0.02	1.1	0.3	3.4	0.04	10.0	1.3	1.2	–	17.7	–	5.0	3.6

FOOTNOTE. The estimates given by Murmansk Centre of Standardization for ^{137}Cs and ^{90}Sr levels are overestimated but indicate the general character of the contamination level.

However, radionuclide concentrations in pelagic fish changed mainly during periods of nuclear explosions and accidents. For example, in 1986-1987, after the Chernobyl accident, ^{137}Cs in Baltic herring and perch fish reached corresponding 10-90 and 9-190 Bq / kg. Concentrations in fish of the short-lived radionuclide ^{134}Cs varied from 3 to 96 Bq / kg (Tables 6.3, 6.4) (Ilus et al. 1993).

Table 6.3. ^{89}Sr, ^{90}Sr, ^{134}Cs, ^{137}Cs in Baltic Sea fish (1986 and 1987) (Ilus et al. 1993), Bq/kg wet weight

Species	Area	1986			1987			
		Date	^{89}Sr	^{90}Sr	Date	^{90}Sr	^{134}Cs	^{137}Cs
Clupea harengus	Hailuoto	14.10	6.4	0.084	09.11	0.11	7.2	22
membras	Vaasa	13.11	0	0.12	21.10	0.087	14.0	41
	Seili	11.10	0	0.067	21.10	0.065	7.0	21
		13.10	0	0.15	20.11	–	6.1	18
	Tvдrminne	14.10	4.8	0.23	26.05	0.074	8.0	28
Esox lucius	Vaasa	08.06	0.015	0.072	18.06	–	9.3	30
	Seili	10.06	4.2	0.044	26.06	0.024	8.2	26
	Tvдrminne	09.07	0.021	0.032	02.06	0.030	4.3	15
Gadus morhua	Tvдrminne	26.06	0.030	0.044	22.06	0.073	5.9	18

Table 6.4. Gamma radionuclides in Baltic Sea fish (1986) (Ilus et al. 1993), Bq/kg wet weight

Species	Area	Date	^{40}K	^{106}Ru	^{131}I	^{134}Cs	^{137}Cs
1	2	3	4	5	6	7	8
Clupea harengus	Loviisa	15.05	110	4.4	13	5.3	11.0
membras	Kallbedagrund	23.07				16.0	35.0
	Porkkala	24.07				31.0	71.0
	Bogskдr	26.07				29.0	57.0
	Gotska Sandцn	27.07				5.1	12.0
	Sдlskдr	31.07				23.0	58.0
	Falkens Grund	03.08				6.7	16.0
	Ulkokalla	05.08				9.2	21.0
	Mдntyluoto	07.08				25.0	64.0
	Isokari	08.08				20.0	52.0
	Loviisa	05.09	120	0	0	44.0	91.0
	Seili	09–13.10				3.7	9.8
	Tvдrminne	14.10				12.0	30.0
	Hailuoto	14.10				6.3	17.0
	Vaasa	13.11				26.0	64.0
Esox lucius	Loviisa	16.05	120	1.4	12	4.4	9.7
		04.06	130	0	0	2.0	7.3
Esox lucius	Vaasa	08.06				1.2	8.9
	Loviisa	02.07	120	0	0	9.5	23.0

Table 6.4. (Cont.)

1	2	3	4	5	6	7	8
	Seili	06–10.07				0	6.4
	Tvдrminne	09.07				3.3	12.0
	Loviisa	05.08	120	0	0	20.0	37.0
		19.09	120	0	0	26.0	56.0
Osmerus eperlanus	Tvдrminne	14.10				2.7	9.6
Abramis brama	Helsinki	22.06				2.4	7.5
	Loviisa	26.06	120	0	0	15.0	28.0
		04.09	110	0	0	11.0	22.0
Rutilus rutilus	Loviisa	12.05	100	1.0	15	6.4	13.0
	Helsinki	22.06				0	8.9
	Loviisa	17.07	110	0	0	23.0	46.0
		31.07	150	0	0	20.0	39.0
		08.08	100	0	0	20.0	42.0
	Helsinki	15.08				0.0	8.9
	Loviisa	03.09	130	0	0	26.0	50.0
		24.09	100	0	0	24.0	50.0
Gadus morhua	Tvдrminne	27.06				1.4	6.9
Lucioperca lucioperca	Loviisa	17.07	120	0	0	5.9	15.0
Perca fluviatilis	Loviisa	12.05	98	1.6	13	3.3	10.0
	Helsinki	22.06				3.1	9.4
	Loviisa	25.06	110	0	0	83.0	160.0
		31.07	120	0	0	73.0	140.0
	Helsinki	15.08				7.7	15.0
	Loviisa	04.09	120	0	0	88.0	170.0
		19.09	120	0	0	96.0	190.0

^{90}Sr was not found in increased concentrations both in seawater and pelagic biota. In the beginning of the 1990s the ^{90}Sr level in polar cod and also in Arctic char inhabiting Stepovoy and Abrosimov Bays on the east coast of Novaya Zemlya, varied from 0.99 to 1.6 Bq / kg (Strand et al. 1994). After the Chernobyl events, Baltic Sea herring contained an insignificant quantity of ^{90}Sr (0.06-0.23 Bq / kg).

$^{239,240}Pu$ concentrations in pelagic fish (herring, polar cod, capelin, sea perch) were below the detection limit - 0.6-2 mBq / kg wet weight (Sickel et al., 1995; Matishov et al. 1996f; Rissanen et al. 1997a) (Fig. 6.3).

Bottom feeding fish feed mainly on invertebrate organisms. As it was noted already, the radioactive contamination of silt-eating marine organisms is higher than in pelagic organisms. For example, 1-45 Bq / kg of ^{137}Cs is contained in the sediments. Large worms, polychaeta, which are fed upon by plaice (*Pleuronectes platessa*, Linnaeus, 1758) and American plaice (*Pleuronectes americanus*, Walbann) contain up to 6 Bq / kg of ^{137}Cs (dry weight).

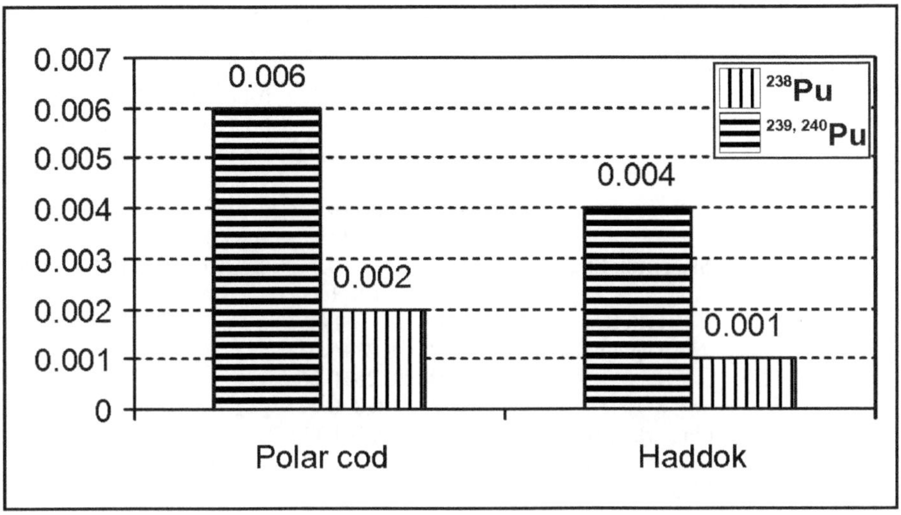

Fig. 6.3. Concentration of plutonium isotopes in muscle tissue of fish from the Barents and Pechora Seas (1993), Bq / kg wet weight

Fish, living near the seabed surface, contain small quantities of ^{137}Cs (0.2-1.3 Bq / kg). In particular, during the 1990s haddock *Melanogrammus aeglefinus* of the Norwegian and Barents Seas concentrated in their organisms about 0.3-0.8 Bq / kg (Table 6.5). Similarly low levels of this nuclide (0.4-1.0 Bq / kg) are noted in American plaice, sea plaice, and catfish.

The White Sea navaga (*Eleginus navaga*, Pallas, 1914) does not migrate over large distances and can serve to a certain extent as an indicator of local radioactive contamination (Christiansen, Bjornstad, 1996). A study of 14 navaga specimens, caught in the Dvina Bay of the White Sea near the Severodvinsk atomic base, contained 2.4-2.7 Bq / kg wet weight (or 8.4-10.7 Bq / kg dry weight) of ^{137}Cs in four fishes with the other fish containing levels below the detection limit.

After the Chernobyl events sea bream (fam. Bramidae) in the Baltic Sea which feed mainly on zoobenthos, accumulated 7-28 Bq / kg of ^{137}Cs and 2-15 Bq / kg of ^{134}Cs (Ilus et al., 1993). In 1999 the sea bream taken from the Azov Sea in Taganrog Bay, had very low levels of ^{137}Cs (0.2 Bq / kg wet weight) (Fig. 6.4).

^{90}Sr *is* not characteristic of bottom feeding fish from northern seas. Examination of organs and tissues from catfish, haddock, plaice and ray, indicate that ^{90}Sr accumulation is very small - from 0.007 to 0.03 Bq / kg. Evidently, the input of ^{90}Sr to fish from seawater and prey items of fish is insignificant. During feeding, benthic fish ingest small quantities of sediment particles containing ^{90}Sr on the order of 0.2-6.0 Bq / kg. However, the small quantities of ingested sediment do not result in significant accumulation in these organisms.

Table 6.5. Artificial radionuclides in fish species (Sickel et al., 1995), Bq/kg wet weight

Species	Area of investigation	Date	^{137}Cs	^{90}Sr	$^{239,240}Pu$
Cod	Barents Sea	11.92	0.7±0.3	0.11±0.05	0.0013±0.0004
		11.92	1.7±0.3	0.28±0.07	0.0012±0.0004
		11.93	0.45±0.02	–	–
		11.93	0.8±0.7	–	<0.002
		1997	0.2–0.5	–	–
			<11	–	–
	Skagerrak Strait	11.92	1.6±0.2	0.06±0.03	0.0010±0.0002
		1997	1.0–1.1	–	–
	North Sea	11.93	0.3±0.02	–	–
	Norwegian Sea	11.93	0.23±0.01	–	–
		09.94	0.23±0.06	–	–
		1997	0.2–0.7	–	–
	Finmarken Bank	10.94	0.7±0.07	–	–
Bone mass	Different areas	11.93	0.28±0.01	0.5±0.1	–
	Skagerrak Strait	01.94	1.09±0.04	–	–
Herring	Salt-Fjord	04.94	0.6±0.08	–	–
Halibut	Andoya	06.94	0.7±0.06	–	–
	Barents Sea	09.94	0.51±0.06	–	–
Haddock	Finmarken Bank	10.94	0.3±0.06	–	–
	Salt-Fjord	10.94	0.7±0.1	–	–
	Norwegian Sea	1997	0.3–0.6	–	–
	North Sea	1997	0.1–0.2	–	–
Coalfish	Barents Sea	1997	0.3–0.4	–	–
	Norwegian Sea	1997	0.4–0.6	–	–
	North Sea	1997	0.1–0.4	–	–
Cusk	Norwegian Sea	1997	0.4–0.8	–	–
Salmon		1997	0.8	–	–
Mackerel	North Sea	1997	0.1–0.4	–	–
Whiting	Skagerrak	1997	1.2	–	–
Sea plaice		1997	0.5	–	–

The highest β-activity (from 15 to 75 Bq / kg) in haddock and catfish was noted during the period of Novaya Zemlya nuclear tests from 1961-1965. Later, ^{90}Sr levels in bottom fish decreased sharply (less than 5 Bq / kg) (Sickel et al. 1995).

The measured values of ^{90}Sr content in Azov Sea sturgeon *Acipencer guldenstadtii* were 0.2-0.8 Bq / kg wet weight and in the stellate sturgeon *Acipencer stellatus* - 0.8 Bq / kg wet weight (Timoshchuk end Sokolova 1970) in 1966.

$^{239,240}Pu$ in bones and meat of haddock, catfish and other bottom fish are below the detection level, i.e. 0.6-2 mBq / kg wet weight. $^{239,240}Pu$ concentrations up to 0.02 Bq / kg are present in the livers of haddock from the Barents Sea. Therefore, it is unclear whether bottom fish result in the transfer of some radionuclides into the pelagic zone for further transfer through trophic food chains.

Piscivorous fish of northern seas, such as cod, coalfish, halibut, ray (*raja radiata)* at present contain low ^{137}Cs concentrations. The levels of this radionuclide vary mainly from 0.2 to 3.2 Bq / kg. In 1993 cod collected from the Barents Sea Central Hollow had 0.5-0.9 Bq / kg of radiocesium in meat and 0.4-0.5 Bq / kg in bones (Table 6.6). In one cod specimen taken from the Skagerrak Strait, ^{134}Cs was found to be 0.03 Bq / kg (Sickel et al. 1995).

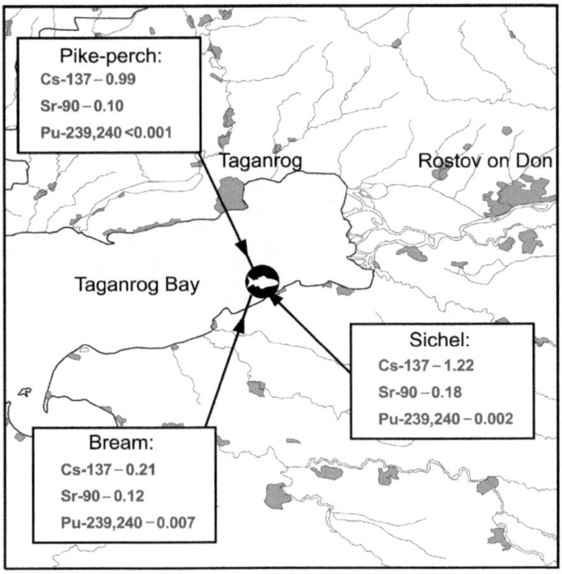

Fig. 6.4. Content of artificial radionuclides in muscle tissue of fish from the Azov Sea (MMBI expedition, June 1999), Bq / kg wet weight

Table 6.6. ^{137}Cs content in Atlantic cod from the Barents Sea central Hollow (Stockman stucture), Bq/kg wet weight

Sample	Meat	Bones
1	0.72	0.44
2	0.54	–
3	0.63	0.39
4	0.51	0.36
5	0.88	0.47
Average	0.66	0.415

Monitoring of radioactivity in carnivorous fish was carried out during the period of 1960-1990s. The ^{137}Cs concentration in Barents Sea coalfish decreased in the following order, Bq / kg wet weight: 1963 - 90, 1968 - 10, 1992-1994 - 0.2-1.0 (Solberg et al. 1995; Foyn end Swaren 1995). During 1980-1990s insignificant oscillations of ^{137}Cs (from 0.2 to 3.5 Bq / kg) were connected with changes in Sellafield discharges (Fig. 6.5, 6.6).

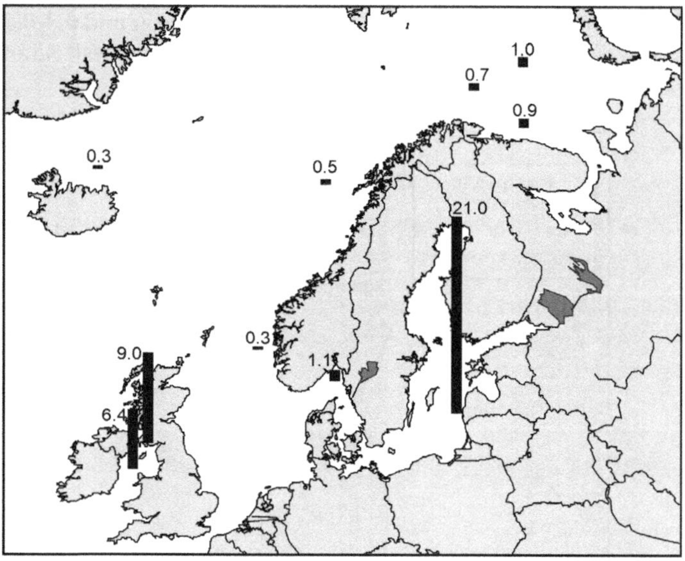

Fig. 6.5. ^{137}Cs concentrations in Atlantic cod from the northern seas during the 1990s, Bq / kg wet weight

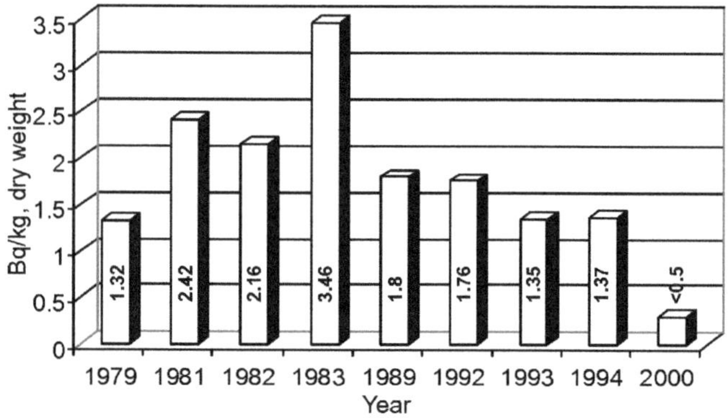

Fig. 6.6. Maximum ^{137}Cs concentrations in cod (Gadus morphua morphua (L.)) from the Barents Sea (1979 to 2000)

At the end of the 1980s and during the 1990s, the highest ^{137}Cs levels were found in Baltic Sea cod (Kanisch end Nagel, 1992; Ilus et al. 1993). The ^{137}Cs concentration in muscle tissues of cod varied from 5 to 21 Bq / kg, and this in liver from 0.8 to 3.0 Bq / kg. The still greater accumulations of this nuclide (6-56 Bq / kg) were noted in the Baltic pike.

^{90}Sr was found in cod, halibut, coalfish and other carnivorous fish from northern seas in small doses from 0.01 to 0.9 Bq / kg during the period of 1980-1990s (Fig. 6.7, Table 6.7). Atlantic cod from the southwest Barents Sea have accumulated about 0.4-0.8 Bq / kg of ^{90}Sr (Matishov D. et al. 1997). Such variations of strontium accumulation by planktivorous predators are the result of feeding behavior.

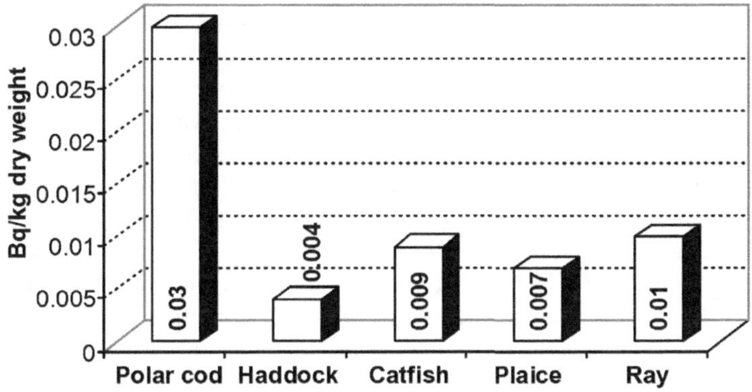

Fig. 6.7. ^{90}Sr concentration in muscle tissue of the Barents and Pechora Seas fish in 1993

Table 6.7. ^{137}Cs, ^{90}Sr, 238Pu, ^{239,240}Pu and ^{40}K concentration in fish from the Barents and Pechora Seas during 1993 (Matishov D. end Rissanen 1993), Bq/kg wet weight

Species	^{137}Cs	^{90}Sr	^{238}Pu	^{239,240}Pu	^{40}K
Cod	0.7	0.030	0.006	0.006	110
Haddock	0.5	0.004	0.002	0.002	110
Catfish	0.4	0.009	–	–	100
Plaice	0.6	0.007	–	–	110
Ray	0.8	0.010	–	–	70
Eelpout	0.0	–	–	–	110
Two-lobed goby	0.3	–	–	–	74

The Institute of Marine Research (Bergen Norway) studied radioactivity in cod and haddock caught in different fishery areas of the Barents Sea before and after the nuclear tests at Novaya Zemlya (Saetre et al., 1992). These data show a relatively rapid decrease in radioactive contamination in fish from 1963. The highest values of β-activity in fish (without activity of ^{40}K) reached 80 Bq / kg (wet weight). Prior to 1986, the level of radioactive contamination in marine fish decreased to 10 Bq / kg wet weight. As it is shown by the observations of Norwegian Institute of Radiation Hygiene (new name Norwegian Radiation Protection Authority) β-activity measured in Barents Sea fish did not change significantly after the Chernobyl accident, remaining low at about 5 Bq / kg.

$^{239,240}Pu$ and also ^{238}Pu are found in a very small quantities (0.001-0.05 Bq / kg) in bones, liver and tissue of cod, coalfish and other omnivorous fish (Fig. 6.8). In bones taken from cod heads, ^{239,240}Pu accumulated to 43 mBq / kg wet weight. A small increase (about 8 mBq / kg) ^{239,240}Pu was noted in the meat of rays. The slight increased accumulation of plutonium in organs and tissues of rays can be explained by the behavior of this fish species. Rays, live near the seafloor and feed on benthic organisms. They often ingest bottom sediments during feeding that in turn leads to the observed higher accumulation of radionuclides in these organisms.

Generally speaking, ^{137}Cs, ^{90}Sr and ^{239,240}Pu concentrations in the marine abiotic environment and in pelagic and bottom fish were relatively low during 1950-1990s. The characteristic feature of this low level contamination of the northern seas is that ^{137}Cs is the dominant source of anthropogenic radioactivity (Fig. 6.9). In fish, this nuclide accumulates mainly in soft tissues and organs.

^{137}Cs concentrations in fish were, in general, very low - within the limits 0.4-0.9 Bq / kg (Table 6.5). The ^{137}Cs level in fish organs and tissues is approximately two times the level in bones. The results of gamma-spectrometric analyses carried out during 15 years has shown that the maximum values of ^{137}Cs in marine fish (herring, cod, perch, ruff) varied from 1 to 3 Bq / kg. In many cases the ^{137}Cs concentration in fish was close to zero. Sometimes the ^{137}Cs level reached 3.5 Bq / kg (cod, 1983) and 5.0 Bq / kg (herring, 1992). By the end of the 1990s similar concentrations of radiocesium (0.7-4.1 Bq / kg) were observed in fish species near Antarctica (Gadidae family and others) (Ross et al. 1993). All these facts, obviously, indicate the important role of ^{137}Cs from global fallout.

The level of ^{90}Sr contamination in marketable fish from the Barents, Norwegian and other Arctic seas is among the lowest in the seas of the Northern Hemisphere. We can note for comparison that fish from the Dnieper and Volga basins contain 1-5 Bq / kg of ^{137}Cs (Kuzmenko et al. 1995). The strontium content in organs and tissues of the Barents Sea fish lies in the interval 0.004-0.03 Bq / kg and is only slightly higher than similar measurements (0.0025-0.02 Bq / kg) taken in Baltic and North Sea fish (Kanisch end Nagel 1992; Ilus et al. 1993).

Strontium accumulates mainly in the bones of fish. Its level in bones occasionally reaches levels that are several times higher than in the organs and tissues of fish. The maximum quantity of ^{90}Sr in fish was noted in 1962-1963 during the period of the most intensive nuclear explosions. During this time, cod, coalfish, and haddock contained levels of 40-75 Bq/ kg.

Maximum 90Sr levels, close to 90 Bq / kg, were detected in Barents Sea coalfish *Pollachiues virens* in 1963 (Foyn end Svaren 1995). Coalfish, being a bathypelagic species, feeds in the sea surface layer, while cod and haddock feed near the sea bottom.

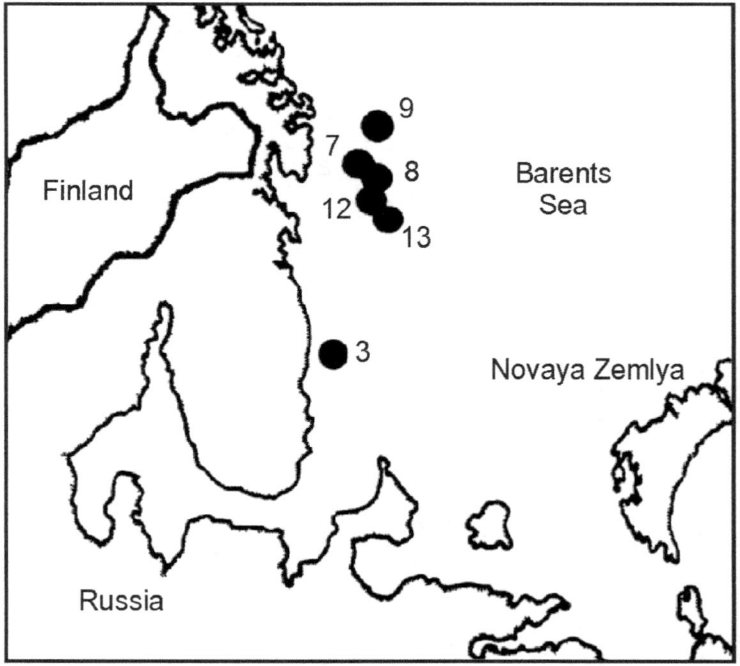

Barents Sea					
Amalgamated sample of n specimens	n	Year	Place of collection	Bq/kg wet weight (я 1 s %)	
				^{238}Pu	^{239,240}Pu
Meat					
Cod (Gadus morhua)	5	1993	12, 13	< 0.0006	<0.0006
	4	1993	3	< 0.002	<0.002
	6	1994	12, 13	< 0.002	<0.002
Haddock (Melanogrammus aeglefinus)	4	1993	3	< 0.002	< 0.002
	3	1994	8	< 0.002	< 0.002
Catfish (Anarhichas lupus)	2	1993	3	< 0.002	< 0.002
Ray (Raja radiata)	3	1994	9	< 0.002	0.0079 (11)
Pollack (Pollachius virens)	3	1994	8	< 0.002	<0.002
Redfish (Sebastes marinus)	3	1994	8	< 0.002	< 0.002
Liver					
Cod	1	1993	12, 13	< 0.002	< 0.002
Cod	4	1994	12, 13	< 0.002	< 0.002
Cod	2	1994	8	<0.005	0.017 (44)
Bones					
Cod	5	1993	12, 13	< 0.009	< 0.003
Cod	4	1993	3	< 0.006	< 0.006
Hard of cod	5	1993	12, 13	< 0.01	0.043 (12)
Haddock	4	1993	3	< 0.006	< 0.006
Catfish	2	1993	3	< 0.01	< 0.01
Ray	2	1993	3	< 0.001	0.0049 (15)
Plaice	7	1993	3	< 0.003	< 0.003
Fishes of the Irish Sea					0.01
Fishes of the North Sea					0.00025

Fig. 6.8. ^{238}Pu and ^{239,240}Pu concentrations in fish from the Barents Sea (1993-1994)

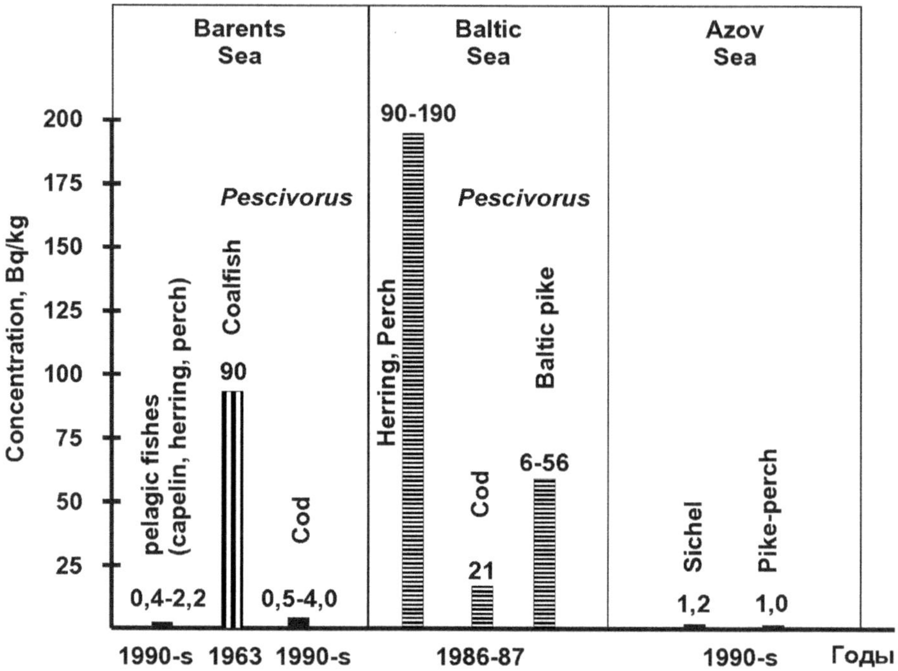

Fig. 6.9. ^{137}Cs concentration in marketable fish species (1960-1990)

During the 1990s, the levels of ^{239,240}Pu in marine fish of the Arctic zone were very low (0.001-0.05 Bq / kg). Such ^{239,240}Pu concentrations are comparable with those in fish from the North Atlantic (Bowen et al., 1971; Aarkrog et al. 1985; Strand et al. 1994). The main conclusion is that no exceptional concentrations of plutonium were found in fish samples collected from the investigated areas.

Atlantic cod, haddock and other bottom fish, caught in Motovsky Bay, the Kildin Island area and the exit from Kola Bay, contained low levels of ^{137}Cs (0.3-1.0 Bq / kg). The study of fish from the Kola Bay, where aerosol releases from maintenance plants for atomic submarines, is of special interest. The long-living radionuclides, such as ^{137}Cs, ^{239,240}Pu and ^{90}Sr, are among those present in the mixture of radionuclides associated with maintenance activities at the plants.

Fish from the Barents Sea contain ten times less radioactive cesium than fish from the Kola Peninsula lakes (Matishov et al., 1994). Recent levels of artificial radionuclides in lake fish are 3-7 times higher than in marine fish. In 1991, lake perch in the Krasnoshchelye settlement area contained ^{137}Cs concentrations from 22.2 to 27.5 Bq / kg. For comparison, in 1982, fish bones, muscles and heads contained correspondingly 10.2, 2.86 and 1.8 Bq / kg of ^{137}Cs. Fish from the Yenisey River in 1973 contained from 15 to 48 Bq / kg wet weight of ^{137}Cs (Table 6.8; Vakulovsky et al., 1994a). Fish taken from adjacent marine waters contained ^{137}Cs and other radionuclides in lower concentrations than in river fish. The process of radionuclide concentration in seawater occurs over a very short time and food sources for marine fish are more varied than river fish species.

Table 6.8. Radionuclide content in Yenisey River fish (August 1973) (Vakulovsky et al. 1994), Bk / kg wet weight

Species	Distance from the source, km	^{32}P	^{65}Zn	^{137}Cs
Dace	190	2800	26	4
Roach	190	3600	37	11
Perch	190	2200	22	7
Perch	330	600	4	7
Dace	600	2800	33	7
Roach	600	1040	56	4
Pike	1230	600	–	15
Roach	1230	600	–	7
Dace	1230	740	–	4
Dace	1460	270	11	48
Roach	1460	20	–	4

Thus the investigation of marine fish during the last decades indicates that the observed low level of radioactive contamination has lead to no observable decrease in biological productivity of the northern seas. The sharp fall in fishery production occurring during 1980-1990s, probably resulted from the superposition of natural cycles of marketable fish yield capacities with over-fishing.

The radioactivity of all fish species of the Barents-Kara region is far less than limits fixed by international standards for the content of artificial radionuclides in food products and is not dangerous for humans. The dose of inner radiation in fishes can be estimated as low, on average from 4 to 24 μSv / year in the Barents Sea and from 6 to 45 μSv / year in the Kara Sea (Fig. 6.10). Consuming 15 kg of fish annually results in an effective dose of about 1-3 μSv / year (Kanisch end Nagel 1992).

6.2 Sea birds

Avifauna, or birds, play a special role in the biogeochemical cycles of substances in Arctic and Sub-arctic marine ecosystems. Birds, owing to their large number, exercise great influence on the functioning of water basins, especially in the coastal areas and in the zones of hydrological fronts of the open seas (Vernadsky 1926, 1929; Ilyenko 1974; Moiseev end Ramzaev 1975; Matishov 1992; Environment ... 1994; Krasnov et al, 1995; Lebedeva 1999). Bird bazaars are an integral part of the Greenland, Norwegian, Barents, White and other seas ecosystems. In the first half of the 20th century, the size of bird populations in the Barents Sea alone reached several tens of millions of specimens. 2-4 million of Brunnich's guillemots, not counting other bird species, nested on the fifty bird bazaars of Novaya Zemlya (Uspensky 1956).

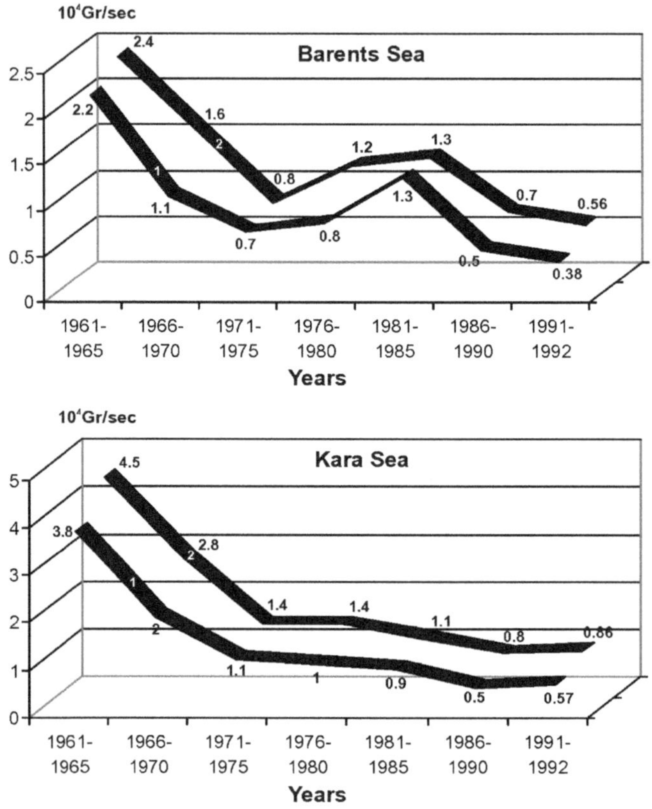

Fig. 6.10. Estimate of radiation doses of fish in Arctic Seas (internal dose) (Foyn and Semenov 1993): **1** - *average*; **2** - *maximum*

As the multi-million bird populations feed on pelagic fishes, plankton and benthic organisms, they play an important role in the cycle of phosphorus and other biogenic elements. In this way, different species of sea avifauna contribute to the increase of biological productivity of the polar seas. At the same time colonial birds perform a key function in the concentration and transfer of anthropogenic radionuclides both within the marine ecosystem and in a transcontinental scale. During the period of nuclear tests in the atmosphere, almost ten million of the Novaya Zemlya population of birds was exposed to direct or indirect radioactive contamination. Although the ^{137}Cs and $^{239,240}Pu$ content in bottom sediments (0-40 Bq / kg) and biota (0-10 Bq / kg) is low, the zones of the Novaya Zemlya nuclear test sites are an exception with concentrations of radionuclides that are considerably higher. Concentrations of artificial radionuclides in bottom sediments of the underwater nuclear test site in Chernaya Bay reach 3-5 thousand Bq / kg. The main bird bazaars are also situated within 50-100 km zones of the atmospheric nuclear explosions. The role of migrating birds, for example guillemots, eiders, geese, and kittiwakes, in transfering radionuclides from Novaya Zemlya to North and West Europe during 1950-1960s remains practically unexplored.

It is important to note that bird bazaars are the most characteristic feature of the Barents Sea ecosystem. The quantity of sea and land birds during the last 3-4 decades has undergone considerable changes due to the influence of climatic and anthropogenic factors. A fall in population sizes is the main trend. A sharp decrease in the number of Novaya Zemlya birds, for example, of guillemots, took place because of overfishing of polar cod - the main food of guillemots (Matishov end Pavlova 1990; Matishov end Denisov 2000).

Widespread radioactive contamination of sea birds has also been hypothesized as a contributing factor to the decreasing trend in bird numbers. Many bird species inhabiting areas subjected to the influence of discharges of the Sellafield nuclear reprocessing facility, have relatively increased concentrations of radiocesium (Solberg et al. 1995). A similar situation existed in the 30-km zone after the Chernobyl accident. The high content of all artificial radionuclides in organs, tissues and excrement of birds indicates the accumulation of radioactive substances in birds (Radioecological sequences .. 1991; Influence ... 1996; Kuzmenko 1998).

However, few investigations have been carried out on the bioaccumulation of ^{137}Cs, ^{90}Sr, ^{60}Co and $^{239,240}Pu$ by birds living on the Arctic archipelagoes and Arctic Seas. Insights can be drawn however from radioecological materials collected during MMBI coastal expeditions between 1991-1997.

In August of 1991 samples of organs and tissues were taken from 10 kittiwakes (*Rissa tridactila*) and little auks (*Alle alle*) from bird bazaars on Guker Island (Franz Josef Land). In August of 1995 13 sea birds were caught in the Ivanovskaya and Drozdovka Bays and on the Nokuev Island (East Murman coast): sea and common gulls (*Larus marinus, Larus canus*), great squa (*Stercorarius skua*), and specimens of other species from which muscle tissue and liver were taken (Table 6.9). In 1996 data on different bird species from the Yarnyshnaya Bay (East Murman) and a number of West Murman bays were analyzed. Additional data were evaluated on birds from the northern part of the Barents Sea obtained in 1991 by Norwegian scientists (Table 6.10) as well as our own data obtained in 2000 (Table 6.11). In total we assembled more than 100 ^{137}Cs determinations in 14 bird species.

It is quite evident that artificial radionuclides are incorporated into birds living in the northern seas mainly by trophic transfer. Colonial and aquatic birds feed on zooplankton, pelagic fish, and some benthic organisms, depending on the geographical situation of nesting, feeding and wintering places, and on migratory patterns.

Together with feeding, an important role in the process of radionuclide accumulation in organs and tissues belongs to the digestion mechanisms of different bird species. It is known, for example, that the quantity of strontium in the skeletons of fish-eating birds is significantly lower than in the bone tissues of their preys. The especially great decrease of ^{90}Sr concentration in this link in the food chain is connected with the fact that gulls regurgitate the skeleton of eaten prey, a sort of «container» with radionuclides (Ilyenko 1974; Krasnov et al. 1995).

Table 6.9. ^{137}Cs concentration in bird tissue and liver samples taken from the Barents Sea and Kola Peninsula, Bq/kg wet weight

Species, date of catching	Ivanovskaya Bay		Nokuev Island		Drozdovka Bay		Yarnyshnaya Bay		Dalnezelenetskaya Bay	
	1	2	1	2	1	2	1	2	1	2
Big sea gull *Larus marinus*										
August 09, 1995	5.6	4.4								
August 16, 1995	2.4	2.2								
Common gull *L. canus*										
August 12, 1995			1.0	1.0						
Great skua *Stercorarius skua*										
August 12, 1995			3.0	2.1						
August 17, 1995					40	2				
Goosander *Mergus merganser*										
August 17, 1995						0.2				
Wood sandpiper *Tringa erythropus*										
August 04, 1995	4.3	4.8								
Little sandpiper *Calidris minuta*										
August 08, 1995	1.5	2.2								
Common eider *Somateria mollissima*										
February 19, 1996									0*	0*
March 09, 1996							0.15	0*		
March 10, 1996							0.17	0.70		
March 10, 1996							0.30	0.29		
Short-billed guillemot *Cepphus grylle*										
March 01, 1996							0.43	0.85		
Purple sandpiper *Calidris maritima*										
February 19-23, 1996									0*	

FOOTNOTE. Concentrations below the detection level are noted by stars; 1 - tissues; 2 - liver.

According to the food composition, all birds, inhabiting the sea or coasts, are divided into ichthyofags, planktofags and benthofags (Krasnov et al. 1995). The following groups are characteristic among the avifauna feeding on fish: gull birds, guillemots, cormorants and sandpipers.: Anatidae (ducks) and guillemots (black guillemots, little auks) feed mainly on plankton and benthic invertebrates. The composition and levels of radionuclides in food items are reflected in the accumulation levels observed in different bird species.

Table 6.10. Radiocesium content in sea birds from islands in the Barents Sea, 1991 (by: Savinova and Gabrielsen 1994)

Number of sample	Date	Area of investigation	Species	Body weight, g	Sex	^{137}Cs, Bk/kg
1	2	3	4	5	6	7
1	11.07	Hornoya Island	Kittiwake	456	♂	<1.3
1a				456	♂	<2.1
2				370	♀	<1.4
2a				370	♀	<3.2
3				452	♀	<1.5
3a				452	♀	<1.5
4				432	♂	<1.2
4a				432	♂	<2.6
5	11.06				♀	<1.1
5a					♀	<2.4
16	11.07		Herring-gull	1128	♂	2.1±0.6
16a				1128	♂	<1.4
17	17.07			860	♀	<1.5
17a				860	♀	<1.3
17б				860	♀	<2.8
18	14.07			1110	juv	<2.3
19				965	juv	Not determined
20	23.06			1200	♀	
21				1380	♂	
32	23.06	Bear Island	Kittiwake	400	♂	<1.0
33				510	♂	<1.3
34				445	♀	1.2±0.5
35					♀	<1.2
36	20.08		Glaucous gull	1430	♀	<1.0
37	21.08			2000	♂	<0.7
38	20.08			1330	♀	<1.1
39	21.08				♀	0.9±0.4
40				1800	♂	<1.2
46	08.07	Ny-Alesund		1515	♀	Not determined
47				1472	juv	<1.1
48				1996	♀	3.3±0.4
49				1579	♀	<0.6
50				1798	♀	<0.7
51			Kittiwake	449	♂	<1.1
52				432	♀	Not determined
52				424	♂	<0.6
54				449	♀	Not determined

Table 6.10. (Cont.)

1	2	3	4	5	6	7
55				453	♂	
62			Eider	1200	♂	
63				1200	♂	
64				1200	♀	<0.6
65				1350	♀	<0.6
66				1949	♀	Not determined
67	10.07			1400	♀	<0.5
41	24.08	Guker Island	Little auk	200	♂	Not determined
42				190	♂	<0.5
43				220	♂	<0.6
44				180	♀	<1.1
45				190	♀	<0.5
61				200	♂	0.6
78	31.08		Kittiwake		juv	<0.6
79					♂	1.3±0.6
80	29.08				♂	1.5±0.5
81	30.08				juv	0.7±0.3

Table 6.11. ^{137}Cs concentration in birds of the Barents Sea coast (Dalnie Zelentsy settlement area) in 2000, Bq/kg dry weight

Number of sample	Species	Sex	Age	Organ	Wet weight, g	Dry weight, g	^{137}Cs	Comment
1	2	3	4	5	6	7	8	9
0075	European cormorant	♂	sad	Musculature	688	174	+	
0076	*Phalacrocorax carbo*	♂	sad	Skeleton	264	124	+	
0067		♂	sad	Skin	356	160	+	
0064	Common eider			Musculature	850	236	+	Composite sample
0066	*Somateria mollissima*			Skeleton	392	144	<0.2	The same
0032		♀	ad			56	1.3±0.2	Found dead
0094		♀	ad			54	1.7±0.2	The same
0095		♀	ad			34	1.6±0.2	”
0199	Great black-backed gull *Larus marinus*	♂	ad	Musculature	916	280	+	Composite sample
0148			ad		930	168	+	The same
0163		♂	ad	Skeleton	276	100	+	

Table 6.11. (Cont.)

1	2	3	4	5	6	7	8	9
0034			juv			94	+	Found dead
0126		♀	ad		232	64	5.0±0.4	
0198		♂	ad		488	264	+	Composite sample
0197		♂	ad	Skin	388	264	4.5±0.3	
0131		♀	ad		488	108	16.8±0.6	
0206		♀	ad		303	206	6.6±0.4	
0155	Black guillemot *Cepphys grylle*		juv	Carcass	304	106	+	
0171	Arctic skua *Stercorarius parasiticus*				654	76	+	Composite sample
0063	Herring-gull	♀	ad	Muscula-ture	252	120	+	
0085	*Larus argentatus*	♂	ad		348	114	2.7±0.3	
0187		♀	ad		288	44	14.7±0.3	
0010		♀	ad		478	130	3.4±0.4	
0065		♀	ad	Skeleton	439	212	+	Composite sample
0009		♀	ad	Skin	232	112	+	
0062		♀	ad		228	120	+	
0084		♂	ad		222	126	+	
0185		♀	ad		184	82	+	
0086		♂	ad	Skeleton	162	80	1.5±0.2	
0186		♀	ad		128	50	2.4±0.3	

Sea gulls - herring-gulls, common gulls (*Larus canus*), kittiwakes and arctic terns are able to obtain their food only from the upper 1 m of ocean depth. The food composition of these birds includes mainly sand eel, capelin, herring, and sometimes polar cod (Table 6.12). Besides cleptoparasitism, feeding near fishing vessels, sea gulls commonly scavenge food remnants from city dumpsites. In the Barents and North Seas the wastes of the fishing industry are an important food source for fulmars (*Fulmarus glacialis*), kittiwakes and great skuas (Krasnov et al. 1995).

The ^{137}Cs concentration in sea gulls of the Barents and Greenland Seas varies from 0.5 to 6 Bq / kg. In particular, kittiwakes from the bird bazaars of Franz-Josef Land and Bear Island contain a small quantity of radiocesium in muscles (0.5-1.5 Bq / kg). Approximately the same accumulation of this nuclide (about 1 Bq / kg) was detected in the muscle and liver of common gull *Larus canus* near Nokuev Island (East Murman).

The ^{137}Cs content in kittiwakes and glaucous gulls from Ny-Alesund and Hornoya Island areas of West Spitsbergen are up to 3.3 Bq / kg (Table 6.10). Cesium levels to 2.8 Bq / kg are found in the hering-gulls of Hornoya Island and big sea gulls (*Larus marinus*) of the Ivanovskaya Bay (East Murman). Probably, the high-

highest ^{137}Cs concentration in muscles (2.4-5.6 Bq / kg) and liver (2.2.-4.4 Bq / kg) of big sea gulls is caused by their specific feeding pattern. In addition to small fish they also consume nestlings (newly hatched birds) and human garbage.

Table 6.12. Frequency of appearance of fish species in the food of colonial birds (by: Belopolsky 1971), %

Fish species	Puffin		Herring-gull		Sea gull	
	I	II	I	II	I	II
Herring	25.4	16.0	32.7	14.3	40.6	23.7
Capelin	12.7	17.0	3.8	10.2	–	9.7
Sand eel	5.4	46.0	–	–	3.1	4.6
Cod	–	2.0	5.8	18.6	31.2	30.9
Total (quantity of stomachs)	55	100	52	322	32	194

FOOTNOTE. I - Ajnovy Islands; II - Seven Islands Archipelago

The radioactivity of **guillemot birds** - puffins (*Fratercula arctica*), razorbills (*Alca torda*), Atlantic murres (*Uria aalge*), Brunnich's guillemot (*Uria lomvia*) and also of European cormorants (*Phalacrocorax carbo*) and shags (*Phalacrocorax aristotelis*) remains practically unstudied. These bird species feed on small pelagic fish - sand eel *Ammodites tobianus* and capelin *Mallotus villosus*, small herring *Chupea harengus*, and in isolated cases, on the fry of large fish including cod and catfish (Golovkin 1982; Krasnov et al. 1995). All of these species extract fish both from the sea surface and from tens of meters below the sea surface, actively moving in water with the aid of wings.

In the case of cormorants and other guillemot birds the contamination of prey is an indirect indication of the presence of ^{137}Cs in their organs. During the 1990s the radiocesium content in pelagic fish of the Barents-Kara region varied from 0.4 to 2.2 Bq / kg. Obviously, the birds themselves contained no lower amounts of radioactive contamination.

The special importance of guillemots, mainly the Brunnich's guillemots, must be noted. Only the Barents Sea guillemot, with several million specimens, annually consumes 200-400 thousand tons of small fish. The role of guillemonts together with other colonial bird species in the biogeochemical and radiogeochemical cycles of matter is especially great during the nesting period. Precisely in this period many thousands of guillemots concentrate into colonies, consuming large quantities of fish and fertilize the water by their excrement near the vicinity of bird bazaars (Uspensky 1956; Golovkin et al. 1972; Golovkin 1982).

Skuas - arctic skua (Stercorarius parasiticus) and great skua (Stercorarius squa) are distinguished by having the highest radioactivity among all of the studied bird species. Skuas obtain fish mainly through cleptoparasitism on other sea bird species (Krasnov et al. 1995). ^{137}Cs accumulation in skua from Droxdovka Bay reaches as high as 40 Bq / kg in muscles and 2 Bq / kg in liver (Table 6.9). Possibly, this form of feeding causes high radiocesium accumulations in skua (Matishov et al. 1996b).

At the same time, one cannot exclude that such high concentrations of radiocesium in skua are the result of their consumption of reindeer remnants (Rissanen et al. 1997a). During the 1990s, the average ^{137}Cs concentration in reindeer meat from the Kola Peninsula coast varied from 200 to 600 Bq / kg. Skua from Nokuev Island contained ^{137}Cs concentrations of only 3 Bq / kg is the result of differences in food items.

Little auks (Alle alle) and black guillemots (Cepphus grylle) contain relatively low levels of ^{137}Cs accumulation (0.4-1.1 Bq / kg). From 40 to 90 % of their food items, depending on season, consist of plankton and littoral invertebrates (crustaceans and molluscs). The prevailing crustacean species in the prey items of birds are euphasiides *Thysanoessa raschi*, *T. longipes*, amphipodes *Parathemisto libellula* (Belopolsky 1957; Golovkin 1982). With the presence of coastal ice and ice fields in the sea, little auks and black guillemots obtain food from under ice (amphipodes *Onisimus glacialis*, polar cods).

The ^{137}Cs concentration of little auks inhabiting the Guker Island (Franz-Josef Land) is 0.4-1.1 Bq / kg (Table 6.10). The black guillemots from the Yarnyshnaya Bay (East Murman) accumulate still less ^{137}Cs both in muscles (0.43 Bq / kg) and in liver (0.85 Bq / kg) (Table 6.9). The radiocesium concentration in these little Barents Sea birds is comparable with levels in zooplankton (to 2 Bq / kg) in the 1990s.

The radioactivity of ducks (Anatidae), such as common eider (*Somateria mollissima*), goosanders, and geese, is known in general terms only. These aquatic birds nest on the coasts of the Barents, White, Greenland and other northern seas. For wintering they migrate to Iceland, the Faroe Islands, and west European coasts. All ducks obtain food in the water or on the seabed. Goosanders feed mainly on fish and partly on invertebrates. Eiders feed on molluscs (periwinkles, mussels) and crustaceans, gathering them from the shallows or in the littoral zone during falling tide.

Organs and tissues of eider from Ny-Alesund (West Spitsbergen) contain 0.5-1.1 and to 3.0 Bq / kg of ^{137}Cs (Table 6.10). The eider, inhabiting the Yarnyshnaya Bay (East Murman) accumulate in muscles and liver still less ^{137}Cs - correspondingly 0.1-0.3 and 0.3-0.7 Bq / kg. The lowest concentrations (0.2 Bq / kg) were noted in goosander from Ivanovskaya Bay (Table 6.9). However, sandpipers living in the same place and feeding mainly on marine invertebrates and insects, had far higher ^{137}Cs levels. Muscles and tissues of sandpiper *Tringa erythropus* contained correspondingly 4.3 and 4.8 Bq / kg wet weight; similar levels are observed in the little stint *Calidris minuta* - 1.5 and 2.2 Bq / kg wet weight.

The dynamics and character of radionuclide accumulation on the Novaya Zemlya coast exemplified using geese (Uspenskaya et al. 1999). During the 1970s, the ^{137}Cs content in tissues was 15-30 Bq / kg, while ^{90}Sr in bones was approximately 16 Bq / kg. By the end of the 1980s geese contained 7-15 Bq / kg of ^{137}Cs (Tables 6.13, 6.14).

Thus, one can conclude that the radioactivity level of colonial and aquatic birds of the Barents and other northern seas is insignificant and not dangerous for the life activity of avifauna. The main source of input of artificial radionuclides, in particular, of ^{137}Cs, to bird species is from small pelagic fish, and, in some cases, planktonic and benthic invertebrates. The level of radionuclides is almost independent of sex and weight (mass) in sea birds. Also there are no direct facts indicating significant accumulations of radionuclides in bird tissues by age that is possibly connected with individual feeding strategies. ^{137}Cs is somewhat elevated in sea gulls, guillemots and skuas. Sea birds have a high rate of metabolism that results in the elimination of cesium and strontium from the birds.

Table 6.13. Radionuclides in biota (Uspenskaya et al. 1999)

Biota	Muscle tissuses			Bone tissues		
	^{137}Cs		^{90}Sr	^{137}Cs	^{90}Sr	
	nKi/kg	Average, Bk/kg	nKi/kg	nKi/kg	nKi/kg	Average, Bq/kg
Reindeers slaughtered in 1976-1978	0.7–1.6	43	–	–	3–5	148
Geese slaughtered in 1976-1978	0.6–1.0	30	–	–	0.25–0.6	16
Chars caught in Nekhvatov Lake and Sakhanin River in 1976-1978	0.06–0.16	4	–	–	0.2–0.5	13

Table 6.14. ^{137}Cs concentration in biological materials (Uspenskaya et al. 1999)

Object	^{137}Cs, Bq/kg				
	1963	1969	1970–1978	1980–1988.	1988–1990
Mosses	222–260	260–370	300–550	220–440	150–180
Lichens	750–1700	1300–2700	750–1500	–	–
Reindeer meat	75–370	80–1100	80–370	80–180	40–75
Fish	2.6–3.7	1.1–1.8	3.0–3.7	2.6–3.7	2.6
Milk	–	0.2	0.56*	0.110*	0.037*
Geese, ducks	–	–	15–22	11–15	7.5–15*

* Average data for USSR

During the 1990s, ^{137}Cs concentrations in muscles of Barents Sea birds varied from 0.2 to 40 Bq / kg. A trend of slightly lower cesium accumulation (from 0.1 to 22 Bq / kg) is characteristic for the bird's livers. There are not so many analytical data at present, in particular, for the Azov Sea (Table 6.15), which would aid in determining the influence of Chernobyl fallout on the dynamics of radioactivity in sea birds.

There are certain species differences in the concentration of radionuclides in birds and their prey inhabiting the same water basin. Barents Sea bird species can be arranged by the level of ^{137}Cs as follows: goosander (0.2 Bq / kg) $\rightarrow$ black guillemot (0.43 Bq / kg) → little auks (0.4-1.1 Bq /kg) → common gull (about 1 Bq / kg) → eider (0.1-3.3 Bq / kg) → kittiwake (0.5-3.3 Bq / kg) → big sea gull (2.4-5.6 Bq / kg) → great skua (3-40 Bq / kg). Such regularity is caused not only by the qualitative composition and radioactive content of the food consumed by birds, but also the place of food gathering. A general descending trend of radioactivity is seen in birds feeding on sea crustaceans and other zooplankton and zoobenthic species.

Table 6.15. Radionuclides in the tissues of European cormorant (Don River delta, channel Malaya Kuterma, Martyshka Island, 47° 11.386'N, 39° 12.988' E, June 19-22, 2000)

Samples	Wet weight, g	Dry weight, g	^{137}Cs, Bq/kg dry weight	^{90}Sr, Bq/kg dry weight	^{239,240}Pu, mBq/kg dry weight
Fish (from stomachs)	1900	476	1.3±0.3	+	2.1±1.0
	3500	508	3.7±1.7	+	8.1±2.0
Skins + feathers	640	450	+	+	1.5±0.4
	800	572	0.5±0.2	+	+
	720	426	+	+	+
	850	498	0.5±0.1	+	–
	820	486	+	+	–
	830	488	+	–	–
	1000	562	+	–	–
	850	500	+	–	–
	730	430	+	–	–
	1070	630	+	–	–
	850	500	+	–	–
	840	494	+	–	–
	880	520	+	–	–
Skeletons	1050	504	0.5±0.16	+	7.0±2.8
	1060	506	0.42±0.16	+	8.9±4.0
	1070	512	<0.2	+	+
Muscles	1024	276	+	+	0.6
	1242	346	+	+	1.2±0.6
	1000	196	0.64±0.17	+	2.3±0.5
	1200	240	<0.2	+	+
	1400	320	0.6±0.2	+	+
	1600	356	+	–	–
	2100	612	+	–	–
	400	172	+	–	–
Skeletons*	1600	740	+	+	–
Excrement	544	298	+	–	+
	808	420	+	–	+
	558	290	+	+	+
	794	412	+	+	–
Skeletons	1100	526	+	+	–
	1500	702	+	+	–
	828	396	+	+	–
	818	352	+	+	–

*One sample is composed of 5 bird specimens

So, the process of pollutant accumulation in sea colonial and coastal aquatic birds depends on many factors: levels of radioactivity in food, food preferences (a function of bird species), contamination of nesting places and migration pathways, and age are important.

The maximum life duration of one generation, for example, of guillemots (Brunnich's guillemot) or Anatidae (eider) varies from 10 to 16-18 years. Consequently, representatives of colonial and aquatic avifauna, which inhabited the polar seas during the 1990s, were subject to the influence of radioactive contamination from Chernobyl discharges to the atmosphere, from the appreciable volume of Sellafield radioactive waste discharges to the sea (1970-1980s) and from local emission of isotopes from the coastal atomic bases. The role of global fallout, connected with the Novaya Zemlya nuclear explosions in the atmosphere, obviously is infinitesimal in the life activity of contemporary birds.

Sea birds usually nest on the shore in colonies, often large (more than 10,000 specimens), where food is readily available. The guillemot birds (puffins, black guillemots) settle their nests on rocks and the turf layer among boulders and driftwood. The nestlings and grown-up specimens, inhabiting bird bazaars, can swallow lichens or berries in which the ^{137}Cs concentration varies from 100 to 900 Bq / kg. Obviously, birds can receive doses of radioactivity during the nesting stage. It must be noted in a comparative aspect that sea birds are contaminated by radiocesium several times more than fish and benthic organisms. The high ^{137}Cs content in great skua (40 Bq / kg) puts forward for the present the task of regular radioecological monitoring of higher trophic level organisms living in Arctic marine and coastal ecosystems. It is necessary to investigate the mechanism of bioaccumulation and transformation of radioactive contamination in these trophic chains.

Additional radiochemical and radioecological investigations in bird colonies of the most prevalent bird species, Brunnich's guillemots and little auks, is needed. The large number of little auks and specific character of their feeding make these birds among the leading vertebrate consumers of zooplankton in ecosystems of the high Arctic Seas. The Brunnich's guillemot, in turn, can be classified among the important fish consumers occupying the highest trophic level in marine ecosystems. The nesting period contributes to the return of significant organic matter by guillemots and little auks from the sea in the form of excrement to the vicinity of bird colonies. Precisely in such places it is necessary to carry out further investigation of radioactivity in all ecosystem components. The population radioecology of sea birds must be investigated concurrently.

6.3 Marine mammals

Sea mammals are of special interest because they are at the top of the food web of the oceans and seas. The most intensive circulation of organic matter occurs in places of feeding, mating and nursery areas. Seals and whales migrate seasonally from the moderate termperature zone to the Arctic Seas and back (Fig. 6.11). They not only accumulate radioactive elements but also transport them, sometimes many thousands of kilometers. Sea mammals are used as food mainly by polar bears and humans.Yet no detailed radioecological investigations of the Barents and other polar seas of pinnipedes and cetaceans have thus far been carried out.

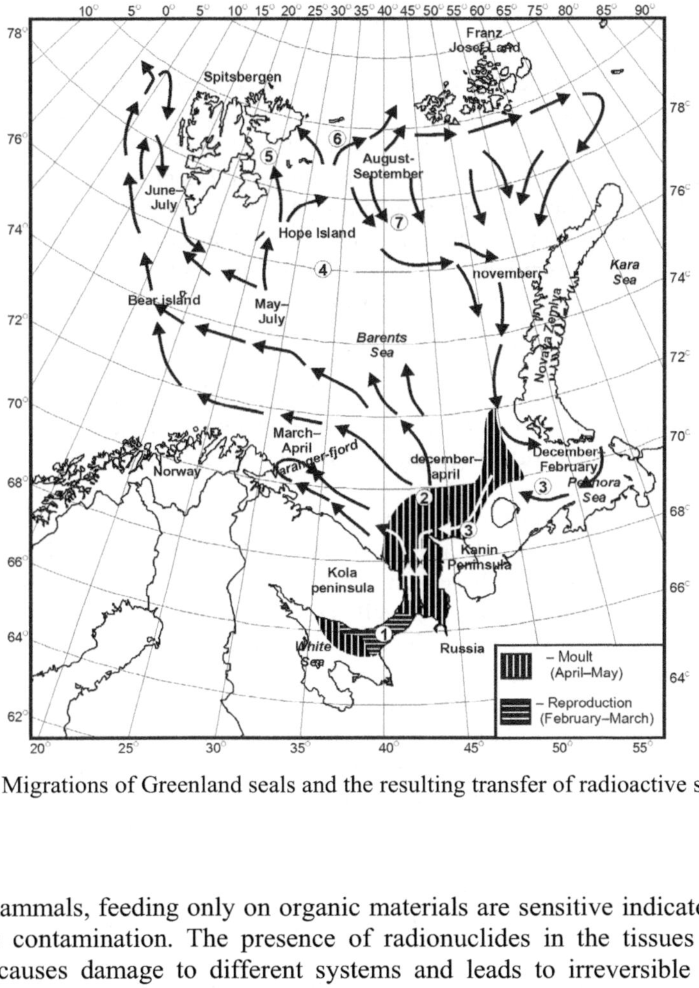

Fig. 6.11. Migrations of Greenland seals and the resulting transfer of radioactive substances by seals

Sea mammals, feeding only on organic materials are sensitive indicators of radioactive contamination. The presence of radionuclides in the tissues of these animals causes damage to different systems and leads to irreversible changes. Although some have hypothesized that the reduced number of seals and whales, the highest level organisms of the pelagic zone, is connected with nuclear tests conducted in the atmosphere and underwater during the 1950-1960s. But this is not the case.

From the 18th to the beginning of the 20th century the Barents Sea was inhabited, by 10-20 million specimens of diverse whale and seal species. The total size of the Greenland seal population decreased because of intensive hunting from 10 million specimens in the middle of the 1950s to 1.7 million speciments in the middle of the 1960s. The hunting of these seals reached 0.5 million during the 1930s and was limited in the 1990s to several tens of thousands (Matishov end Denisov 2000). Similar dynamics were characteristic for the hunting of small whales. In 1938-1964 Norwegian hunters killed about 80 thousand specimens of blue whales and in the 1990s only several hundred. Ultimately the excessive extermination of Greenland whale, Atlantic walrus, and Greenland seal led to the prohibition of hunting on these species.

It is important to note that the pinnipeds and cetaceans have a very low reproduction rate. The reproductive cycle of such species as walrus, Greenland whale and others is not less than 3-4 years and the total life duration - 20-40 years. The life of seals, unlike cetaceans, depends to a greater extent on the dynamics of hydrometeorological factors. Pinnipeds during the reproduction period, without fail, form large colonies on the drifting ice, shores of islands and continental coasts.

The most typical pinniped species of the Barents, White, Kara, Greenland and other northern seas are Greenland seal *Pagophilus groenlandica*, Atlantic walrus *Odobenus rosmarus*, grey seal, ringed seal, and sea hare. The cetaceans are represented mainly by Greenland whale *Balaena mysticetus*, finner *Balaenoptera physalus*, blue whale *Balaenoptera musculus*, sei whale, white whale, little piked whale *Balaenopterus acurostrata* and Minke whale. All whales are social, forming different groups containing from several to several hundred and even thousand specimens.

Radioactive contamination accumulates in the organs and tissues of marine mammals. Whales and seals in the course of feeding swallow both prey and seawater. Their feeding is based on pelagic fish and invertebrates and for some species - on zoobenthic species as well. The composition and amount of food are determined by size and mass of the animal. Greenland seals weighing up to 100-200 kg, consume 5-10 kg of fish daily. The Greenland whale with a body length 14 m and mass of 50 tons can consume during the summer feeding season about 24 tons of copepods (Sokolov end Arsenjev 1994). To eat only this biomass of Calanus, the whale must filter 30-40 million m^3 of water. The daily absorption by Minke whale is approximately 3-4 % of body weight.

Marine mammals of the northern seas are divided by their food composition into planktivorous, piscivorous and benthivorous. The planktivorous group includes whalebone whales - Greenland, blue, finners, sei whales and others. These animals, filtering the pelagic waters, eat mainly crustaceans - copepods, gammarids, hyperiids, euphausiids (Calanus finmarchicus, Temora longicornis, Thysanoessa raschi, Calanus hyperboreus, Parathemisto libellula, Gammarus zaddachi and others).

The ^{137}Cs concentration in these crustaceans is from 1 to 5 Bq / kg that is far less than the ^{90}Sr concentration. In cases when the quantity of plankton is insufficient, whalebone whales feed on capelin, herring and other small fish and also on squid. Other species of plankton invertebrates and fish are consumed, depending on the area of migrations. So, for example, finner whales migrate in the Barents Sea, near the Murmansk coast and in the north-eastern part of the sea reaching the

Novaya Zemlya coast. It passes to the Kara Sea rarely but has been observed in the Voronka and Gorlo areas of the White Sea, and in the southern part of the Baltic Sea (Sokolov end Arsenjev 1994).

The group of piscivorous whales includes many pinnipeds (Greenland seal, grey seal and others) and cetaceans (killer whale, Minke whale, partly sperm whale). Their main food consists of shoal fish species, crustaceans and, to a small degree, Cephalopoda molluscs. The composition of prey items for sea mammals differs depending on the area of inhabitance and season of year. Herring, mackerel, and capelin, all rich in calories, are preferred. These fish accumulate little ^{137}Cs (0.4-2.2 Bq / kg) and ^{90}Sr (0.1-1.6 Bq / kg). The spectrum of feeding, for example, of the Greenland seal includes 53 fish and 54 invertebrate species, i.e. seals are, as a matter of fact, omnivorous. A similar picture was determined in the course of study of little pike whales feeding in the waters near the Westerolen, Lofoten, Bear, Spitsbergen Islands and Greenland (Sokolov end Arsenjev, 1994). Their food consisted of fish (to 51 %) and crustaceans (31 %).

The small group of benthivorous species includes Atlantic walrus and sea hare. Walrus inhabits the coastal waters of polar seas and feeds mainly (to 70-80 %) on molluscs. For example, in the Franz-Josef Land coastal waters, walrus feeds on 20 mollusc species (*Hiatella arctica*, *Macoma calcarea* and others). These large animals (mass to 1-2 tons) need up to 30-50 kg of food daily. Such molluscs usually contain very few ^{137}Cs (0.2-1.5 Bq / kg), ^{90}Sr (0.01-0.3 Bq / kg) and ^{239,240}Pu (0.001-0.02 Bq / kg).

The data on radioactivity of little pike whales (48 grown-up animals) were obtained in 1988 and 1992 (Sickel et al. 1992; Gordon et al. 1999). The whales of 5-8 m length were hunted in the coastal waters of Northern Norway (1988) and to the north from the Kanin and Kola Peninsulas (1992). The cesium activity was determined in samples of muscle tissue, strontium - in bones, and plutonium - in liver of Minke whales. No other information can be found in the literature.

^{137}Cs in organs and tissues of sea mammals ranges from 0.2 to 6 Bq / kg. The ^{137}Cs concentration in muscle tissues of grey seals caught at the end of the 1980s in the North Sea ranged from 6.4 to 27.5 Bq / kg (Andersen et al., 19900. The radiocesium level in seal milk was still lower (to 3 Bq / kg). The quantity of radionuclides was close to the same in fish consumed by grey seals.

The content of radionuclides in marine mammals remain poorly studied. Muscle tissue was taken from Greenland seal pups collected from the East Murman coast by MMBI specialists in summer 1993. In March 1995 samples of muscle tissue, liver, brain, spleen, kidneys and bones were taken from two pups of Greenland seal from the southeastern coast of the Kola Peninsula (Fig. 6.12; Table 6.16). In 1996 the organs and tissues of seven newborn Greenland seals in the Zolotitsa area of the White Sea were investigated (Fig. 6.13) (Matishov et al. 1996e; Rissaneen et al. 1997a,b, 1998). These pups only fed on mother's milk before the sampling.

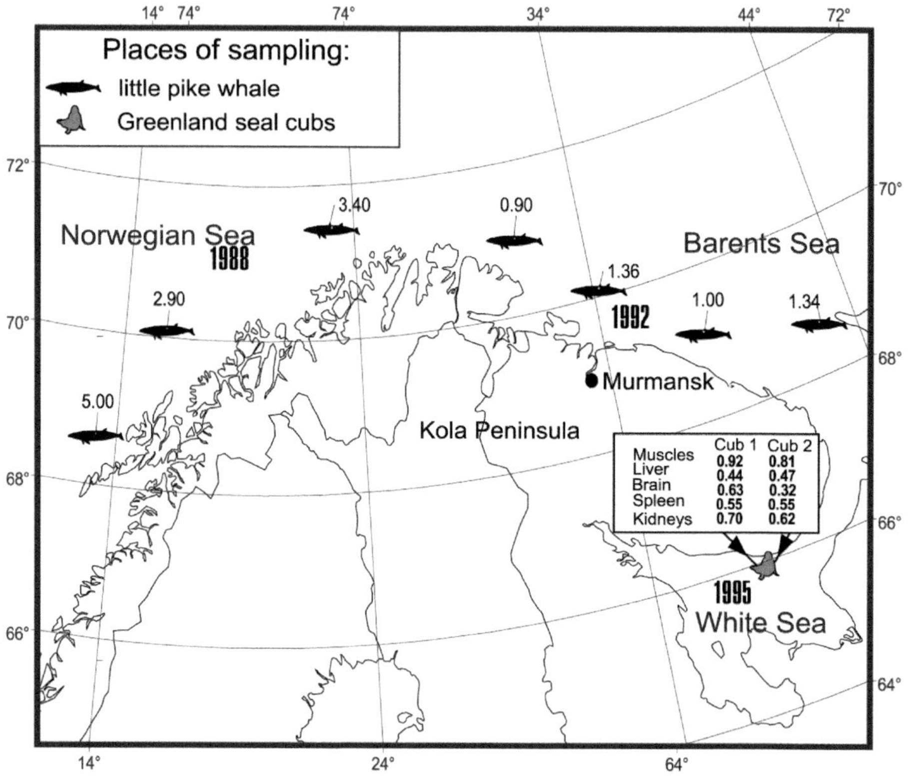

Fig. 6.12. ^{137}Cs content in muscle tissues of little pike whale (1988, 1992, by: Gordon et al. 1999) and in the tissues of Greenland seal cubs (1995, MMBI data), Bq / kg wet weight

Table 6.16. ^{137}Cs content in organs and tissues of Greenland seal pups in 1995, Bk/kg wet weight

Organ, tissues	Cub 1 66°06'N 39°00'E	Cub 2 66°05'N 38°51'E
Liver	0.44	0.47
Muscles	0.92	0.81
Brain	0.63	0.32
Spleen	0.55	0.55
Kidneys	0.70	0.62
Bones	Not determined	Not determined

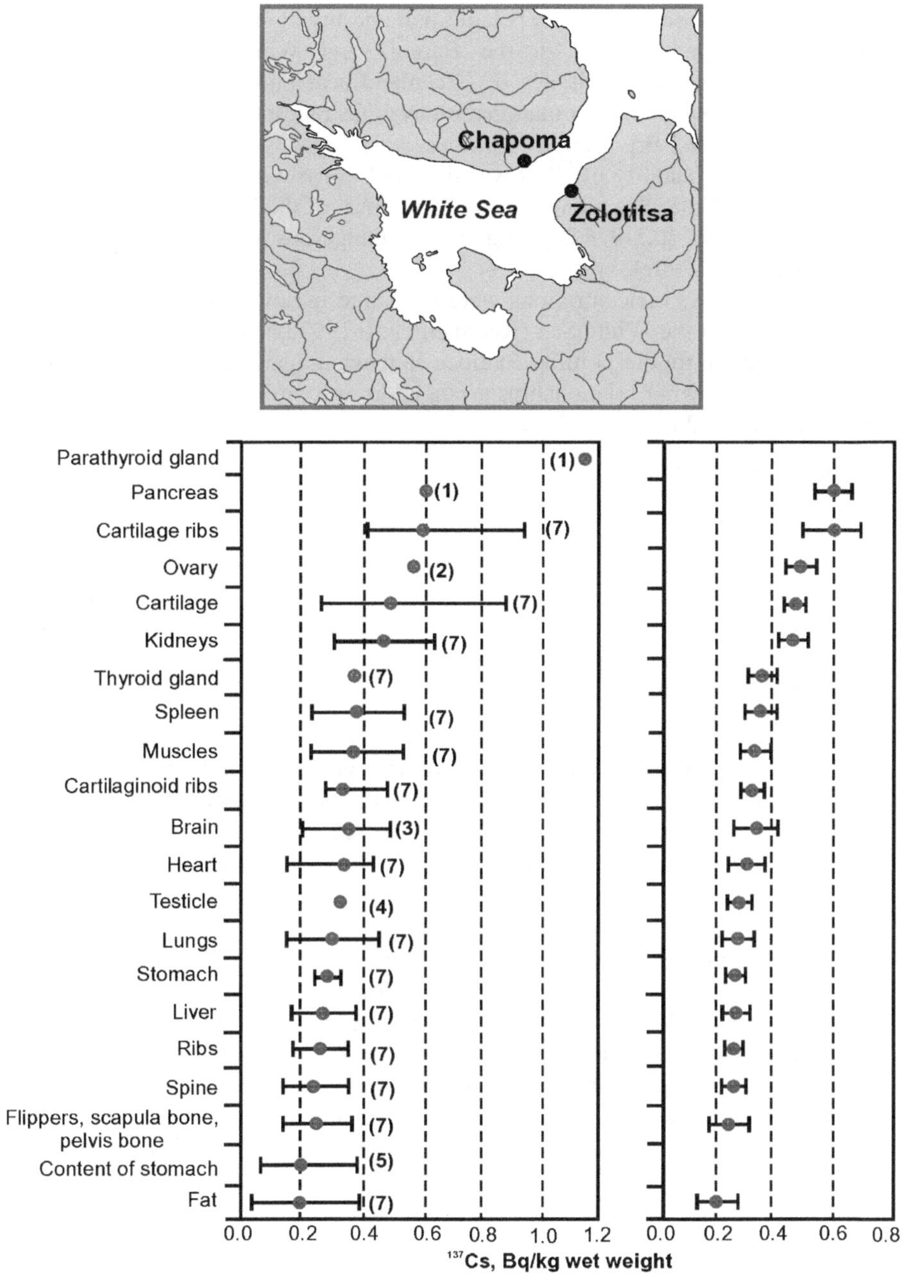

Fig. 6.13. ^{137}Cs concentration in different organs and tissues of Greenland seal cubs in the White Sea

The Greenland seal, feeds on pelagic fish and invertebrates, is one of the most prevalent pinniped species in the Barents and White Seas. The gamma-spectrometric analysis of tissues of Greenland seal pups (1993) have shown a comparatively low level of contamination by ^{134}Cs radionuclide (2.84 Bq / kg) and almost no ^{137}Cs (0.2 Bq / kg).

A very small quantity of ^{137}Cs was detected in two Greenland seal pups taken in 1995 in the White Sea along the southeast coast of the Kola Peninsula. The low radiocesium level (0.3-0.9 Bq / kg) is not dangerous for their health and life (Matishov et al. 1996e).

Still lower ^{137}Cs concentrations were measured in newborn seal pups from the Zolotitsa area on the White Sea (Rissanen et al. 1997a,b, 1999b). These newborn pups fed only on mother's milk. Radiocesium accumulation in organs and tissues exhibits the following relationships arranged in order of decreasing concentration (Bq / kg wet weight):

Cartilage ribs and pancreas	0.6-0.95
Ovaries	0.56
Cartilage and kidneys	0.47-0.49
Heart, muscles, brain, gall bladder, thyroid gland, cartilage coccyx	0,34-0.37
Liver, lungs, stomach and testicles	0.29-0.31
Consolidated samples of bones (limbs, shoulder, pelvis), vertebral column and ribs	0.24-0.26
Fat	0.20

The concentration in the fat component of stomach varies from 0.07-0.21 Bq / kg. The higher level is registered in samples with blood traces.

The ^{137}Cs content in tissues of White Sea seal pups (0.36±0.12 Bq / kg) is similar to the concentration (0.4±0.12) Bq / kg determined in seals from the Greenland area in 1990-1994 (Aarkrog 1997a,b). In 1960-1964 the ^{137}Cs level was 5 times higher - (2.2±0.13 Bq / kg).

A similar radioecological picture is characteristic for the pinnipeds of the coastal zone of Antarctica. The ^{137}Cs concentration in bodies of sea leopard (Hydrurda leptonyx), Weddell seal and other species varies from 0.2 to 2.5 Bq / kg dry weight (Table 6.17) (Ross et al., 1993). The formation of such levels of radioactive contamination, undoubtedly, is related to global fallout.

Baltic seals, hunted in the Bothnia Bay, contained up to 50 Bq / kg of ^{137}Cs (Rissanen et al. 1999b). In addition, ^{99}Tc, connected with discharges to the sea from Sellafield during the second half of the1990s, was also detected in the seals.

About 3-6 Bq / kg of ^{137}Cs was found in the muscles of little pike whale, hunted in the Norwegian Sea in 1988. The level of this nuclide in little pike whales on the south of the Barents Sea in 1992 was noticeably lower (0.9-1.4 Bq / kg) (Gordon et al., 1999). The increasing trend of radioactivity levels in Minke whales toward west European sources of contamination is quite evident (Fig. 6.12; Table 6.18). At the end of the 1980s and beginning of the 1990s, marine waters of the Norwegian Sea pelagic zone (the Lofoten Islands area) contained ^{137}Cs from 10-20 and ^{90}Sr - 4-12 Bq / m^3. As already noted, in the course of feeding, whales filter large volumes of seawater containing this level of radioactive contamination.

Table 6.17. ^{137}Cs, ^{210}Po and ^{210}Pb content in pinnipeds of the Antarctic coastal zone (by: Ross et al. 1993), Bq/kg dry weight

Species	Organ	^{137}Cs	^{210}Po	^{210}Pb
Sea leopard	Muscles	N.D.	5.0±0.3	0.2±0.3
	Liver	N.D.	46±1	2.6±0.6
	Kidney	08.±0.5	72±2	N.D.
Crabeater seal	Muscles	2.5±0.8	4.1±0.6	0.1±0.1
	Liver	0.3±0.1	90±13	8±3
	Kidney	0.2±0.1	52±11	0.7±0.3
Weddell seal	Muscles	N.D.	16±8	0.7±0.1
	Liver	0.9±0.4	195±90	1.4±0.7
	Kidney	N.D.	150±53	1.1±0.1

Table 6.18. Concentration of radionuclides in little pike whale, Bq/kg wet weight

Area of investigation	Year of sampling	^{137}Cs (muscles)	^{90}Sr (bones)	$^{239,240}Pu$ (liver)
Norwegian Sea	1988	4.4±1.7 (29)	5.7±1.0 (5)	0.010–0.008 (5)
	1992	3.4±2.5 (4)	–	–
Barents Sea	1992	1.2±0.3 (17)	0.9±0.5 (5)	0.004±0.001 (4)

^{90}Sr, being absorbed like calcium in nature, concentrates mainly in the bones of seals and whales (Rissanen et al., 1997a; Gordon et al., 19990. So, for example, ^{90}Sr accumulations in White Sea newborn seals were, Bq / kg wet weight: 0.18 - in ribs; 0.14 - in vertebrate column; 0.10 - in other bones and 0.03 in cartilage ribs. In the area of Chapoma settlement the concentration was less than 0.01 Bq / kg in the muscle of one seal. The interpolated ^{90}Sr content for the period 1990-1994 in Greenland waters (Aarkrog 1997a,b) was 0.0013 Bq / kg wet weight of seal muscles.

During the 1960s in Knipovich Polar Research Institute of Marine Fisheries and Oceanography the ^{90}Sr content was determined in seal bones (Crested seal and Greenland seal) hunted in the Greenland Sea (Petrov end Fedorov 1967). Depending on the feeding pattern and age (years of birth 1946-1953), the quantity of ^{90}Sr in the bones varied from 7.4 to 22.2 Bq / kg and in crested seals - from 1.85 to 3.7 Bq / kg dry weight. Evidently, the feeding of sea mammals by large amounts of plankton crustaceans and fish contributes to the high accumulation of radioactive substances.

The ^{90}Sr accumulation in little pike whales occurs mainly in bones. The analysis of bones of several Minke whales, hunted in 1988 in the Norwegian Sea, have shown levels to 6 Bq / kg. These values are relatively high.

Strontium is accumulated mainly in fish bone. Its level in bones (0.3-0.6 Bq / kg) is in some cases several times higher than in organs and tissues of fish. The maximum quantity of 90Sr in fish was noted in 1961-1962 during the period of the most intensive nuclear explosions. Cod, catfish, and haddock contained this radionuclide levels of 40-75 Bq / kg. The accumulation of radionuclides by fish depends to a certain degree on species. Lower concentrations of ^{90}Sr (0.9±0.5 Bq / kg) were found in Barents Sea whales in 1992.

^{239,240}*Pu* in sea mammals has been documented in several cases. In grey seals from the White Sea this nuclide was found to be at very low concentration levels (0.001-0.05 Bq / kg) (Andersen et al., 1990). The ^{239,240}Pu concentration in White Sea newborn seals from Zolotitsa settlement was below the detection level (Rissanen et al., 1997a). These values were of the following order, Bq / kg wet weight:

bones	< 0.034
ribs and cartilage ribs	< 0.008
vertebrate column	< 0.005
muscle tissue, liver and gall bladder	< 0.003
kidney and lungs	< 0.002

The level of plutonium in leg bones of newborn seals from the Chapoma settlement area was less than 0.013 Bq / kg wet weight. ^{239,240}Pu in the liver of Minke whales is 0.004-0.01 Bq / kg.

In spite of the limited volume of information, it is clear that ^{137}Cs accumulation takes place mainly in muscle tissues of whales and seals. Its level varies from 0.2 to 6 Bq / kg. In some cases, when the local sources of contamination exist in habitats, ^{137}Cs accumulation can reach 20-50 Bq / kg. ^{90}Sr is more noticeable in bones of newborn Greenland seals and Minke whales. The strontium level in White Sea seal pups feeding on the mother's milk is 0.2 Bq / kg. In whale bones it can reach 1-6 Bq / kg.

Generally speaking, sea mammals contain very low concentrations of artificial radionuclides. It is characteristic that the radionuclide concentration in fish, plankton and other prey items differ little from the concentrations in seals and whales. The observed activity is several orders of magnitude lower than the level which could arouse anxiety from the viewpoint of the influence of radiation on humans.

Toothed and whalebone whales and also pinnipeds, being the main consumers of pelagic crustaceans and fish, top the trophic food web in the ecosystems of northern seas. The role of sea mammals in the biogeochemical cycles of matter is rather great, taking into account their number and total biomass. Whalebone whales, which filter huge volumes of sea water in the process of separation of small crustaceans, are of special importance.

Thus, during the period of time from 1960-1990s, the level of radioactivity in the Arctic marine ecosystem has decreased by an order of magnitude owing to the processes of self-purification of the basins, natural decay of ^{137}Cs and ^{90}Sr, and reductions in discharges of artificial radionuclides. No changes of species, population and trophic structure of marine biota, caused by the action of low (background) radioactivity levels, were detected. The contemporary concentrations of artificial radionuclides in marketable bioresources are not dangerous for humans.

7 Transfer and assimilation of radionuclides in marine ecosystems

7.1 Natural purification of marine systems

7.1.1 Conditions of radionuclides assimilation

How rates of biological and physico-chemical factors influence self-purification processes of marine ecosystems is a key aspect in understanding the problem of Arctic Ocean and North Atlantic anthropogenic radioactivity. This complex radioecological problem has previously been outlined from a theoretical standpoint by many radioecologists (Vernadsky 1926, 1929; Peredelsky 1957; Shvedov end Patin 1968; Kuznetsov 1971; Gromov end Spitsyn 1975; Alexakhin 1982; Gromov et al. 1985; Izrael end Tsyban 1989; Golubchikov 1993; Braginsky et al. 1994; Kuzmenko 1998). As a result, a conceptual model of assimilation capacity exists today. The model is used to describe the main processes and to determine quantitative estimates for the output and transformation of contaminants in the marine environment (Izrael end Tsyban 1989; Izrael 1998). According to this approach, every marine ecosystem contains a determinable volume of material (e.g. seawater, sediment, biota) and, therefore, the assimilation capacity of each ecosystem is also finite.

The assimilation capacity of marine ecosystems is defined as the maximum amount of pollutants that can be transformed or transferred outside the given basins without the loss of ecosystem function (Izrael end Tsyban, 1989; Braginsky et al. 1994). Based on these concepts, one can derive theoretical estimates for the development of the radioecological situation for example, in the semi-enclosed Novaya Zemlya basin Chernaya Bay after the nuclear explosion.

Throughout the entire nuclear era, biological communities of the northern seas have not been subjected to high radiation loads. The functioning of a broad spectrum of radiological and oceanographic processes has lead to a high degree of self-purification for the northern seas.

Information on the composition and levels of ^{137}Cs, ^{90}Sr, ^{239,240}Pu and other radionuclides and also available information on radioactive discharge volumes from major sources allows us to formulate the following principles. During the 1950-1990s, the role and input of direct discharges from atmospheric and underwater atomic explosions decreased over time. During these decades the ^{137}Cs and ^{90}Sr content of the Arctic and Sub-Arctic seawater decreased correspondingly from 50-400 to 2-20 Bq / m^3 and from 10-100 to 1-7 Bq / m^3. The concentration of these radionuclides in the organs and tissues of marine organisms has decreased as well (Matishov et al. 1994a,b,c, 1995a, 1996b,f,g, 1997b,c,f, 1999d). The maximum levels of radioactive products, including 90Sr, are detected in the vicinity of atomic nuclear test sites (for example, in the waters of Chernaya Bay (up to 140 Bq / m^3 for ^{90}Sr).

At the end of the 1980s the influence on the physical environment and biota of radioactivity originating from Western Europe and Siberia, on the one hand, and global fallout, on the other hand, are similar in magnitude within the northern seas where the levels of cesium and strontium reach background values. The relative input and role of radioactive sources in the Arctic Seas during the 1990s began to change as a result of the penetration far northward and eastward (to the Laptev Sea) of Chernobyl fallout radionuclides.

The fallout from the Chernobyl accident was transported to the Barents-Kara Basin both as a result of the advection of relatively contaminated Baltic seawater and because of wash-off from the vast catchment basins of the northern rivers. The ratio between ^{134}Cs and ^{137}Cs isotopes in surface waters of the Skagerrak and Kattegat Straits varied in November 1996, 1997 and in 1998 within the limits 0.023, 0.017 and 0.013 (Brungot et al. 1999a,b). These ratio values were 0.009-0.014 in the coastal waters to the south-east of Norway in 1996-1997, indicating the presence of Baltic pollutants of Chernobyl origin,

At the end of the 1990s global radioactive fallout was the main source of ^{137}Cs. Also at this time the contribution of global fallout to the total radioactivity of marine ecosystems began to diminish. Radionuclide levels diminished over time through radioactive decay. As a result the radioecological importance of various nuclides in the dose loads to organisms also changed. In the coming decades the primary danger for sea flora and fauna will be caused by exposure to radiation in small doses. Over time, long-lived radionuclides such as $^{239,240}Pu$ will become more important as a component of radiological doses to marine biota. At the same time the contemporary presence of short-lived radionuclides such as ^{60}Co and ^{134}Cs requires monitoring in the coastal zone to remain knowledgeable of impacts to local and regional sources.

During the period between 1980-1990s the $^{239,240}Pu$ concentration in seawater, macrophytes, bottom sediments, and the organs, tissues and bones of fish was low, often below instrument detection limits (Matishov et al. 1996f, 1997a,b,d,e,f, 1998b,c, 1999c,d,e,g,h,i; Matishov D. et al. 1996). It can be noted that $^{239,240}Pu$ levels in seawater, sediment and fish varied correspondingly from 2 to 70 mBq / m^3 (up to 0.3 Bq / m^3 in some locations), from 0.1 to 6 Bq / kg (in some bays to 10-20 Bq / kg) and from 0.001 to 0.05 Bq / kg. Such concentrations reflected mainly global radioactive fallout sources. Considering the main contamination sources, the input to bottom sediments of α-emitting radionuclides from the European processing plants relative to global fallout is 20-55 % (Stepanov et al. 1999).

The analysis of the ^{238}Pu:$^{239,240}Pu$ ratio in bottom sediments of Arctic shelves has shown that this ratio varied from 0.009 to 0.065, 0.034 on average. The comparison of these ratios with published data on known sources of plutonium indicates that there is little input of $^{239,240}Pu$ from Western European discharges and sunken atomic reactors in the Barents-Kara region.

Owing to the ecosystem characteristics of the polar basins, the minimal influence of the existing radiation levels on seawater and biota in the Barents-Kara region is noted. Taking into account the relatively low levels of ^{137}Cs, ^{90}Sr

and $^{239,240}Pu$ in the environment and biota of the Arctic and Sub-Arctic seas during the 1950-1990s, one can suppose that self-purification processes in the basins is rather significant (Matishov et al. 1994c, 1995a,b, 1996g, 1997b,f, 1998b,c, 1999d, 2000b). The major radionuclides remain in the marine ecosystem over long timescales (from tens to hundreds of years and even more). Therefore long-lived radionuclides can display their chemical properties to a greater extent than short-lived elements (^{134}Cs, ^{60}Co) derived from local fallout sources. All of the mentioned radionuclides are not only subject to dilution in the sea environment, but also are involved in the general biological cycle of matter, including the transfer of radionuclides through trophic food webs (Fig. 7.1).

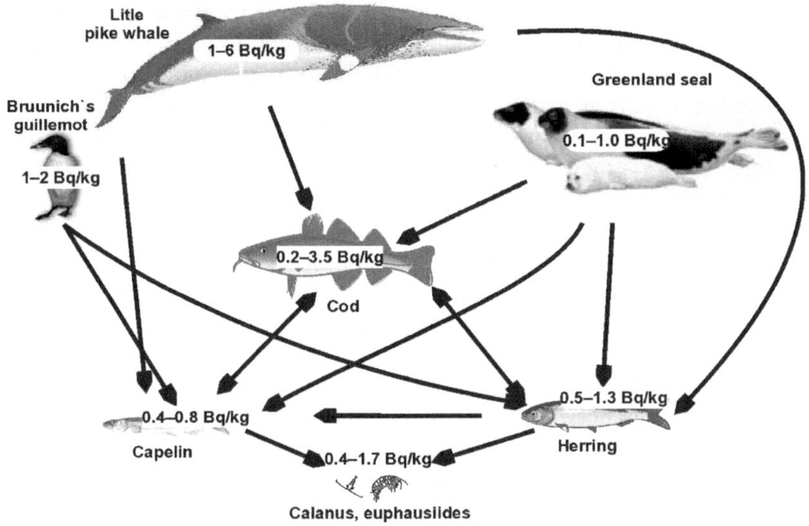

Fig. 7.1. ^{137}Cs transfer by the trophic food web of the Barents Sea

The analysis of the radiological situation during the period of nuclear tests and after the Chernobyl accident shows that the marine environment is less contaminated by radioactive substances than freshwater and land environments. The Barents Sea fish collected during the Novaya Zemlya atmospheric tests had radioactivity levels not more than 80 Bq / kg wet weight; the Chernobyl contamination didn't impact the Barents Sea fishery grounds. At the same time, owing to radioactive Chernobyl fallout, the average values of ^{134}Cs and ^{137}Cs in Arctic char and sea trout from high latitude lakes exceeded 1600 Bq / kg. Radiocesium concentrations in reindeer muscle varied from 10000 to 50000 Bq / kg wet weight. The modern ^{137}Cs content in lichens from the Kola coast presently reaches 800 Bq / kg, though levels do not exceed several Becquerels per kilogram in Phaeophyceae algae from the adjacent littoral zone.

Low concentrations of artificial radionuclides in the oceans and coastal seas reflect the great assimilation capacity of marine systems as compared to freshwater systems. Indeed, the large volumes of these water bodies contribute to the effective dilution of radionuclides to their observed low concentrations.

The marine ecosystems express their buffer capacities through the interaction of numerous oceanographic processes. Ecological and physico-chemical conditions are leading controllers on the processes of radionuclide migration, accumulation and redistribution in the oceans. The variety of these processes leads to significant variability of radionuclide accumulation coefficients among different environmenal compartments and biota (Polikarpov 1964a,b; 1971; Gromov end Spitsyn 1975; Gromov et al. 1985; Polikarpov end Egorov 1986; Nikitin et al. 1991; Kuzmenko et al. 1995; Duursma and Carroll, 1996; Carroll et al. 1997; Kuzmenko 1998; Chebotina end Kulikov 1998; Carroll et al. 1999).

The fate of radioactive admixtures received by the Ocean depends also on the forms of their introduction to the marine environment. In particular, global fallout of ^{90}Sr and ^{137}Cs from aerosols occurs mainly as water-soluble forms (Kuznetsov 1971; Izrael end Tsyban 1989). Radioactive elements present as dissolved forms in seawater, neither form compound complexes, or colloids and exist in the form of simple ions. Some sea organisms are able to absorb and accumulate radionuclides even when the dissolved concentrations in seawater are extremely low. As a result, it makes sense that detailed investigations of the structure of marine ecosystems and related biogeochemical cycles is warranted. Efficient bioaccumulation of radionuclides is observed in certain marine species, including several species inhabiting the littoral zone, polar hydrological fronts and bird bazaars.

To understand the differentiation and distribution of anthropogenic radioactive substances that have been transported from the atmosphere or introduced directly to the sea, one must consider not only the transport pathways but also the life cycles of organisms inhabiting the affected areas. Biological and biogeochemical phenomena influence the distribution and transport of radionuclides by sea currents. The influence of the warm, saline waters of the North Atlantic Current system in the Barents Sea is traced by the position of the Arctic polar hydrological front which is characterized by rich bioproductive zones. Planktonic larvae of Atlantic cod, boreal Arctic foraminifera and other zooplankton species are carried periodically to the western part of the Barents Sea from the Norwegian Sea. The migration pathways of zooplankton change every year in accordance with climatic cycles.

Marine organisms play an enormous role in the energy and matter transfer within the Gulf Stream. Atlantic cod, herring, salmon, sea colonial birds, Atlantic walruses, seals and whales which contain artificial radionuclides in their bodies, migrate every year from Novaya Zemlya to Western Europe and Iceland after reproduction and spawning

At present we have only a general understanding of the patterns of radioactive particle deposition in relation to regional oceanographic circulation processes. In the Barents and Kara Seas the process of marine sedimentation of radioactive substances must accelerate sharply during the period of biological production in response to the growth of plankton species and the filtration of the surface ocean layer by organisms. Macrophytes and benthic organisms accumulate radionuclides by filtering in the coastal zone. Absorption of radionuclides in the open sea is accomplished by colonies of molluscs, mainly the Iceland scallop, and polychaetes.

The self-purification of marine systems is largely due to a few important factors. In the first place, the assimilation capacity is determined by the interaction of geomorphological, hydrodynamic, hydrochemical and biogeochemical factors. Through our own research and other published materials we will attempt to broaden and synthesize current knowledge on the factors of self-purification and radioecological capacity of the Arctic and Sub-Arctic marine ecosystems. The behavior of artificial radionuclides as chemical constituents in the marine environment is emphasized including their likely forms and transfer and assimilation mechanisms in the northern seas.

7.1.2 Geomorphological and hydrodynamic factors

Geomorphological factors, which include the elements of orography and bottom topography, determine the degree of exchange across ecosystem boundaries, the volume of water in the basin, the current circulation pattern and other features that influence the rate of removal of pollutants from the basin. For example, the border and transverse troughs of glacial shelves play the same role in the migration of radionuclides as river valleys on land. Both channel pollutants within the dominant direction of water flow.

Marine sedimentation is determined by a number of conditions: depth and bottom topography, velocity and direction of steady currents, ice conditions, drift of sea ice and icebergs, water salinity and stratification, river runoff and the general water balance, distance from the source of wash-down, and rates of sediment accumulation. Radionuclide migration depends to a large extent on the development of the thermocline and picnocline in the sea and also on the processes of biofiltration, bioturbidity, trophic structure and bioproductivity of the basin.

However, bottom relief is one of the most important elements controlling sedimentation on the shelf and in the open ocean. Geomorphology determines to a large extent the type of bottom sediments and hence rates of sediment accumulation. Anthropogenic radioactivity levels in bottom sediments of the Barents, Pechora and Kara Seas are low. The ^{137}Cs concentration in bottom sediments, depending on sediment type, varies from 0.1 to 20 Bq / kg dry weight. However within some bays and shelf depressions sediments contain from 30 to 100 Bq / kg of ^{137}Cs. The maximum levels of radionuclides are associated with sediments deposited near the Novaya Zemlya nuclear test sites. The radionuclide 60Co (2-6 Bq / kg) is detected in bottom sediments of Abrosimov and Stepovoy Bays and indicates that regional leakage has occurred of radionuclides from nuclear waste burial sites.

Hydrodynamic factors are responsible for the vertical and horizontal circulation of water masses and therefore play an important role in the fate of pollutants in the marine environment from the time they are first released into the sea. The self-purification capacity of a marine system depends on the cross-boundary water exchange, velocity of currents, turbulence and water temperature. The importance of river runoff is self evident given that river runoff results in a distinct density stratification of waters in the Arctic Ocean and its marginal seas. The influence of water transfer and mixing processes varies in the coastal zone, pelagic zone, deep-water and bottom areas. These processes control the velocity and direction of motion for small radioactive particles (μm size) as well as particle residence times in the active (to 200 m) surface layer of northern seas. Many regional features of hydrodynamic phenomena, influence the transport of ^{137}Cs and other dissolved radionuclides, as discussed in previous chapters.

Tidal currents are particularly effective in the Barents Sea fjords, on the shelves of British Islands, of the Gorlo part of the White Sea and of the North Sea (Matishov 1987). The direction of transport for resuspended sediments carrying radionuclides in the coastal zone will depend on the wave direction. Rip currents, associated with waves, also play an important controlling role in transporting resuspended matter. These currents are capable of removing to the open sea large masses of resuspended sand and silt containing sorbed radionuclides.

The mass transport and differentiation of radioactive substances deposited deep below the sediment-water interface can occur as a result of turbid flows, underwater slumping, flow or avalanche of bottom sediments and also as a result of bottom density currents. In all these cases anthropogenic radionuclides are displaced from shallow to deeper depths.

Hydrochemical factors play a major role in the self-purification process. The buffering capacity of seawater is provided by the presence of humic substances and fulvic acids, organic compounds, amino acids, carbon dioxide, phosphates, hydrocarbons and other compounds (Lisitsyn 1978). The factor of greatest importance for the buffer capacity of radionuclides is the content of potassium, sodium, magnesium, calcium, sulfate, carbonate (carbonate balance) ions, general mineral content of water, and its alkalinity or acidity. Radiocesium participates in the metabolism of marine biota in the same metabolic reactions as $K^+ + Na^+$; radiostrontium participates in the same reactions as calcium and magnesium.

As previously discussed, the minimum and maximum concentrations of cesium in Barents Sea bottom sediments vary. The small range of cesium concentrations in sediments can be explained by the fact that cesium in seawater, forms the simple ionic form Cs^+. It tends to remain in a dissolved form and to be only weakly sorbed by suspension and accumulated in bottom sediments (Gromov et al. 1985). The existence of cesium in water solution, mainly in the ionic state, determines its migration in the sea which almost exclusively depends on hydrological factors. Low concentrations of cesium in bottom sediments are characteristic for the Barents Sea because practically all of the cesium remains in seawater.

Water temperature and light penetration are important ecological factors in the marine environment. They determine the rate of physiological and biochemical processes in organisms, the solubility of salts, the stability of colloid forms of elements in solutions etc. As temperature increases, the rate of ion migration through membranes increases and, consequently, the passive exchange of dissolved elements in marine organisms becomes higher (Khlebovich 1974).

The alkaline and acidic conditions of the environment are an important physico-chemical feature of seawater and bottom sediments. They determine the state of hydrolyzing elements in solution and their sorptive properties. The pH values in seawater undergo seasonal and inter-annual variations within narrow limits around pH=8 (Hydrometeorology, ...; Matishov and Pavlova 1990).

The salinity is one of the most important abiotic environmental factors. It is relatively constant in the open sea (34-35 ‰) and undergoes substantial changes in the coastal zone, in estuaries and some other parts with variable hydrological conditions. The amplitude and duration of salinity variations depends on the amount of precipitation and continental runoff, ice melt, currents and a number of other conditions.

7.1.3 Influence of the marginal filter

The filtering capacity of marginal seas which we denote as the marginal filter results in the natural purification of northern seas from pollutants. One of the main functions of the marginal filter is performed within the littoral zone characterized by tidal currents and a rim of macrophyte algae. The dominant algae species (e.g. *fucus* and *laminaria* species) absorb dissolved radionuclides transported by coastal currents (Matishov et al. 1994b,c, 1995a, 1996f,g, 1997b,c,f, 1999d; Matishov D. 1994). Macrophyte stocks in the Barents Sea have an average biomass of 3-15 kg / m^2 or 5-40 tons per hectare.

As described earlier, during 1980-1990s these macrophytes contained ^{137}Cs, ^{90}Sr and ^{239,240}Pu at levels of 0.4-5.0, 0.3-3.0. and 0.02-0.3 Bq / kg respectively. Close to radioactive release points, for example, the «Atomflot» nuclear reprocessing facility, or as a result of the Chernobyl accident, ^{137}Cs concentrations are observed to increase dramatically for a short time to 0.3-5.0 Bq / kg. The efficient filtering capacity of macrophytes make macrophytes useful sentinels of radioactive leakage.

This powerful marginal filter exists from the sea- side of many estuaries, deltas, and bays, into which large rivers flow. Sedimentation rates in the deltas and estuarine zones of major rivers (e.g. Lena, Don, Yenisey, Ob, and Pechora) are many orders of magnitude higher than in the open sea (Safjanov 1987; Khrustalev 1989; Lisitsyn 1994). In particular, the sedimentation of colloidal matter and fine-grained clayey sediments takes place during mixing of fresh and seawater. Mixing results in the coagulation and mass sedimentation of fine suspended particles. During active sedimentation ^{137}Cs, ^{60}Co, ^{239,240}Pu accumulate in seabed sediments. ^{90}Sr also accumulates but to a lesser degree. ^{90}Sr is sufficiently stable in the water phase, so that the effective accumulation from seawater suspensions to bottom sediments is less efficient (Kuznetsov et al. 1994).

The self-purification process in the marginal filter zone is connected with natural sorbents such as iron oxyhydroxides, organic matter, clayey materials, silicic and carbonate remnants of organisms. All these elements contribute to the sedimentation of suspended material, including the finest suspensions and colloid particles, which are the main carriers in the water phase of such hydrolyzing radionuclides as $^{239,240}Pu$ (Kuznetsov et al. 1994). The finest clay particles which are present in suspension, are able to accumulate significant quantities of ^{137}Cs because of the stable fixation of ^{137}Cs in the interplanar intervals of clay minerals.

These processes lead to the formation of active sedimentation areas containing higher concentrations of $^{239,240}Pu$ and ^{137}Cs in these regions. The direct sorption of radionuclides to bottom sediments also serves as a filtering mechanism for the removal of radionuclides from estuarine waters. As a result, the highest levels of ^{137}Cs, to 50-100 Bq / kg, are observed for seabed sediments of the Taganrog, Dvina and Yenisey Bays. Plutonium releases from nuclear reprocessing plants to the basins of the Yenisey and Ob Rivers are fixed to silty particles in the estuaries resulting in the accumulation of $^{239,240}Pu$ (to 2-6 Bq / kg), ^{60}Co (to 1-6 Bq / kg), ^{90}Sr (to 10-16 Bq / kg).

Dissolved radionuclides in river freshwater also undergo the process of biological sorption. The accumulation of radionuclides to river bottom sediments is also facilitated by biofilters. Coastal vegetation, from green vesicular algae to macrophytes, actively remove radionuclides from seawater through filtering. After death, radionuclides that previously accumulated in plant tissues, are transferred to riverbed sediments.

Bottom sediments in marginal filter zones have many times higher radioactive contamination than shelf sediments. The assimilation and sorptive capacities of the Barents, White and Kara Seas serve to efficiently remove large quantities of radioactive isotopes transferred to coastal margins from rivers. Similarly, the biogeogeochemical barrier zone of the Arctic hydrological front also acts as a kind of pelagic marginal filter. The frontal zones are especially well-developed in the shelf troughs where warm Atlantic waters are dispersed. In the polar and subpolar frontal zones, in particular, near the ice edge and icebergs, intensive chemical and biological processes controlling biological productivity persist during the long polar day. Phytoplankton and zooplankton organisms absorb dissolved constituents including contaminants, directly from seawater. Radionuclides are largely introduced at the sea surface either from the atmosphere or as a result of radioactive waste dumping. As a result, biofiltration is of great importance for the removal and transfer of suspended and dissolved radionuclides to bottom sediments.

7.1.4 Significance of biological composition

The biological filtration processes resulting in the natural purification properties of marine ecosystems includes photosynthesis, microbiological decomposition, biofiltration and bioaccumulation. Photosynthesis is accompanied by the oxidation of organic matter which leads to a decrease in the contamination of basins (Polikarpov 1971; Izrael end Tsyban, 1989; Braginsky et al. 1994). Anthropogenic radionuclides together with iron, manganese, organic matter and phosphorus are absorbed by plankton organisms into the chemical composition of their protoplasm. Quantitative measures of the importance of these bioassimilation processes are the primary production and biomass of phytoplankton. After the death of plankton organisms the organic matter and remnants of hard parts (chitin, silica, carbonates) are selective sorbents for a number of heavy metals and radionuclides (Lisitsyn 1994).

Plankton organisms convert dissolved substances into suspension mainly during the period of spring biological production. The most intensive absorption of radionuclides from the pelagic zone takes place in areas of phytoplankton growth, where biomass reaches 5-10 g or more (wet weight) per 1 m^3 of seawater. Bacteria-plankton and bacteria-benthos decompose and mineralize organic matter to simple precursor compounds (carbon dioxide, ammonia and others). The nuclides, fixed in tissues and skeletons, either return to the dissolved phase or descend to the bottom and accumulate in sediments (Fig. 7.2). Consequently, the quantitative assessment of this process is important to the analysis of the assimilation capacity of marine ecosystems.

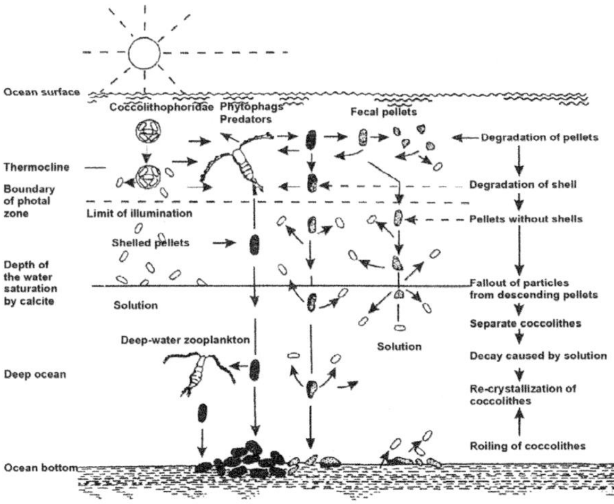

Fig. 7.2. Formation of zooplankton fecal pellets, contributing to the transfer of radionuclides to the sea bottom

Biofiltration and bioaccumulation processes contribute actively to the removal of radionuclides from seawater. Zooplankton and zoobenthic organisms and also many colonial birds and sea mammals, filter seawater containing contaminants resulting in the incorporation of contaminants into the biotic cycle and transfer through trophic chains (Fig. 7.3). Bioaccumulation is defined as the accumulation of dissolved radionuclides by macrophytes or other sea organisms from seawater.

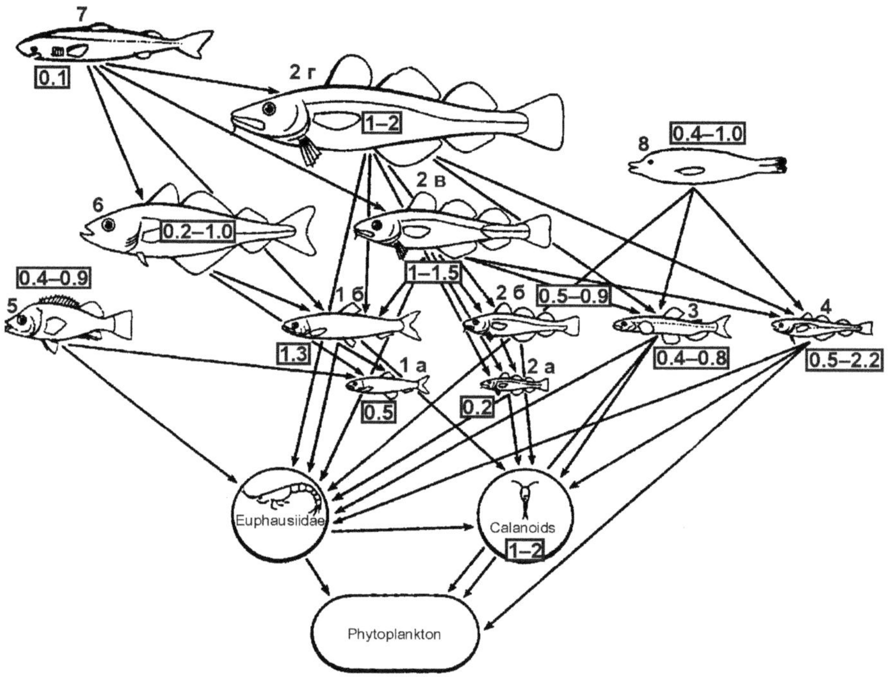

Fig. 7.3. Probable scheme of ^{137}Cs migration and accumulation (Bq / kg wet weight) over the trophic food web of the Barents Sea planctivorous and carnivorous fish species (1990s): **1a, 1b** - *little and big herring*; **2a, 2b, 2c, 2d** - *little and big cod*; **3** - *capelin*; **4** - *polar cod*; **5** - *sea perch*; **6** - *coalfish*; **7** - *Greenland shark*; **8** - *seal*

The biological transfer pathway is through metabolism, growth and death of different groups of marine organisms. At the same time, in spite of the high metabolism level in colonial birds which feed mainly on pelagic fish, there is no noticeable accumulation of radionuclides in these organisms.

In general, the biota of the northern seas is characterized by relatively small concentrations of radioactive contamination. ^{137}Cs, for example, is observed in organisms within the narrow limits 0.1-4.0 Bq / kg. However, some species of molluscs, polychaetes, and brown algae are distinguished by high sorption of ^{60}Co and ^{137}Cs. On the other hand, it is also noted that marine invertebrates accumulate ^{90}Sr in relatively small quantities (from 0.01 to 2 Bq / kg wet weight).

The migration of radionuclides across marine food webs is described using bioaccumulation coefficients (Polikarpov 1964a,b, 1971; Iljenko 1974; Krivolutsky 1983, 1999; Izrael end Tsyban 1989; Merezhko et al. 1996; Pavlyutin end Babitskaya, 1996; Kuzmenko 1998; Chebotina end Kulikov 1998). These coefficients are the ratio between the concentration of a radionuclide in an organism and the concentration in seawater, bottom sediments or the previous level of the trophic structure.

The concentration of radionuclides in seawater is the key factor determining accumulation in plankton, macrophyte and nekton organisms, and in bottom sediments for the benthos and bottom fish. Investigations have shown that the accumulation of ^{137}Cs and ^{90}Sr in marine organisms is directly proportional to the density of contamination in the surrounding environment (Fig. 7.4).

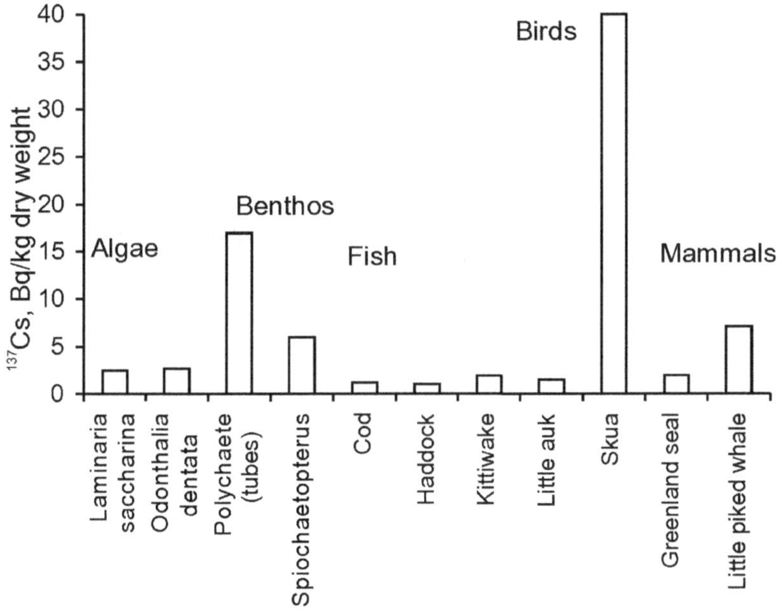

Fig. 7.4. ^{137}Cs distribution for the biotic component of the marine ecosystem

7.1.5 Salinity and radionuclide transfer

The radioecological significance of salinity to the functioning of marine ecosystems is quite obvious, but relationships of radionuclide sorption to salinity is poorly investigated. Salinity clearly influences the adsorption capacity of clayey particles. The main sedimentation mechanism for particles of small size is the salinity flocculation which is typically most intensive when seawater salinities reach a value of 10-12 ‰ (Lisitsyn 1994).

The dependence of artificial radionuclide concentration on salinity (Fig. 7.5, 7.6) and on the volume of terrigenous suspended matter is also quite obvious. The properties of seawater, salinity, and water-salt exchange processes are of a primary importance in the functioning of the shelf and oceanic ecosystems. Considering the migration and transformation of artificial radionuclides, the concentration of radionuclides decreases in sea basins much more rapidly than in lakes and rivers (Fig. 7.7). Also, the high mineral content of seawater contributes to a lower accumulation of artificial radionuclides by marine biota (Chebotina and Kulikov 1998).

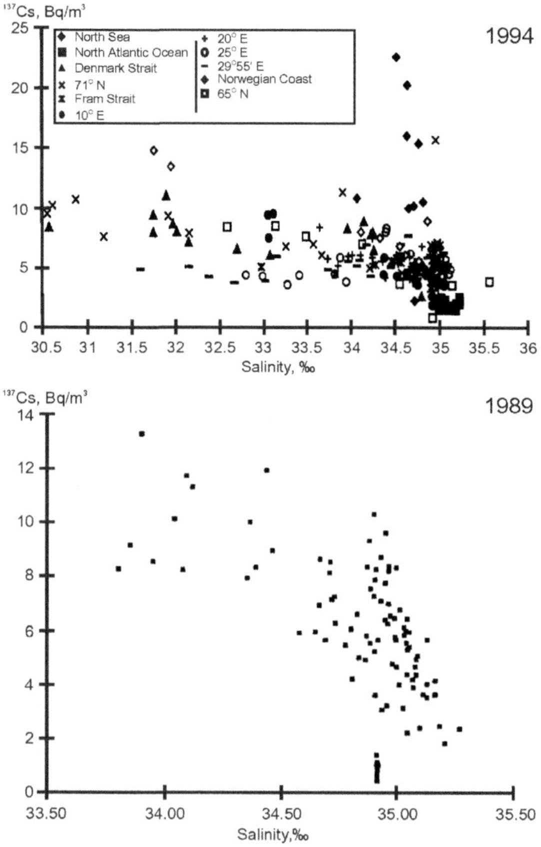

Fig. 7.5. General trend of higher ^{137}Cs concentrations in the northern seas with decreasing salinity and temperature (Kershaw 1997)

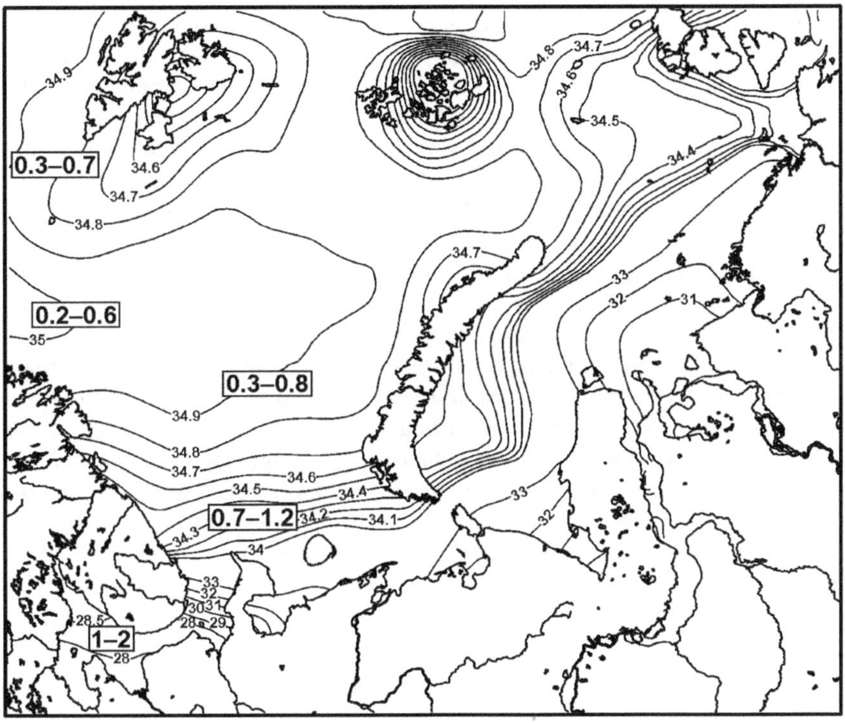

Fig. 7.6. Variations of ^{137}Cs concentration in cod fish in relation to salinity, Bq/kg wet weight

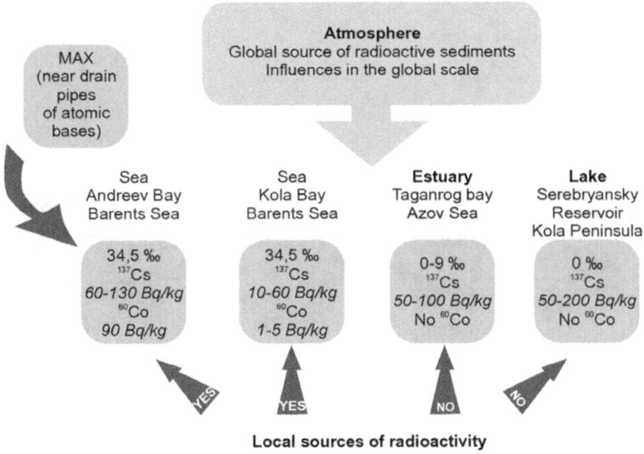

Fig. 7.7. Comparative accumulations of ^{137}Cs and ^{60}Co in basins with different salinity regimes

As a result, the transfer of artificial radionuclides through the food web is inefficient. This is particularly true for the biota of open areas of the Norwegian-Greenland Basin and the Barents Sea. One explanation for this trend is dilution by large water masses coming from the North Atlantic and also that competition for adsorption sites between salts and stable isotopes of cesium and strontium and radionuclides. In the Arctic, the lowest cesium and strontium concentrations are observed in marine organisms and humans, living mainly on marine food sources (Strand 1998).

Concentrations of radionuclides in fish, seals and whales in the Barents, Greenland and other polar seas from the beginning of 1990s are very low. This is in contrast to ^{137}Cs and ^{90}Sr concentrations in reindeer and caribou. It is thought that salts in seawater hinder the absorption of nuclides by marine organisms (Matishov et al. 1994c, e, 1997; Strand 1998). Seawater differs from freshwater in the ratios among different ions. Significant freshening of seawater as well as the salinization of freshwater will simultaneously change the osmotic and salinity conditions that organisms living in these transition zones are exposed to. The relationship among ions in seawater is constant only at salinities higher than about 5-8 ‰. At lower salinities, the proportion of bivalent cations in seawater increases significantly (Khlebovich 1974; Berger 1986). The work of osmotic and ionic regulation mechanisms is affected as well (Fig. 7.8).

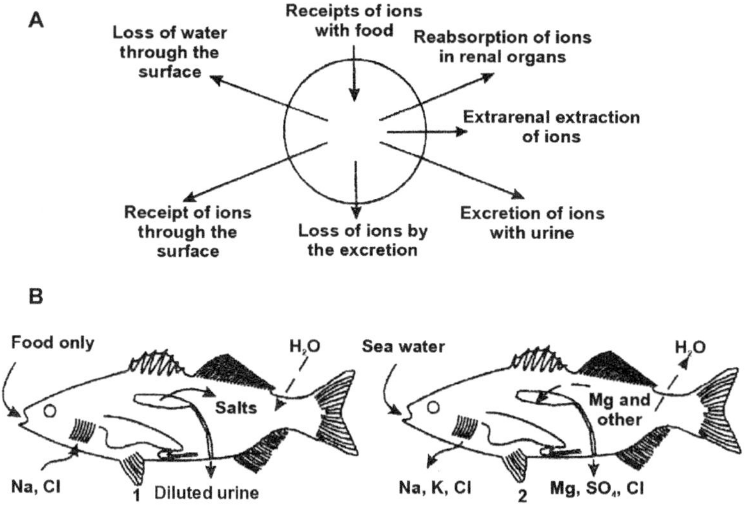

Fig. 7.8. (**A**) Direction of processes of osmotic regulation in marine invertebrates. (**B**) Main pathways of ions and water movement during the process of osmotic regulation in *freshwater* (**1**) and *sea fish* (**2**). In (**B**) the solid arrows represent active movement and dotted arrows represent passive movement (by: Berger 1986)

In general, the salinity range 5-8 ‰ demarcates the extreme boundary for the distribution of marine (>5-8 ‰) and freshwater (<5-8 ‰) organisms. This ecological barrier is named «critical salinity». It is important to note that the isolating reaction of «critical salinity» influences biological distributions differently in different seas: the Azov and Black, Caspian, White, Baltic, North and others, and also for different faunistic groups. As V.V. Khlebovich (1974) has shown, the number of marine species is reduced sharply for salinity lower than 5-8 ‰.

The nearly constant composition of ion concentrations in marine organisms is attained mainly by means of ionic regulation. Ionic transport is especially active in specialized cells covering the body, e.g. gills, the walls of excretion organs, intestines etc. (Fig. 7.8). These media remove surplus salts from the marine fish and sorb ions from the environment into freshwater fish.

The ionic coefficient, the ratio between $K^+ + Na^+$ and $Ca^{++} + Mg^{++}$ - is of great significance for marine organisms. The ionic coefficient decreases with decreasing salinity because the calcium concentration increases relative to sodium.

The analysis of all available data on ^{137}Cs in fish inhabiting fresh, brackish and salt waters of polar seas provides a general view of the effect of salinity on radionuclide filtration by organisms (Table 7.1). As a rule, salinity is negatively correlated with ^{137}Cs content in fish (to 20-80 Bq / kg) (Fig. 7.5, 7.6). A similar relationship is seen for ^{90}Sr accumulation in fish. Large fresh water releases from Siberian rivers to the Kara Sea during spring and summer cause salinity values to be reduced significantly from those in the Barents Sea. An even greater degree of freshening is observed in the White Sea and, in particular, in Onega Bay. Given the same ^{137}Cs level in the water, we expect that concentration levels in fish will be higher in sea areas more highly influenced by freshwater releases (Baklanov et al. 1996).

Table 7.1. Average concentration of ^{137}Cs in muscles and of ^{90}Sr in bones of fish from the coastal waters of Northern Norway, Barents and White Seas (Miretsky et al. 1993; Solberg et al. 1995), Bq/kg wet weight

Species	Coastal waters of Northern Norway and the Barents Sea		White Sea			
			Sea basin		Onega Bay	
	^{137}Cs	^{90}Sr	^{137}Cs	^{90}Sr	^{137}Cs	^{90}Sr
Cod	0.5	–	8	8	18	10
Herring	0.4	–	–	–	–	–
Smelt	–	–	80	10	21	23

7.1.6 Radionuclide sorption processes

Radioactive substances entering into continental sea ecosystems undergo physico-chemical sorption reactions. This phenomenon is key to the natural purification process of the marine environment and biota (Kuznetsov 1971; Lisitsyn 1978;

Izrael end Tsyban 1989; Matishov et al. 1994c; Alexeenko 2000). The concentration of many chemical elements, including toxic ones, is controlled for the most part by sorption processes, both adsorption and absorption. The sorption of ions and molecules depends on the composition of natural sorbents. The most important natural sorbents are colloids of Mn oxide (IV), hydrooxides of Fe (III), silica, and dolomite (Lisitsyn 1978; Alexeenko 2000).

The elaboration of the principal mechanisms of radionuclide concentration by sorption processes in the ecosystems of northern seas is an important task. Here we describe only in general terms the mechanics of sorption in the marine environment. Chemosorption, for example, the formation of iron-manganese concretions is widespread in the marine environment but poorly understood.

Ionic exchange processes in seawater, i.e. the ability of mineral adsorbents to exchange ions with liquid and gaseous electrolytes is a key mechanism (Lisitsyn 1994). The ionic exchange properties of natural sorbents are determined by the exchange (sorptive) capacity - quantity of exchange ions per 1 g of sorbent. The absorption of radionuclides usually occurs on colloid particle surfaces. The absorption of contaminants from aerosols is characteristic for the ocean-atmosphere boundary. In this case the absorption of gas or vapor occurs throughout the entire volume of dispersed particles.

Micro-quantities of strontium, cesium, and plutonium are received by the Arctic Ocean mainly as dissolved constituents remaining in ionic form. Probably, the majority of ^{239,240}Pu in ocean waters is present in seawater in an ion exchangable form_(Champ et al. 1994). The sorptive properties of natural sorbents are determined by a number of physico-chemical characteristics of the marine environment. The following ones are important: specific surface area of sorbents, particle charge, pH, water temperature, season, bottom sediment type, behavioral characteristics of different organisms, ecosystem trophic status and also the concentration of radionuclides in seawater and sediments.

The main qualities controlling the filtration capacity of radioactive contamination from marine ecosystems are lithodynamic and biological adsorption and also, to a lesser degree, chemo-sorption. The biological extraction of ^{90}Sr, ^{137}Cs, ^{239,240}Pu particles is moderated by the loss of material from the pelagic zone and extraction by particles in suspension, mainly iron and manganese dioxides.*Lithodynamic absorption* is the adsorption of radionuclides by the surfaces of sedimenting particles in the course of their transport, separation and descent to the sea bottom. The high degree of ion adsorption by sedimenting particles determines the high buffer capacity of the bottom boundary layer at the sediment-water interface. This situation leads to radionuclide sorption to bottom sediments and uptake by benthic organisms.

Several components are of main importance in the lithodynamic sorptive system: clayey minerals, iron, organic carbon, manganese and aluminum. The greatest ability to extract, for example, cesium from seawater is associated with clayey minerals, especially illite, for which distribution coefficients are highest among the different clay minerals (ratio of the concentration in hard phase to the same in solution).

The specific surface, for example, of manganese hydroxides is 70-350 m^2/g (Lisitsyn 1978). If the average suspended sediment concentration in the sea is assumed to be 1 g/m^3, then the total surface area available for sorption to particle surfaces in the pelagic water column (1000 m depth) and with a base of 1 m^2 is equal to 10-40 thousand m^2. Particle exchange reactions, adsorptive accumulation of ions and atoms, isomorphic substitution and other physico-chemical processes continuously occur at the interfaces between particles and solutions.

Bottom sediments serve a dual geochemical role because they can both retain radionuclides and consequently contribute to the removal of radionuclides from seawater and to desorb them and so to worsen the water quality.

Biological sorption of radioactive elements takes place continuously during the process of salt exchange between marine organisms and the environment. Active sorption is caused by the permeability for ions of the surfaces of flora and fauna species. Biological absorption ends through transfer to other organisms in the food web or by loss to bottom sediments. The rate of sorption and absorption is determined by the individual characteristics of organisms: morphological and physiological features. Differences in the absorption capacity of body surfaces also play an important role in this connection.

8 Radioactive monitoring principles for marine ecosystems

During the era of nuclear weapons testing, monitoring of artificial radionuclides in the environment and biota of the Arctic and Sub-arctic Seas was not conducted on a routine basis. Institutes of the Former Soviet Union carried out some observations in the Barents and Kara Seas. Other countries sponsored periodic surveys in the Norwegian, North, Irish and Baltic Seas. These surveys provided time series observations on the dynamics of radioactive contamination in seawater. At the time, little emphasis was placed on protecting the quality of living and nonliving resources of the northern seas.

Today national and international authorities are placing much greater emphasis on the protection of natural resources from radiological damage. As part of this effort, MMBI has participated in the development of scientific principles for carrying out routine radioecological monitoring of the polar seas (Matishov D. 1992b, 1993a, 1994, 1998; Matishov et al. 1994a,b,c, 1995a,b, 1996c,f,g, 1997b,c,f, 1998b,c, 1999d, 2000b). The current system of radiological monitoring is based on methodological and theoretical considerations of the functioning of ecosystems. To the degree possible, monitoring includes organisms from different trophic levels of the marine food web associated with Arctic continental margins. Data collection and reporting are carried out within a systematized framework in order to ensure both accuracy and reproducibility. In this chapter we present fundamental requirements for monitoring in Arctic areas and develop some practical guidelines for carrying out integrated cross-border, national and local monitoring of these areas.

8.1 Important considerations in radioecological monitoring

A system for monitoring radioactivity in the Arctic marine environment (seawater, suspended particles, bottom sediments) and biota (plankton, macrophytes, zoobenthos, ichthyo- and avifauna, sea mammals) has not been fully realized at present. It is impractical to conduct routine radioecological surveys over numerous critical geographical areas, species, and all components of the ecosystems of the northern seas. At the same time, identifying, analyzing and forecasting possible changes to the environment and biota as a result of the presence of artificial radionuclides amidst a background of natural oceanographic processes, remains an important problem. This challenge must be solved by two interdependent methods. The first is to implement and operate to their full capacity, regional and cross-border monitoring systems. The second, parallel, method is to conduct systematic investigations of the phenomena and factors that are essential in radioecological monitoring.

This complex approach, in spite of the fragmented nature of current radiation surveys, will significantly increase the quality of estimates and forecasts that necessarily are based on extrapolation of the present situation and the general patterns observed in investigated areas. Such phenomena and processes, as lithodynamic and biological sorption of radionuclides, bioaccumulation, biosedimentation, bioturbation, have been discussed in previous chapters. Here we shall dwell additionally upon the main properties and features of these factors in regard to the aims of monitoring in the northern seas.

8.1.1 Marine sediments as an indicator of radioactive contamination

The dynamics of marine ecosystem contamination can be partly obtained through the study of nuclide concentration levels in sediments. It was noted (Chapter 4), that bottom sediments accumulate the majority of radionuclides received by the basins of the northern seas. The degree of sediment contamination will depend upon the distance from the radiological source. Once deposited on the seabed, radioactive particles undergo transformations leading to the transfer of radionuclides to groundwater, interactions with benthic organisms, organic matter etc. The main mechanism of radionuclide migration is through diffusion, including sorption-desorption on sedimented particles (Kuznetsov, 1971; Lisitsyn, 1994; Kuznetsov et al. 1997b).

Estimation of radionuclide migrations in marine ecosystems is predicated on knowledge of the mechanism of radionuclide transfer between seawater and sediments. The sorptive capacity of sediment is characterized by the dimensionless parameter known as the distribution coefficient (K_dS). This parameter is determined as the ratio of the radionuclide content in sediment to its content in water (Sediment 1985). Values of distribution coefficients determined for specific areas have been shown to vary depending on many environmental factors (Duursma end Carroll, 1995). Paramount among these factors is the type of bottom sediments deposited along the continental margins (Gurevich 1992; Matishov et al. 1992c, 1994b,c, 1998; Kuznetsov et al. 1994; Domanov et al. 1996; Namyatov 1998). Distribution coefficients are widely used in marine radiological risk assessments. As a result, the dependency of radionuclide sorption (^{137}Cs) with sediment lithology has been greatly studied (e.g. Smith et al. 1995a,b; Matishov et al. 1998b,c, 1999d). The sorption of ^{239,240}Pu, ^{137}Cs, ^{90}Sr is largely determined by the surface area of particles, content of organic matter, ion exchange capacity of cations, and the granulometric composition of sediments.

Studies have shown that ^{137}Cs and ^{90}Sr are minimally fixed by sand and gravel; these radionuclides are easily removed from large particles by bottom currents leaving only a small fraction (approx. 5%) attached. Aleurite sediments scavenge the majority (approx. 95%) of radioactivity that is removed from the larger particles. Nuclides, especially ^{90}Sr, are fixed more permanently in fine, clayey sediments.

For example, in several bays of the Barents and Kara Seas, ^{137}Cs activity changes from 1 Bq / kg dry weight in areas dominated by sand (sand fraction = 86 %) to 30 Bq / kg dry weight in areas dominated by aleurite-clay sediments (clay fraction = 90-95 %). The correlation coefficients determined for the different areas ranged from 0.48 – 0.91(Fig. 8.1). In a similar study conducted in the Sea of Azov, the maximum in ^{137}Cs concentration corresponds to the maximum quantities of pelite silts (< 0.01 mm), and the dependence between these factors is linear (Fig. 8.3) (Matishov et al. 1998d). The average distribution coefficient of ^{137}Cs was 210 for sand and 7500 for sediments composed of < 90% pelite. The relationship between grain size and radionuclide content varies among different areas and even with individual seas.

As for ^{137}Cs, ^{239,240}Pu concentrations are highest in the fine aleurite and clayey sediments that are typically deposited in shelf depressions and along continental margins. Maximum levels of ^{239,240}Pu, up to 300 thousand Bq / m2, are found in silts of the (1) Chernaya Bay as a result of the underwater nuclear weapon explosions conducted at the site, (2) Bailot Strait to the west of Greenland as a result of the Thule accident, and (3) on the Irish Sea bottom in the zone of direct influence of Sellafield releases ((Matishov et al. 1994a,b;c; Smith et al. 1995a,b). The transfer of fine- dispersed particles of plutonium from Chernaya Bay to the Pechora Sea is observed within a 50 km range of the Bay. Regular observation of the area would be beneficial toward assessing temporal trends in the transfer of radionuclides from this contamination source into the adjacent sea areas.

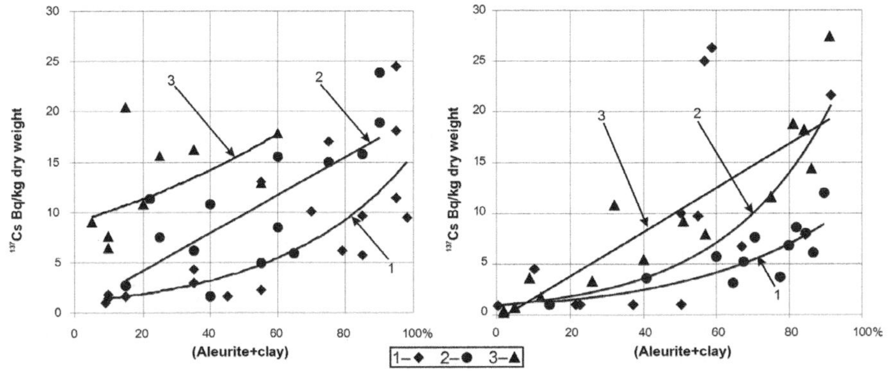

Fig. 8.1. Dependence of ^{137}Cs concentration with the content of aleurite + clay fraction in the upper layer of bottom sediments: a - Kola and Motovsky Bays: **1** - *Motovsky Bay and northern elbow of Kola Bay, r = 0.74;* **2** - *middle elbow of the Kola Bay, r = 0.74;* **3** - *southern elbow of Kola Bay, r = 0.48; b - Ob Bay, (1, r = 0.80), Pechenga Bay (2, r = 0.51), Pechora Sea (3, r = 0.80)*

Clay minerals, with their lattic structure, also influence substantially the capacity of particles to fix radionuclides and the subsequent mobility of radionclides in seabed sediments. The sorption potential of clayey minerals is unequal and depends on the content of illite, smectite, chlorite and kaolinite in association with the smallest particle sizes (<0.001 cm size fraction). Overall, the lithodynamic absorption of ions by suspended particles and silts is responsible for the large buffering capacity of the «water-silt» barrier zone.

A historical perspective of the accumulation process and dynamics of radioactive contamination of northern seas is often essential to understand the timing and sources of contamination events. We have utilized this approach to investigate the Kola Bay (Tables 8.1-8.3). The input dynamics for anthropogenic radionuclides was reconstructed by ^{210}Pb geochronology. At one location (Station 14) the presence of ^{152}Eu, ^{154}Eu, ^{60}Co was discovered below the surface layer and is likely the result of neutron activation of construction materials. Maximum concentrations of ^{60}Co and ^{137}Cs were also discovered during 1966 and 1972 (Fig. 8.2).

The ^{239,240}Pu content in a column of sediment collected at Station 6 varied from 1 to 2.4 Bq / kg. However, in the 2-3 cm layer the concentration was 5.7 Bq / kg. The peak in plutonium concentration occurred during the period 1990-1993 and is obviously the result of activities conducted by the Northern Fleet at one of the numerous bases situated along the periphery of the Kola Bay. The need for local monitoring of inlets adjacent to smilitary facilities is clear.

Table 8.1 Vertical distribution of radionuclides in Kola Bay bottom sediments (by γ-spectrometric analysis)

Depth of section, cm	Specific activity of nuclides, Bk/kg					
	^{137}Cs	^{60}Co	^{228}Ra	^{228}Th	^{226}Ra	^{40}K
			Station 1			
0–1	10±2	–	35±7	19±6	12±4	570±60
1–2	7±2	–	46±8	24±4	11±3	670±80
2–3	8±1	–	40±4	14±3	10±2	580±90
3–4	2±1	–	41±5	12±3	9±3	650±90
4–5	5±1	–	36±6	24±5	12±4	690±90
5–6	4±1	–	41±8	22±4	16±2	650±80
6–8	3±1	–	43±4	23±5	12±5	720±90
8–10	18±2	–	45±5	25±4	12±3	700±90
10–12	13±2	–	44±7	26±4	11±2	670±90
			Station 6			
0–1	13±2	–	72±28	38±7	9±2	740±150
1–2	2±1	–	61±22	22±7	10±2	580±130
2–3	10±2	5±2	35±8	42±13	17±3	750±160
3–4	16±2	2±1	37±7	46±12	19±3	550±150
4–5	14±2	1.0±0.5	40±7	31±9	10±3	620±170
5–6	14±2	5±2	35±8	16±6	9±2	780±150
6–8	14±2	1.0±0.5	42±7	30±7	11±3	660±140
8–10	16±2	4±2	36±8	26±6	10±2	570±190
10–12	27±3	–	50±10	28±7	13±3	720±160
12–14	16±2	–	55±12	28±7	10±2	660±170
14–16	8±2	–	45±10	25±6	11±2	520±190

Table 8.2 Vertical distribution of total ^{210}Pb and unsupported ^{210}Pb in Kola Bay bottom sediments

Sediment layer, cm	Total specific activity of ^{210}Pb, Bq/kg		Unsupported ^{210}Pb activitiy, Bq/kg	
	St.1	St. 6	St. 1	St. 6
0–1	94.3±12.3	246±19	12±4	9±2
1–2	62.4±7.0	146±12	11±3	10±2
3–4	48.3±6.7	156±17	9±3	19±3
6–8	24.7±3.8	108±8	12±5	11±3
8–10	19.6±0.9	–	12±3	–
10–12	16.2±2.1	104±17	11±2	13±3
14–16	–	56±7	–	11±2

Table 8.3 Average age of sediment depth intervals for Kola Bay bottom sediments based on ^{210}Pb geochronology

Depth of the level, cm	Age, years	
	St. 1	St. 6
0–1	6±0.6	1.5±1
1–2	12±1	4.4±0.3
2–3	20±2	7.4±0.4
3–4	29±3	10.3±0.6
4–5	37±4	13.2±0.8
5–6	45±5	16.2±1.0
6–8	58±6	20.6±1.2
8–10	75±8	26.5±1.6
10–12	92±9	32.4±1.9
12–14	–	38.2±2.2
14–16	–	44.1±2.6

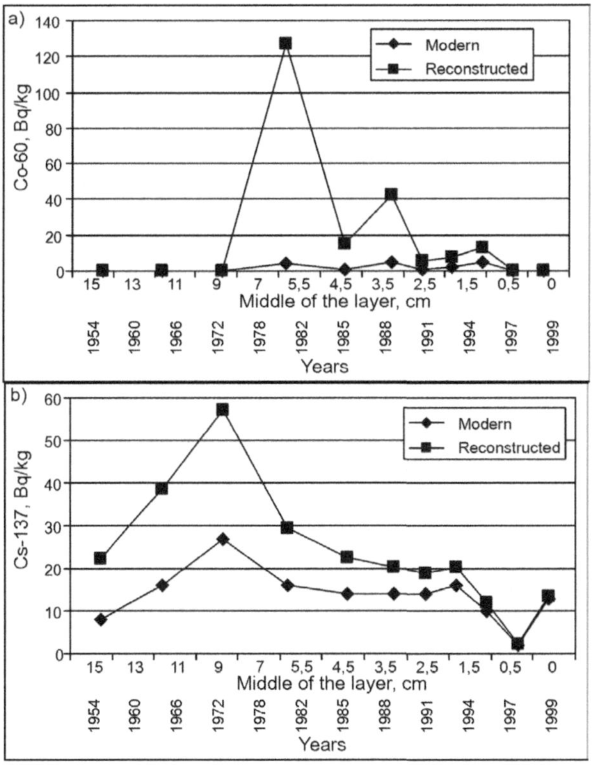

Fig. 8.2. Modern and reconstructed concentrations of ^{60}Co (**a**) and ^{137}Cs (**b**) in Kola Bay bottom sediments

The determination of sorption dependencies with lithology is important for minimizing the uncertainty in the estimation of radionuclide storage capacities of different marine systems. This information is especially importance for coastal areas where the anthropogenic impact on the marine ecosystem is the greatest, and the probability of accidents is far higher than for the open water areas.
Clay-rich sediments, deposited in the depressions and troughs of the underwater continental border, are the most important integral indicator of the state of contamination of marine ecosystems. The essential dependence between the concentration of artificial nuclides and the quantity of pelite fraction (< 0.01 mm) in clayey deposits has been demonstrated. Consequently, the easily dispersed, fine particles, i.e. modern clays of shelf troughs and coastal bays, must be a focal point of radioecological monitoring in northern seas.

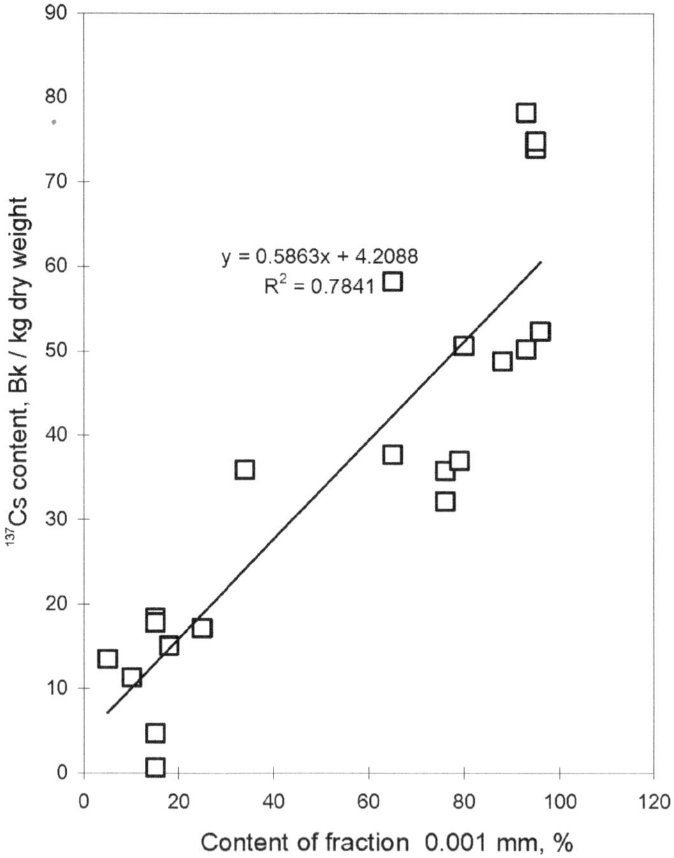

Fig. 8.3. ^{137}Cs concentrations in pelite fraction of bottom sediments

8.1.2 Secondary radiation contamination of the benthic zone

As part of the normal functioning of benthic ecosystems, radioactive elements are brought to the sediment-water interface through processes such as biological mixing, accumulation of sediments, and transport by currents. As it was noted already, the modern sedimentary cover of the shelf, continental foot and abyssal plains of the Arctic Ocean and North Atlantic together with zoobenthos and bottom fish, fix rather quickly and remove the majority of ^{137}Cs, ^{90}Sr, $^{239,240}Pu$ and other artificial radionuclides.

The seabed, depending on its geomorphological features, is subject to the influence of various physical and biological processes. Secondary redistribution of radionuclides occurs as a result of water circulation, desorption, bioturbation and other factors which lead to the disturbance and mixing of bottom sediments. The radiobiogeochemical cycle of the layer just above the seafloor and the role of this cycle in the secondary contamination of the sea environment has only been addressed in a very limited capacity. It is essential that monitoring programmes address this secondary source of contamination.

Radioecological observations are especially necessary in the mixing zones of aleurite and clayey sediment deposits. Several mechanisms of radionuclide transfer act simultaneously here: diffusion of free and adsorbed ions; advective transfer via porewater, transport with colloid particles; intermixing by bottom fauna and also the massive disturbance of silts caused by trawling and other anthropogenic impacts on the sea bottom.

Desorption phenomena (including cation exchange) occurs when there is a misbalance of radionuclide concentrations between water and bottom sediments causing polluted sediments to release radionuclides back into the water. As a result, the quality of seawater is reduced even after the removal of a radiation source. This situation has been detected in a few of the Novaya Zemlya bays, in particular, Chernaya, Stepovoy, Abrosimov Bays, in the Kola and Motovsky Bays, and in the sea environment adjoining Sellafield (Matishov et al. 1994a,b,c, 1996f, 1998b,c, 1999d). The concentrations of artificial radionuclides, existing at these locations, are much higher than background values, making these areas potential sources of radioactive pollution. In particular, regular and long term monitoring of $^{239,240}Pu$ exchanges is needed.

The «hot particles» are highly radioactive (on the order of 10^2-10^3 Bq / kg) materials which are incorporated into sediments and bottom biota as a result of emissions from nuclear facilities and seabed burial sites. The presence of «hot particles» has been identified in close proximity to a number of radioactive objects (Egorov et al. 1993; Braginsky et al. 1994; Namyatov 1998; Strand 1998; Nikitin et al. 1999; Rissanen et al. 1999a).

The extreme radioactivity of sediments in bottom depressions of some of the bays along the Kola Peninsula, located in close proximity to nuclear facilities, is the result of the wash-off of ^{137}Cs, $^{239,240}Pu$, and ^{60}Co-containing sediments, «hot particles», and the tissues of decaying macroalgae. High and low intertidal macrophytes are exposed to different dose loads due to the presence of irradiation sources in bottom sediments. The absorbed dose to macrophytes is also partly caused by internal irradiation from radionuclides, this dose is also related to the presence of radioactive elements in bottom sediments.

The migration of radionuclides in sediment porewater within the zone inhabited by benthic organisms must be addressed in radioecological monitoring. As it is known, bottom sediments having a moisture content of 1-40 %, gradually exchange radionuclides to porewaters. Radioactive substances are transferred to porewater by diffusion processes (Klenova 1960; Belov end Lapina 1961; Khrustalev 1989; Matishov end Pavlova 1990). The highly porous silt deposits on the bottom of troughs sometimes reach thicknesses of 1-2 m.

Obviously, radionuclides in sediment and porewater are accessible for silt eating organisms both of zoobenthic species and fish of the plaice family. Investigations are needed to estimate the importance of porewaters as the source of long term radioactive contamination of bottom waters.

Biological mixing and bioturbation are of great importance in the process of migration and redistribution of the radioactive contamination of bottom sediments (Matishov et al. 1994b,c, 1998b,c; Smith et al. 1995a,b). The transfer of radionuclides from bottom sediments is facilitated by different bottom animals: crustaceans, echinoderms, polychaets, mollusks. Detritus eaters, silt eaters and sand inhabitants, which live in enormous communities and quickly reproduce, are the most important organisms responsible for transformation of a contaminated environment and the subsequent impact on the ecosystem, especially in the most productive shelf habitats. The accumulation of pollutants in these organisms depends not only on the level of pollutants in sediments but also on the organism's metabolism. Obviously, during the 1990s, in spite of the process of radioactive decay, radionuclides in the sediments of the northern seas were accessible to biota of all trophic levels after initial incorporation into the benthos.

In summary, monitoring of the radioactivity in zoobenthos and macrophytes must be the focal point of radioecological surveys. Bottom organisms take an active role in the self-purification processes of sea basins and are indicators of water and sediment contamination. Anthropogenic mixing of shelf and continental slope bottom sediments is also an important factor contributing to the secondary redistribution of radionuclides. Fish trawling of the North, Barents, Azov and other European Seas, has resulted in the large-scale ploughing of sediments and dispersion of previously deposited radionuclides. In fact, there are no pristine habitats in commercial fishing areas today. During monitoring it is reasonable to investigate periodically the influence of trawling on the process of secondary accumulation of ^{137}Cs, ^{90}Sr and especially of the long-lived radionuclide, $^{239,240}Pu$.

8.2. Role of bioindicators in radiological monitoring

The high biological productivity of the pelagic, benthic and coastal zones of the Barents, Norwegian, North and other high-latitude seas determines to a great extent the processes of self-purification of these basins from radioactive contamination. The distribution patterns and concentration of ^{137}Cs, ^{90}Sr, $^{239,240}Pu$ and other radionuclides in the flora and fauna of the seas along with the biogeochemical processes operating in marine ecosystems must be taken into account when radio-ecological monitoring is to be carried out.

For the present, our investigations confirm unequivocally that macrophytes, zoobenthos, fish, birds and sea mammals of northern seas contain low concentrations of ^{137}Cs, ^{90}Sr and $^{239,240}Pu$ (Matishov et al. 1992c, 1993b, 1994a,b,c, 1995a, 1996f,g, 1997b,c,f, 1999d; Matishov D. 1992b, 1993a, 1994, 1997; Matishov D. et al. 1993, 1995b, 1996, 1997). The level of plutonium in fish is within the limits of instrument detection. Radionuclide concentrations are higher in the intercontinental seas - White, Baltic and Azov. Fresh traces of ^{60}Co are observed only in macrophytes and zoobenthos of the Kola and Kandaslaksha Bays, as well as the Oslo Fjord. Despite the risks, potential sources of radioactive contamination (e.g. dumped objects in the Kara Sea) have not resulted in widespread contamination of the marine food web.

The sustainable self-purification of northern sea ecosystems from radioactivity is connected in a number of cases with the functional activity of large populations and communities of organisms. Some representatives of marine biota that accumulate anomalous quantities of nuclides in their bodies must be noted in this connection. It is also important to consider that many species of fish, birds, seals and whales are marketable. So the maximum allowable concentrations of radionuclides that are not dangerous for human consumption are well-known for sea products. The maximum allowable contents of ^{137}Cs and ^{134}Cs, adopted by European countries, is 600 Bq / kg in food products and 370 Bq / kg in children's nourishment. The cesium limit adopted in Norway for salmon filets and reindeer meat is 3000 Bq / kg. Similar standards for the consumption of fish exist in the Former Soviet Union.

Marine organisms that are used as indicators of radioactivity belong, as a rule, to species that are widespread geographically and abundant (Fig. 8.4). The annual production of the Barents Sea zooplankton species, *calanus*, for example, is several tens of millions of tons. It is important to consider that highly productive populations of species such as *calanus*, also facilitate the large-scale biosorption of radionuclides. The active sorption of ions by plankton is carried out by excretion through cell membranes in connection with the processes of water and salt regulation and through morphological and behavioral adaptations of the organisms. Fish, invertebrates and other sea animals absorb dissolved radionuclides through the body surface and gills, and ingest radionuclides through the intake of water and food. The final by-products of metabolism are subsequently removed by excretion through the bowels.

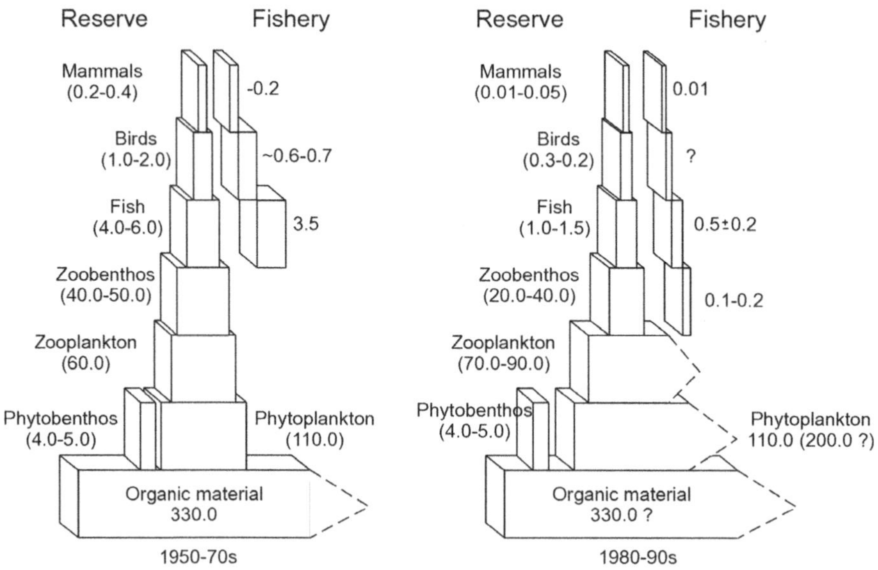

Fig. 8.4. Annual production (mln tons) of main communities and amount taken by the commercial fishing industry in the Barents Sea and adjoining basins

Osmotic regulation and salt exchange of sea animals (Khlebovich 1974; Berger 1986) is related to the phenomena of biological sorption and extraction of artificial radionuclides. It is important to note that ionic regulation mechanisms can function differently in different marine organisms. For example, the transport system of salmon and other andromous fish is polarized in such a way that sodium ions are removed to seawater. The osmotic defense, including the accumulation of radionuclides, is further supported by the formation of slime on the body surface of fish and many other organisms.

The transfer of radionuclides through trophic food webs is parameterized using bioaccumulation coefficients (K_a). These coefficients are calculated as the ratio of nuclide level in the given organism to the concentration in sediment and water or in the previous link of the food chain (Kilezhenko end Podymachin 1964; Polikarpov 1964a,b, 1971; Iljenko 1974; Braginsky et al. 1994; Namyatov 1998). Thus it is the radionuclide content of water and sediments that are the main factors used to predict radionuclide accumulation in all other ecosystem components. Sedentary species of marine biota sorb radionuclides more than organisms undergoing daily and seasonal migrations (Matishov et al., 1994c, 1996f,g, 1998c). In spite of the relatively small quantity of radionuclides accumulated by organisms, their role in the migration and distribution of radionuclides in the ecosystems of polar seas is important.

Based on our experience of radioecological observations, representatives of macrophytes, zooplankton, zoobenthos, fish and birds can be recommended as organism-indicators of the radiation state of ecosystems of northern seas. It is necessary to use for the comparative analyses the species inhabiting both pure and contaminated biosystems.

Among the macrophytes, brown algae: fucoids (*Fucus vesiculosis*) and laminarias (*Laminaria saccharina*) are the most sensitive to radioactive contamination, are found in significant quantities, and are characterized by cross-border spreading. In particular, they can serve as excellent indicators of the toxicant level in fjord ecosystems, especially near nuclear facilities and military bases (Matishov et al. 1996b, 1997a,b,c, 1998b,c). In the 1980s and 1990s the concentration of ^{137}Cs in these macrophytes varied from 13 to 6 and from 1 to 4 Bq / kg dry weight for the Oslo Fjord and the Kola Bay respectively (Matishov et al. 1994b,c, 1996f,g; Rissanen et al. 1997a, 1999a; Brungot et al. 1999a,b).

For plankton, species of crustaceans are useful as bioindicators. As it was noted already, the fecal pellets, produced by medium-sized copepods such as *Calanus finmarchicus*, are cylindrical (of dimensions 600 x 70 μm) and descend to the sea at speeds of two hundred meters per day (Lisitsyn 1978, 1994; Lisitsyn end Vinogradov 1982). The short-lived plankton organisms, which filter and sorb radionuclides from water, serve as the main biological filterers of the basin. During the spring bloom in the northern seas (from the end of March until the beginning of June), the rapid growth of plankton results in large-scale bioaccumulation and biosedimentation of artificial nuclides. Analysis of radionuclides in plankton will help to reveal the daily and monthly cycles of ^{137}Cs and in part $^{239,240}Pu$ concentrations in seawater. Microalgae and zooplankton, having no calcified components, accumulate almost no ^{90}Sr.

The various zoobenthic species differ significantly in their capacity to accumulate ^{137}Cs, ^{60}Co and other nuclides (Matishov et al, 1994b,c, 1995a, 1996f,g, 1999d). However, benthic organisms of northern seas in general contain relatively little ^{137}Cs and ^{60}Co (from 0.1 to 6 Bq / kg). The concentration of ^{90}Sr and $^{239,240}Pu$ in benthic organisms is even less. This is characteristic for mollusks, crustaceans, starfish, sea urchins, sea cucumbers. There is practically no strontium (0.01-0.34 Bq / kg) in benthic organisms (Foraminifera, Asteroidea, Ophiuroidea). These findings include observations taken from the area of the sunken submarine, Komsomolets (Kuznetsov et al. 1993; Kuznetsov et al. 1997a).

Some filter- and silt- eating organisms can serve as bioindicators among the benthic organisms. Some polychaeta ingest large quantities of organic-rich, fine sediments making them also efficient accumulators of radionuclides (^{137}Cs, ^{60}Co). We have found the presence of ^{137}Cs from 6 to 17 Bq / kg in polychaetes from several parts of the shelf. Bivalve mollusks contain the largest quantities of ^{137}Cs among filter-feeders; blue mussel (*Mutilus edulis*) and scallop (*Chlamys islandica*), in particular. ^{90}Sr is also more perceptible in the shells of these organisms.

The highest levels of anthropogenic radionuclides in commercial fish species was detected in the vicinity of Novaya Zemlya during the early 1960s (Saetre et al. 1992; Foyn end Svaren 1995). Monitoring during the 1950-60s indicated a rapid transfer of radionuclides to fish after atmospheric nuclear tests and a pronounced seasonal pattern in fish contamination levels. The absorption and transfer of radionuclides through the pelagic food web would be initiated through consumption of phytoplankton by zooplankton. This bioconcentration would be most efficient during the summer-time peak in secondary production in response to a strong spring-time phytoplankton bloom. However, even during the period of the most intensive nuclear tests, when the Barents Sea received direct atmospheric fallout, the maximum value of β-activity in fish did not exceed 90 Bq / kg.

Among marketable fish species, the Atlantic cod *Gadus morhua* and Pollack *Pollachius virens* should receive priority in a monitoring plan. These species have a wide geographical distribution and reflect the movement of critical radionuclides through the marine food web.

The influence of radioactive contamination on colonial bird populations of the Arctic Seas can be effectively traced through some of the avifauna. Birds are calcifiers, a condition that is very important for the control of ecosystems contaminated by ^{90}Sr. As with all vertebrates having developed muscles, birds also accumulate ^{137}Cs (Lebedeva 1999). The accumulation of radionuclides in some birds, is higher than in many other sea animals. This is caused by the intensive consumption and metabolism of food per unit of body mass in birds. The big sea gull *Larus marinus* and great skua should be classified as indicator species. These species have comparatively high ^{137}Cs levels (from 2 to 40 Bq / kg).

Clearly it is advisable to control the rate of supply of radionuclides into the environment and biota, so that the levels of artificial nuclides accumulating in organisms remain low. The elaboration of bioindication methods is an important component of a protection strategy because the method allows the prompt detection of radiation events and the contamination source can be traced even when the source is hundreds or thousands of kilometers away.

The use of bioindicator methods certainly must be applied in radioecological programmes conducted in the Barents-Kara Sea region. The current understanding of the content and distribution of radionuclides in these polar seas testifies to the necessity of creating a routine radiation control system for the region. The control must be based not only on the traditional determination of abiotic ecosystem components (sea and fresh waters, air, soils), but also on biota, especially those known to bioconcentrate radionuclides and those that are key components in Arctic food webs.

8.3 Human radiological dose estimates for consumption of sea food

The main reason for monitoring is to be able to properly assess radiation doses for humans. The international recommendations on radiation protection, nuclear safety and the treatment of nuclear wastes oblige countries to carry out risk estimations, including the calculation of long-term doses from possible leaks from potential sources within the Arctic. In many countries, the highest allowable annual dose for personnel working with radioactive materials is 20 mSv (Strand

1988). In the post-Chernobyl period, ^{137}Cs and ^{90}Sr became the artificial radionuclides of greatest biological significance in the northern seas. However over time as the levels of ^{137}Cs and ^{90}Sr decrease through radioactive decay and ocean dilution, the ecological significance of $^{239,240}Pu$ will certainly increase.

Radiation influences every level of life organization and is modified by a complicated tangle of abiotic and biotic factors. But the expression of radiobiological effects is apparent at the organism level (Polikarpov 1964a,b; Polikarpov end Egorov 1986; Krivolutsky 1983; Radiosensitivity ... 1989; Abaturov et al. 1996; Impact ... 1996; Pozolotina 1996; Moiseenko end Kudryavtseva 1999). The uneven spatial and temporal distribution of radionuclides in marine ecosystems results in different accumulation patterns not only by representatives of different trophic levels but by organisms of the same species. Certainly, low concentrations of ^{137}Cs, ^{90}Sr and $^{239,240}Pu$ in the ecosystems of pelagic and benthic zones have no toxic influence on the functioning of communities of marine organisms. Benthic fauna are subject to relatively higher background radiation doses than pelagic fauna due to the continuous presence of a weak radiation dose in clayey silts.

The relationship between levels and the doses influencing human health is complex. However, investigations clearly show that populations consuming mainly sea products receive the lowest radiation doses (Foyn end Semenov 1993; Strand 1998; Matishov et al. 1999d). Those eating food products mainly obtained from land and freshwater ecosystems receive the highest irradiation doses both from natural and athropogenic radionuclides. In particular, the ^{137}Cs level in populations of Saami who live in northern scandinavia and consume reindeer meat, increased after the Chernobyl accident from 3-4 to 10-40 thousand Bq / kg.

During the 1990s, a number of assessments were conducted on the radiation dose for humans through the food chain water → fish → human (Fig. 8.5). The values of effective doses, received from consumption of Barents Sea fish were very low, on the order of 0.6 μSv / year and 4 μSv / year for fishermen. In the marine ecosystem of the Kola and Motovsky (Zapadnaya Litsa) Bays, the maximum effective dose for the trophic chain water → fish → men is 6.7 μSv / year based on a yearly consumption of 100 kg of fish (Fig. 8.6). (Matishov et al. 1999d, 2000). The main component of the dose (about 90 %) is from ^{137}Cs; the dose from plutonium is insignificant.

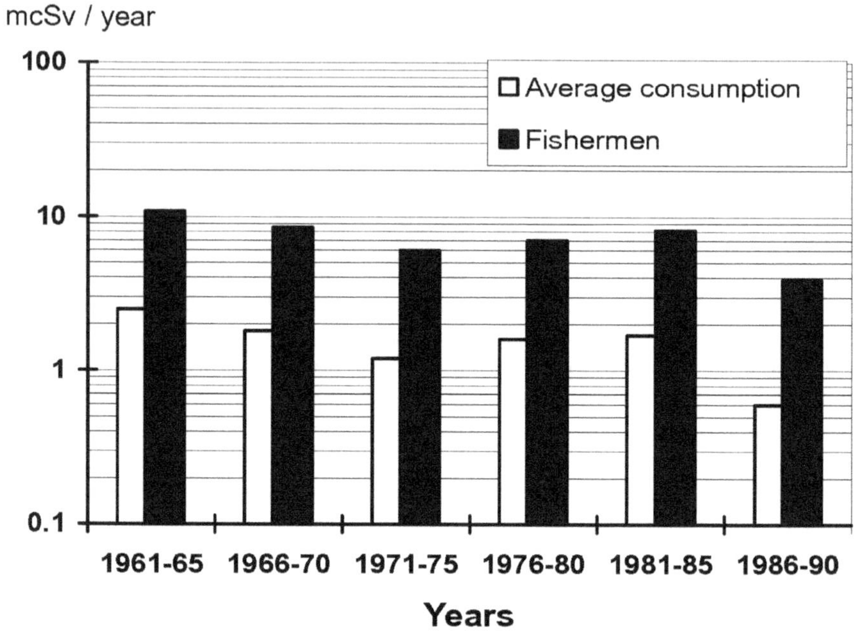

Fig. 8.5. Effective doses (axis of ordinates) received by two groups of populations consuming fish from the Barents Sea (Dumping ... 1996)

8.4 Modeling as a tool in radioecological monitoring

Forecasting the consequences of potential emissions from underwater nuclear dumping sites and coastal nuclear bases occupies an important place in the framework of radioecological monitoring. Different models and scenarios have been designed to calculate and forecast the hydrodynamic and hydrobiological processes that control the assimilation of radioactive substances by the northern seas (Fig. 8.7-8.9). We shall note only some important points without describing fully the various approaches being employed (Paluszhkevicz et al. 1995; Matishov et al. 1996g, 1997f, 1999d; Dumping ... 1996; Namjatov 1998; Nies et al. 1998; Goncharov end Klementjeva 1999).

As noted previously, radioecological investigations conducted at radioactive waste sites in the Novaya Zemlya bays, near the sunken atomic submarine «Komsomolets» and near release points for atomic bases in the Kola and Motovsky Bays have indicated only localized impacts on the marine ecosystems. The limited presence of contamination in bottom sediments near radioactive waste containers and submarines indicates that these objects are stable, at least temporarily.

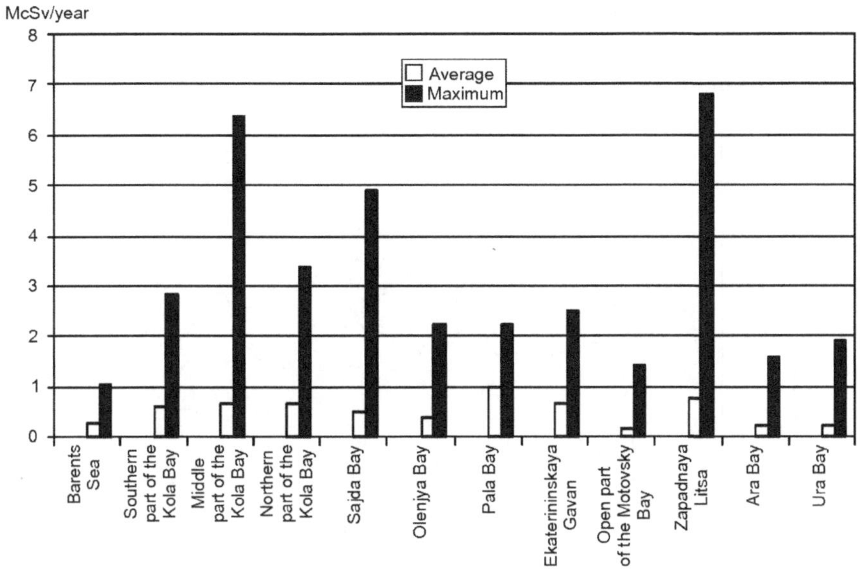

Fig. 8.6. Average and maximum potential effective doses from consumption of 100 kg/yr of fish calculated from different areas of the Kola and Motovsky Bays

Modeling of hypothetical scenarios of instantaneous radionuclide releases from sunken submarine reactors near Novaya Zemlya and the «Komsomolets», predicts that the volume of contaminated seawater would be very low, representing only several hundredths of one per cent of the water volume of the Kara or Norwegian Seas (Fig. 8.9). These accidents potentially are not able to cause damage to marine biological resources. The radioactive contamination of the pelagic zone would not be perceptible in 5-7 years (JRNEG 1993; Blindheim et al. 1994; Saetre 1994; Hamilton et al. 1994; Foyn end Svaren 1995; AMAP 1997; Pfirman et al. 1997; Baxter et al. 1998).

Emergency discharges into the Kola and Motovsky bays does pose a potential radiation danger for the marine ecosystems and population of the region (Matishov et al. 1995a, 1996b,f,g, 1997c,f, 1998a,c, 1999d,h). The circulation of surface and bottom currents is the leading factor controlling the direction of pollutant transfer during controlled releases. About 89-90 % of ^{239,240}Pu and ^{60}Co is usually deposed in the bottom sediments of bays near the place of discharge. However, the fine dispersed fractions of suspension are able to migrate long distances. In particular, cobalt is found up to 100 km from the Kola Bay.

During spring-time in the Kola and partly the Motovsky Bays a pronounced halocline is formed in the pelagic zone of bays (to 15-20 m). The velocity of surface freshwater (melted and from rivers) can reach 1-4 km / hour. Radionuclides, brought by rivers or reaching the bay basins during routine releases are partly transported to the Barents Sea coastal zone (Matishov et al. 1997c). The radioecological risk for living organisms would be significantly lower and the situation less dangerous, if the planned releases to the waters of Kola and Motovsky Bays were conducted in phase with the falling tides and during periods of maximum freshwater runoff.

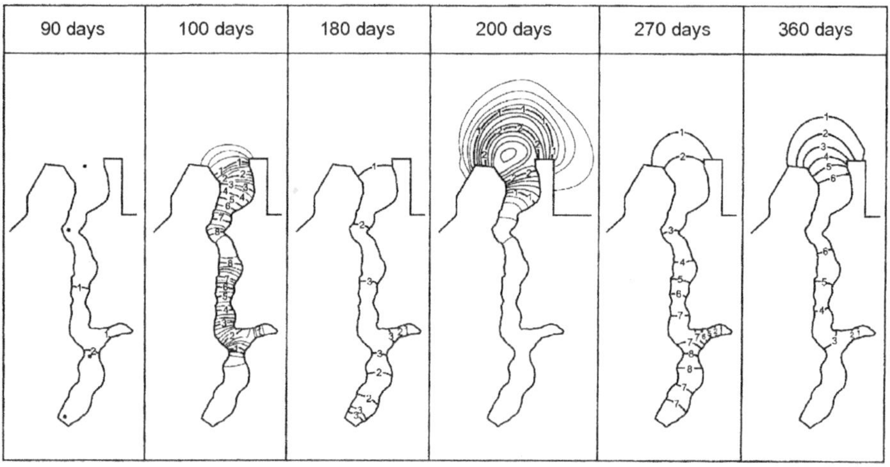

Fig. 8.7. Model determinations of ^{90}Sr concentration (Bq / m^3) and cumulative discharge (TBq) for 1985. The release of radionuclides to the Techa River from the reservoir at the «Mayak» facility (Paluszkiewicz et al. 1995). Contour lines show the ^{90}Sr concentra-tion in the Ob mouth on the 100th and 200th days after the release, Bq / m^3

Modeling of possible scenarios of radioactive discharges in the Kola Bay have shown the following (Matishov et al. 1997c). Consider a 100 m^3 release of inert radioactive liquid into the surface layer of the Kola Bay during a falling tide. Within the northern section of the Kola Bay, the concentration will be diluted by 1 million times over a distance of 15 km in 2 days (Fig. 8.10) and in the southern section – over a distance of 11 km in 16 hours (Fig. 8.11). If the spill were to occur during a rising tide, the point source would spread over a distance of 8 and 7 km within the northern and southern sections respectively. We next consider instantaneous releases from the «Atomflot» facility. Approximately 5-7 % of the release will be exchanged into the Barents Sea every day from the 2nd to 10th day after the release (Namjatov 1998). Kola Bay bottom sediments will accumulate approximately 40 % of the released ^{241}Am, 17 % of ^{60}Co, 9 % of ^{63}Ni and ^{239}Pu, 0.3 % of ^{137}Cs, 0.1 % of ^{90}Sr and 0.002 % of ^{129}I.

The liquid radioactive waste purification system at «Atomflot» was upgraded in 1998. As a result, radioactive discharges to the Kola Bay have increased by nearly an order of magnitude. To consider the impact of this new situation, we modeled the increased release volumes assuming the same radionuclide composition as previous to the modernization (Table 8.4). The resulting environmental concentrations are not dangerous for the fjord ecosystem.

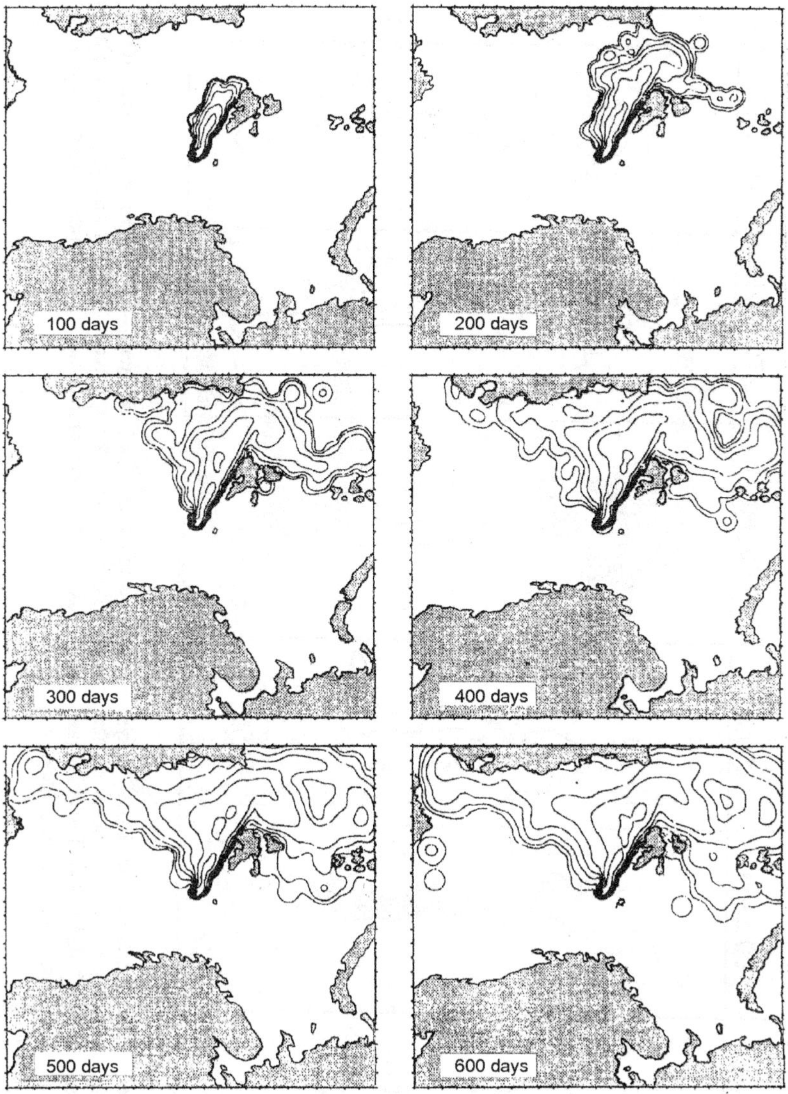

Fig. 8.8. Temporal dynamics of the vertical complex concentration (Adlandsvik and Martinsen 1996)

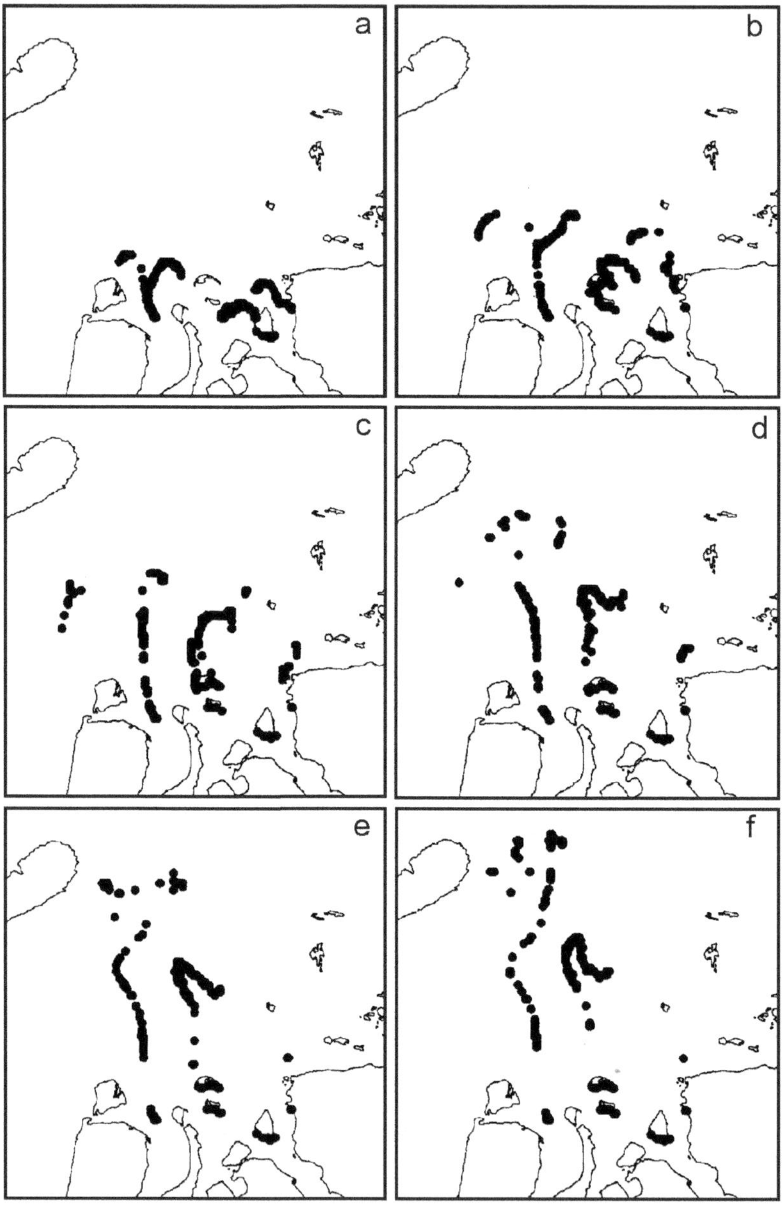

Fig. 8.9. Predicted transfer of radionuclides with drifting ice calculated with a 4-day time step (Johannessen et al. 1996): **a** - *16th day*, **b** - *32nd day*, **c** - *48th day*, **d** - *64th day*, **e** - *80th day*, **f** - *96th day*

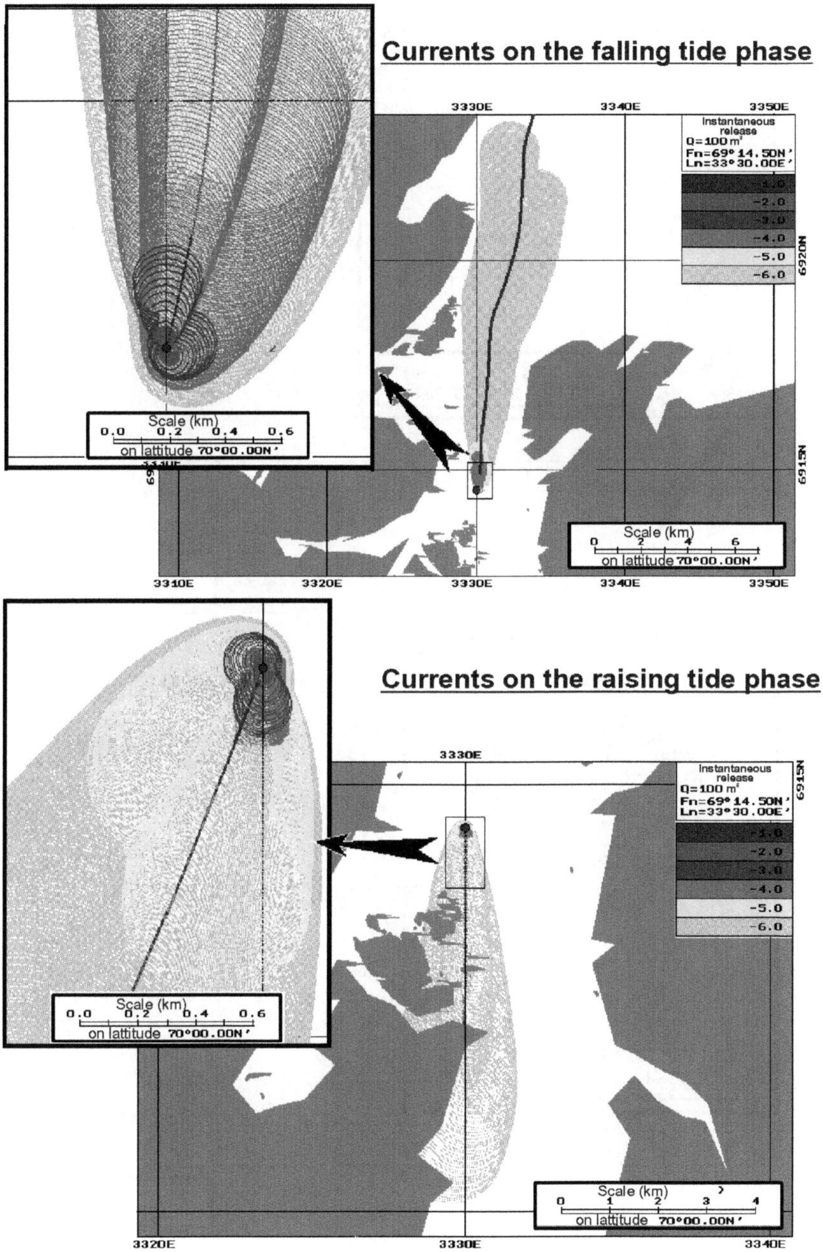

Fig. 8.10. Estimate of the distribution of dissolved tracer in the northern section of the Kola Bay

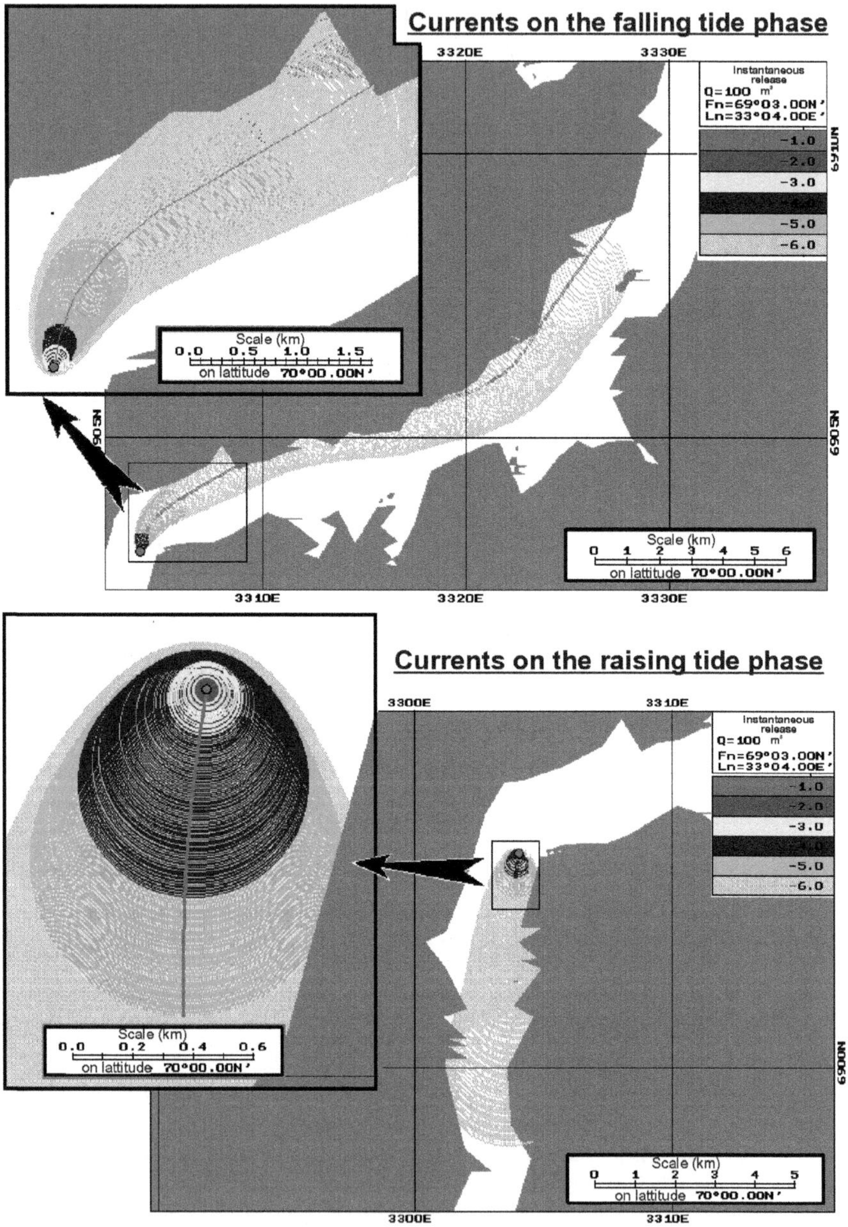

Fig. 8.11. Estimate of the distribution of dissolved tracer in the sourthern section of the Kola Bay

Table 8.4 Calculated radionuclide concentrations in Kola Bay waters resulting from discharges from «Atomflot» (Namjatov 1998), Bq / m^3

Nuclide	Southern elbow C_1		Middle elbow C_2		Northern elbow C_3	
	average discharge	maximum discharge	average discharge	maximum discharge	average discharge	maximum discharge
^{90}Sr	$3.62 \cdot 10^{-5}$	$8.52 \cdot 10^{-5}$	$8.67 \cdot 10^{-5}$	$2.04 \cdot 10^{-4}$	$2.56 \cdot 10^{-5}$	$6.02 \cdot 10^{-5}$
^{137}Cs	$1.75 \cdot 10^{-4}$	$4.81 \cdot 10^{-4}$	$4.20 \cdot 10^{-4}$	$1.15 \cdot 10^{-3}$	$1.24 \cdot 10^{-4}$	$3.40 \cdot 10^{-4}$
^{134}Cs	$5.34 \cdot 10^{-5}$	$1.42 \cdot 10^{-4}$	$1.28 \cdot 10^{-4}$	$3.41 \cdot 10^{-4}$	$3.77 \cdot 10^{-5}$	$1.01 \cdot 10^{-4}$
^{125}Sb	$1.58 \cdot 10^{-4}$	$4.26 \cdot 10^{-4}$	$3.79 \cdot 10^{-4}$	$1.02 \cdot 10^{-3}$	$1.12 \cdot 10^{-4}$	$3.01 \cdot 10^{-4}$
^{60}Co	$1.42 \cdot 10^{-4}$	$2.99 \cdot 10^{-4}$	$3.40 \cdot 10^{-4}$	$7.18 \cdot 10^{-4}$	$1.00 \cdot 10^{-4}$	$2.12 \cdot 10^{-4}$
^{154}Eu	$2.06 \cdot 10^{-4}$	$3.69 \cdot 10^{-4}$	$4.93 \cdot 10^{-4}$	$8.84 \cdot 10^{-4}$	$1.45 \cdot 10^{-4}$	$2.61 \cdot 10^{-4}$
^{152}Eu	$1.15 \cdot 10^{-4}$	$2.39 \cdot 10^{-4}$	$2.75 \cdot 10^{-4}$	$5.72 \cdot 10^{-4}$	$8.12 \cdot 10^{-5}$	$1.69 \cdot 10^{-4}$
^{95}Zr	$5.66 \cdot 10^{-5}$	$1.17 \cdot 10^{-4}$	$1.36 \cdot 10^{-4}$	$2.82 \cdot 10^{-4}$	$4.00 \cdot 10^{-5}$	$8.32 \cdot 10^{-5}$
^{95}Nb	$3.72 \cdot 10^{-5}$	$6.96 \cdot 10^{-5}$	$8.91 \cdot 10^{-5}$	$1.67 \cdot 10^{-4}$	$2.63 \cdot 10^{-5}$	$4.93 \cdot 10^{-5}$
^{54}Mn	$4.52 \cdot 10^{-5}$	$7.91 \cdot 10^{-5}$	$1.08 \cdot 10^{-4}$	$1.90 \cdot 10^{-4}$	$3.19 \cdot 10^{-5}$	$5.60 \cdot 10^{-5}$
^{144}Ce	$1.77 \cdot 10^{-4}$	$3.62 \cdot 10^{-4}$	$4.25 \cdot 10^{-4}$	$8.73 \cdot 10^{-4}$	$1.25 \cdot 10^{-4}$	$2.57 \cdot 10^{-4}$
^{58}Co	$4.13 \cdot 10^{-5}$	$7.49 \cdot 10^{-5}$	$9.89 \cdot 10^{-5}$	$1.79 \cdot 10^{-4}$	$2.92 \cdot 10^{-5}$	$5.29 \cdot 10^{-5}$
^{106}Ru	$3.17 \cdot 10^{-4}$	$6.24 \cdot 10^{-4}$	$7.58 \cdot 10^{-4}$	$1.49 \cdot 10^{-3}$	$2.24 \cdot 10^{-4}$	$4.41 \cdot 10^{-4}$
^{103}Ru	$2.98 \cdot 10^{-5}$	$5.28 \cdot 10^{-5}$	$7.14 \cdot 10^{-5}$	$1.27 \cdot 10^{-4}$	$2.11 \cdot 10^{-5}$	$3.73 \cdot 10^{-5}$

In 1995-1998 specialists from the German Hydrographic Institute, University of Hamburg in collaboration with MMBI investigated transport scenarios of ^{137}Cs in the Arctic Ocean (Nies et al., 1998) using a 3-dimensional numerical model and expedition observations. The transport of dissolved radioactive substances is based on the advective-diffusion equation. At the present time, Sellefield is the largest source of radioactive contamination in the Arctic Seas. Therefore the model was used to simulate the transport of radionuclides discharged from the Sellafield nuclear reprocessing plant from 1965 to the middle of 1990s.

^{137}Cs transported from the Irish Sea enters the southern and eastern parts of the Barents Sea after approximately 6 years. The maximum ^{137}Cs content in water was 30-60 Bq / kg. The main part of the Sellefield ^{137}Cs signature is transported via Atlantic waters to the Kara Sea between Franz-Josef Land and Novaya Zemlya. To a lesser extent ^{137}Cs is transported through the Karskie Vorota Strait (Fig. 8.12). The calculated maximum ^{137}Cs concentration in the strait in 1983 was about 40 Bq / m^3.

Modeling was also carried out on discharges from underwater nuclear objects. The radionuclide ^{137}Cs is also characteristic for the contents of sunken containers in the Novaya Zemlya bays of the Kara Sea. The hypothetical emissions scenarios were compared with the real situation of radionuclide transfer from Sellafield. A 2-5 Bq / m^3 correction was made for the 1990s concentration data to account for global fallout ^{137}Cs from atomic weapons explosions. Fallout from the Chernobyl accident added 50 Bq / m^3 in 1986 and 20 Bq / m^3 in 1995 to the values of ^{137}Cs concentrations in the Norwegian Current (Fig. 165) (Nies et al. 1998).

The modeling of hypothetical leakages of ^{137}Cs from underwater objects in the Abrosimov Bay (1 TBq / year) was carried out assuming both gradual and instantaneous release scenarios. The advection of contamination from the bay occurs in the surface water layer and the peak dispersion pathway in the Kara Sea occurs in 5-6 years (Fig. 8.13). For an instantaneous release, the ^{137}Cs concentration in the water area around the containers will be approximately 2000-3000 Bq / m^3 (Nies et al. 1998). ^{137}Cs will reach the Fram Strait in approximately 8 years. The dissolved concentration of ^{137}Cs decreases along the transport pathway from 1 to 0.1 Bq / m^3 in the Kara Sea to 0.01 Bq / m^3 in the East Greenland Current.

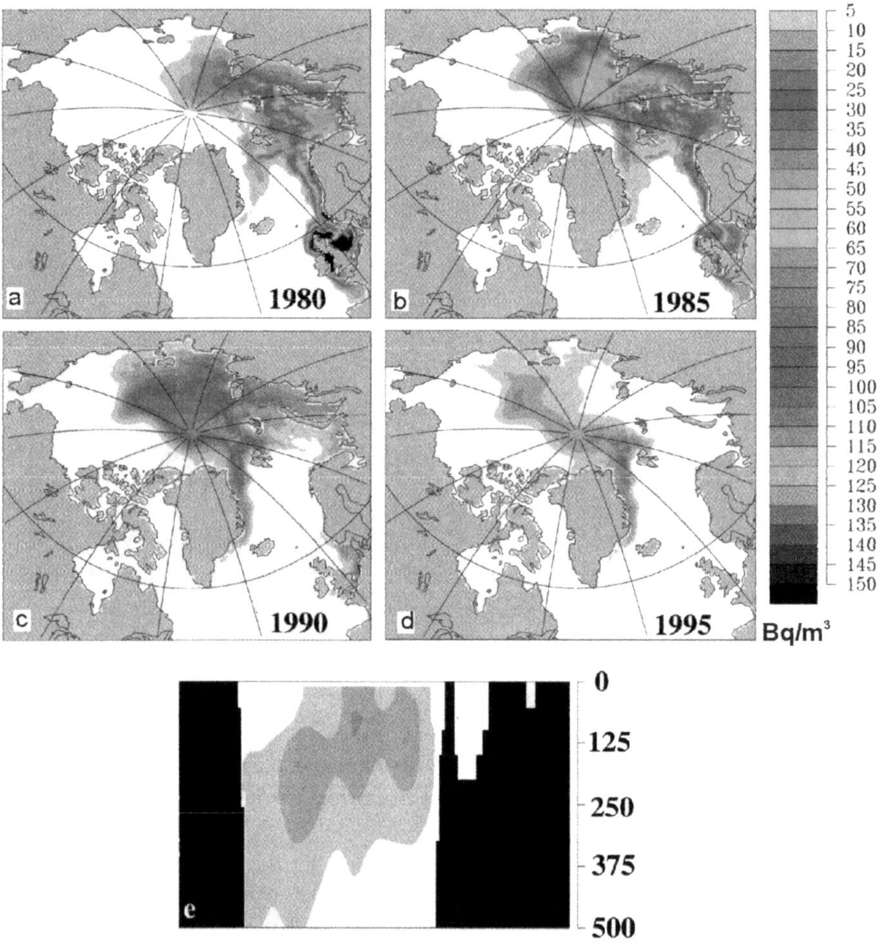

Fig. 8.12. Calculated concentrations of ^{137}Cs, resulting from Sellafield discharges (Nies et al. 1998)

The results briefly described here indicate that, the gradual release of ^{137}Cs from dumped objects will result in a concentration in the Kara Sea that is two-three orders lower than the actual maximum value of ^{137}Cs in 1983. Model scenarios conducted for $^{239,240}Pu$ indicate a similar finding. In the case of an instantaneous release, the ^{137}Cs content in the Kara Sea will be a little higher than the levels which existed in this basin during the period of maximum Sellafield releases.

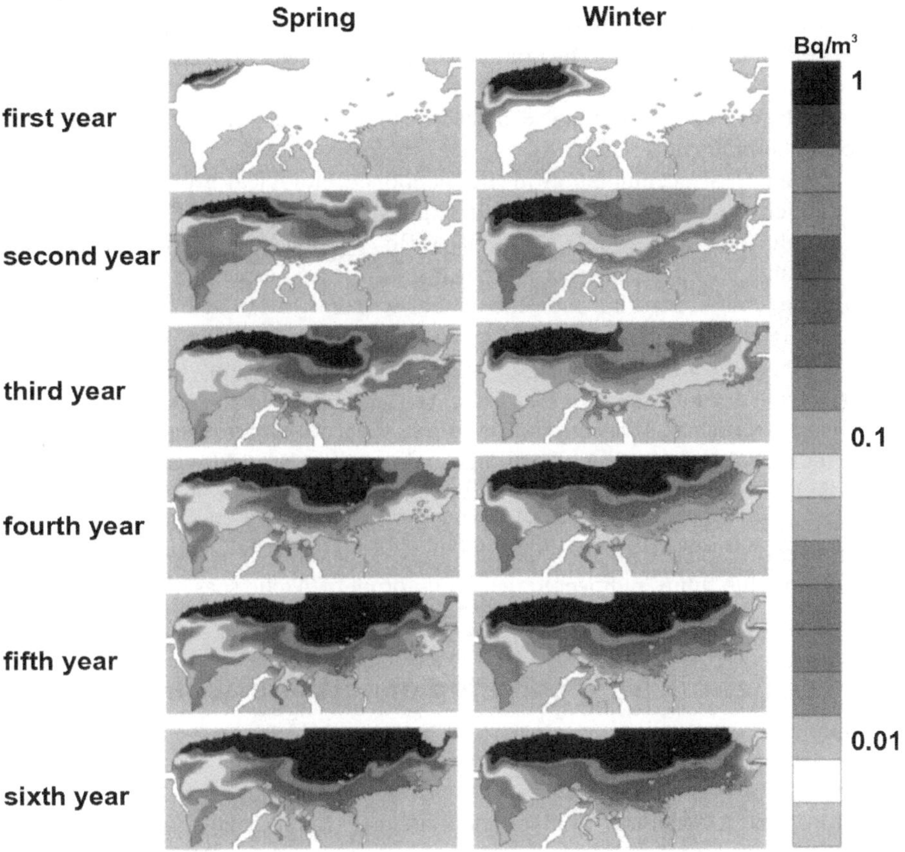

Fig. 8.13. Predicted concentrations of ^{137}Cs resulting from the continuous release of 2 TBq / year of ^{137}Cs in the Abrosimov Bay (Nies et al. 1998)

So, the large-scale contamination of the Barents Sea or Arctic Ocean as a result of dumped objects in the Novaya Zemlya bays seems scarcely probable when considering realistic release scenarios.

8.5 Withdrawal of commercial bioresources

The biological production of the Barents and Norwegian Seas - algae, mollusks, crustaceans, fishes, birds, sea mammals - is 4-5 million tons annually. During the

1960s, 2 million tons of Atlantic herring and about 1 million tons of cod were harvested. During the 1980s, 3 million tons of capelin and about 120 thousand tons of northern shrimp were harvested. In the mid-1990s , as much as 12 thousand tons of scallops were harvested. As a whole, during the 40-year period of fish harvesting activities, approximately 200 million tons of bioresources have been extracted from these seas. The amount of radionuclides removed from the marine ecosystem with the harvesting of sea products during the past 40 years may be tentatively estimated from these data on harvesting activities and radionuclide levels in commercial species.

^{137}Cs and ^{90}Sr accumulation in the mentioned biological groups of the Arctic Seas varied from 0.4-5.0 to 10-60 Bq / kg (during the period of atmospheric nuclear explosions) and from 0.2-4 to 10-70 Bq / kg otherwise. If the concentration of ^{137}Cs and ^{90}Sr in commercial organisms will be assumed for calculations to be on the order of 1 Bq / kg, then the annual withdrawal of each of these nuclides from the Barents and Norwegian Seas was about $1 x 10^9$ Bq. In total, during the whole atomic history, $4 x 10^{10}$ Bq of ^{137}Cs and the same quantity of ^{90}Sr could have been removed from these two seas only.

Two important conclusions follow. First, current scenarios and models of the transfer and transformation of artificial radionuclides should account for the annual decrease of radioactivity within the northern seas related to the harvesting of bioresources. Second, the commercial fishing industry is a relevant component of the process of self-purification and this must be taken into account in the context of planning the future radioecological monitoring of Arctic seas.

8.6 Framework for monitoring in northern seas

The ideal monitoring system for radioactive contamination of the marine environment (seawater, suspended particles, bottom sediments) and biota (plankton, macrophytes, zoobenthos, ichthyo- and avifauna, sea mammals) has yet to be realized. It is currently impractical to carry out routine radioecological surveys over all key geographical areas including numerous indicator species and other parameters that contribute to the dynamic state of the ecosystem.

Proceeding from these observations and the experience gained through our research in the northern seas, a framework for conducting monitoring is as follows. The objective of the system must be to use a combination of methods and approaches that will most reliably characterize the current radiation levels and impact on the Arctic environment and its biota. This system must also support the modeling of trends and scenarios associated with extreme cases of radiation contamination in northern seas. Two interrelated principles are taken as the basis of the system: first, the system must integrate cross-border, regional and local radioecological monitoring systems; second, the system must account for a core set of radiologically significant phenomena and factors (i.e. bioindicator organisms, clayey sediments, and secondary radioactive contamination). In addition to considering human doses the transfer and assimilation of radioactivity by marine organisms must also be an integral component of radioecological monitoring.

Mrine organisms perform an important biological function as purification systems for seawater toxicants. Based on our previous estimates of the amount of radioactivity removed by the harvesting of sea products between 1960s-90s the scale of this purification process is highly significant to the mass balance of radioactivity in northern areas. So far this factor has not been accounted for in the current radioecological framework. The removal of bioresources must be taken into account if we are to achieve a clear understanding of radiobiogeochemical phenomena and to control the balance between the influx of radionuclides to the basin and their assimilation into living organisms.

The challenge of detecting, analyzing, and forecasting radioecological changes must be solved using two interconnected approaches. First, regional and cross-border monitoring systems must be developed and maintained. Second, systematic investigations must be carried out for the essential phenomena and factors in radioecological assessments. Such phenomena and processes include lithodynamic and biological sorption of isotopes, their bioaccumulation, biosedimentation, and bioturbation, and these processes were already characterized above.

This complex approach, despite the patchy nature of radiological investigations in Arctic areas today, will increase significantly the quality of the coupled radiobiological and oceanographic estimates needed to describe general patterns in the investigated areas and for improving our forecasting capabilities.

On a practical level, the structure and organization of the radiation monitoring of biota and the environment of northern seas must focus on characterizing the complexities of the ecosystem. Observations must be carried out with sufficient regularity to identify radiochemical and radioecological trends. It is particularly necessary to routinely collect observations in order to reliably forecast changes in the condition of dangerous radiation objects. Routine procedures must be established for all aspects of the programme including sampling and processing procedures and the preparation and reporting of results. A programme may be designated as cross-border (international), regional (federal) or local based on the geographic scale and systematization level of the monitoring activities and procedures. It is also critical that all countries, which contaminate the northern seas participate in a programme of cross-border radioecological monitoring.

Cross-border monitoring surveys should be conducted across a network of standard hydrological sections every 5-7 years. The aim is to ascertain the patterns of cross-border transfer of pollutants in the pelagic and benthic ecosystems and also to predict the potential sequence of events in cases of emission from dumped objects on the shelf and in the bays. Regional monitoring programmes should focus on obtaining spatial observations in coastal zones and polar fronts. Within this context, it is important to monitor the background level of radiation related to the aggregate influence of local sources and releases of large northern rivers. The frequency of data collection must be no more than 2-3 years.

Local radioecological monitoring must be practiced every year within limited areas with special attention to ^{134}Cs, ^{60}Co and other short-lived radionuclides.

There are many objects of radiation danger in the coastal waters of the Barents and Kara Seas and, consequently, the probability of accidents, leading to the input of artificial radionuclides into the ecosystem, is higher than in other sea areas.

As a result of the operation of nuclear facilities in the inlets of the Kola and Motovsky Bays, small amounts of radiation are released in the form of aerosols and liquid discharges. Accidents with uncontrolled discharges of radioactive substances into adjacent waterways and into the atmosphere also occur in some years (Nielsen et al. 1996). In addition, local contamination near sunken atomic submarines (e.g. Komsomolets and Kursk) must be detected and controlled.

The bays on the eastern coast of the Novaya Zemlya, containing solid radioactive waste and several reactors with nuclear fuel, are also potentially dangerous areas. The potential also exists for artificial radionuclides to enter the Ob and Yenisey Bays from sources located in the upper parts of the rivers and their tributaries. These local problems, exemplify the need for local-scale radiation monitoring in some areas.

Conclusion

Radioactive contamination of marine ecosystems is considered one of the most dangerous consequences of the technological use of nuclear materials. Sea mammals and fish occupying the upper trophic levels are key organisms for investigating radio-biogeochemical processes in the northern seas and for estimating cumulative effects of environmental radiation. Through our own prolonged research on the ecological impacts from environmental radiation combined with an analysis of diversified literature we have developed new theoretical principles and conclusions on the behavior of marine organisms, ecological mechanisms of their purification in connection with exposure to global fallout from nuclear tests, the Chernobyl accident, and discharges of radioactive wastes.

The main conclusions of the work are the following.

1. Using our compiled database on the composition, concentrations and distribution of artificial radionuclides in northern seas we have created detailed maps of ^{137}Cs, ^{90}Sr, ^{60}Co, ^{239,240}Pu spreading in the Arctic and Subarctic environment and biota, including areas containing high contamination levels. From this information we have examined many classic principles of radioecology and have identified those without a factual basis.

2. Synthesizing knowledge from the fields of biology, radioecology and oceanography in order to achieve a true interdisciplinary understanding of the structure and function of the ecosystem of the northern seas, we have formulated a new paradigm for examining the role of radiation in the natural marine environment.

3. The input of global, regional, and local sources of radioactivity has been represented in a more complete way through analysis of sources and pathways of artificial radionuclides. In particular, radionuclide migration through deep-water troughs, which act as arteries for the transfer of contaminants across Arctic shelves is presented for the first time on the map «Levels and pathways of radionuclides in the Barents and Kara Seas». Discharges of plutonium from weapons facilities on the Yenisey and Ob Rivers are fixed in estuarine silts by high accumulation of ^{239,240}Pu (2-6 Bq / kg), ^{60}Co (1-6 Bq / kg) and ^{90}Sr (10-16 Bq / kg). Chernobyl fallout has been the main source of ^{137}Cs and ^{90}Sr for the Barents and Kara Seas during the 1990s. These radionuclides were brought both by advection of relatively contaminated Baltic waters and by wash-off from the catchment areas of large northern rivers.

4. During 1960-1990s the primary transformation of radioactive contamination leading to a decrease of nearly an order of magnitude has been the decay of ^{137}Cs and ^{90}Sr, reduces discharges, biofiltration, bioaccumulation and other natural purification processes. During the last 10 years ^{137}Cs concentrations in the upper water layers of northern seas has changed from 70 to 5 Bq / m^3. During the period of maximum discharges from Sellafield, the ^{137}Cs level in Scotland's coastal waters increased to 200-400 Bq / m^3 leading to ^{137}Cs accumulation in planktonic crustaceans to 70-150 Bq / kg. The zone of the Barents Sea polar hydrological front is

an important pelagic biofilter. It is here that several hundred million tons of new biomass is formed annually.

5. The coastal ecosystem is serves as a large filter and accumulator of anthropogenic radionuclides. Maximum levels of ^{137}Cs, ^{90}Sr, and ^{239,240}Pu are found in bottom sediments and zoobenthos of coastal bays. Radionuclide concentrations are higher by one-two orders of magnitude compared to shelf troughs. The macrophyte belt bordering the Arctic sub-littoral and littoral zone serves as a filtration system for coastal areas. Macrophytes taken from the Baltic Sea after Chernobyl contained fallout up to 1-5 thousand Bq / kg dry weight of ^{137}Cs. Local emissions from nuclear bases and dumping sites is partly removed by incorporation into macrophytes and sediments.

The classification of coastal ecosystems based on contamination levels and mechanisms of radionuclide fixation in coastal bays is presented for the first time. Bottom sediments of the first three types are characterized by the absence of ^{60}Co, and the quantity of ^{137}Cs, ^{90}Sr and ^{239,240}Pu depends on the size of the catchment area from which the radioactive global fallout originated. Types four to seven consist of bays (Yenisey, Chernaya, Zapadnaya Litsa, Stepovoy and others), where 60Co is present, along with high levels of ^{137}Cs, ^{90}Sr, ^{239,240}Pu and where radionuclide accumulation is a function of emission rates from local radiation sources.

6. Investigation of radionuclide accumulation in diverse benthic ecosystems has shown that ^{137}Cs, ^{90}Sr, ^{239,240}Pu in sediments (respectively 1-10, 0.2-3.0, 0.1-3.0 Bq / kg) and in zoobenthos (0.1-3.0, 0.01-0.40, 0.01-0.05 Bq / kg) reflects the dynamics of global fallout. Radionuclide concentrations in bottom sediments increase with decreasing distance to the Novaya Zemlya nuclear test sites and to West Siberian and Baltic rivers to 30-45, 4-6 and 5-15 Bq / kg. A similar trend is observed for benthic fauna. ^{137}Cs and ^{239,240}Pu concentrations correlate with sediment type increasing to maximum levels in clayey silts deposited in shelf troughs. The radioactive contamination of benthic organisms can be related to the feeding mode and food composition. The relatively high accumulation of ^{137}Cs by Polychaete worms illustrates the point.

7. The accumulation of radionuclides by different species is dependent on habitat, behavioral strategies. The efficiency of marine algae to accumulate ^{137}Cs increases in the following order: *Laminaria saccarina* → *Fucus vesiculosus* → *Fucus inflatus* → *Ascophyllum nodosum* → *Palmaria palmata* → *Fucus* Sp. → *Chorda filum*. ^{239,240}Pu accumulation is highest in the red algae *Ptilota plumosa*. Arranging organisms in a similar manner based on 137Cs accumulation, Bq / kg, is as follows: zoobenthos - Annelida (higher worms) polychaetes 4.7-6.2, chitin tubes - 10-20; → Bivalvia (bivalve mollusks) - 1-4 and to 9 Bq /kg; → Decapoda (decapode crustaceans) - 0.3-1.0 and to 8 (in shrimps) → Amphipoda - 2-3; → Echinodermata - 0.3-2.2; → Gastropoda - 0.3-1.1; Barents seabirds - Goosander (0.2) → black guillemot (0.43) → little auks (0.4-1.1) → common sea gull (about 1) → kittiwake (0.5-3.3) → big seagull (2.4-5.6) → great skua (3-40). Recent concentrations of ^{137}Cs in organs and tissues of newborn seals from the White Sea

are as follows, Bq / kg wet weight: in cartilage ribs and pancreas 0.6-1.0; in ovaries - 0.6; in cartilage and kidneys - 0.5; in heart, muscles, brain, gall bladder - 0.4; in liver, lungs, stomach and testicles - 0.3; in bones of limbs, vertebral column, ribs and fat - 0.2.

8. During 1980-1990s the level of radionuclide concentration in marine vertebrate fauna reflected global fallout patterns and was low. Fish, birds, seals and whales in northern seas had accumulated from 0.2 to 0.6 Bq / kg of ^{137}Cs and from 0.001 to 0.05 Bq / kg of $^{239,240}Pu$ (wet weight). The ^{137}Cs level in organs and tissues of fish and marine mammals is approximately double the concentrations found in their bones. In cases of local contamination sources, ^{137}Cs accumulation can reach 20-50 Bq / kg. ^{90}Sr is accumulated mainly in the bones of fish and seals (0.2-0.6 Bq / kg). It can reach 1-6 Bq / kg in the bones of Minke whales. A maximum value of ^{90}Sr (90 Bq / kg) was observed in 1963 in coalfish *Pollachius virens* from the Barents Sea.

The degree of radionuclide concentration by marine fauna is closely connected functionally with the species specificity of trophic chains, feeding behavior and metabolism but there are also interpopulation and geographical differences. Based on the database and conclusions of the body of work presented in this monograph, a tropho-dynamic model of artificial radionuclide assimilation in marine ecosystem food webs is formulated.

9. The mechanisms of transfer and accumulation of radionuclides in marine ecosystems is described conceptually. The essence of the model is that in the northern seas there exists a favorable combination of high biological activity, transboundary water exchange, thermohaline circulation and hydrodynamic conditions. As a result, these processes acting within the basins facilitate natural purification of radioactive contamination. The most important functions of this natural purification system are related to the lithological conditions and the large capacity for biological absorption. Salinity gradients along coastal margins are a dominant feature of these areas. The critical salinity for biological processes (osmotic regulation, biodiversity) and lythological processes (salinity flocculation) are correspondingly 5-8 and 10-12 ‰. Low salinity, for example, in estuaries, is related to higher ^{137}Cs content in sea fish (to 20-80 Bq / kg).

In general, low concentrations of ^{137}Cs, ^{90}Sr, $^{239,240}Pu$ in the environment and biota indicate a higher assimilation capacity of marine ecosystems in comparison with freshwater ecosystems. Factors such as the presence of clayey minerals, iron, and organic carbon, which accelerate the adsorption of ions to suspended sediments, are of main importance in the lythodynamic sorption system. The rate of biological sorption, which consists of the absorption of dissolved radionuclides through permeable body surfaces or through direct ingestion, depends on species behavior. Zooplankton and zoobenthos biomass serves as a quantitative indicator of the capacity of the biological filtration system. These organisms together with colonial birds and sea mammals incorporate contaminants into their life cycle through trophic transfer through food webs leading to humans.

10. A system for the radiation monitoring of marine ecosystems is developed, the essence of which consists in the synchronization of systems of transboundary,

regional and local control of changes of contamination concentrations in environment and biota and regular observations of phenomena that are needed to determine ecological safety and to solve the problems of radioactive waste disposal. Monitoring must include the study of clayey sediments (sorbing fraction less than 0.01 mm), and secondary contamination resulting from desorption and bioturbation, groundwater migration, and large-scale dredging of the benthic zone by trawling. The determination of radionuclide transfer from marine ecosystems to land through marketable biological resources is also important. About $4 \cdot 10^{10}$ Bq of ^{137}Cs and the same quantity of ^{90}Sr could be removed from the Norwegian and Barents Seas through harvesting. Useful bioindicator organisms are macrophytes (*Fucus vesiculosus*) and laminarias (*Laminaria saccarina*), calanuses (*Calanus finmarchicus*) and polychaetes (Annelida), cod (*Gadus morphua morphua*) and coalfish (*Pollachius virens*), big seagull (*Larus marinus*) and great skua. The most immediate task is the monitoring of ^{239,240}Pu since this is a long-lived radionuclide that accumulates in marketable bio-resources and marine ecosystems.

11. Low concentrations of ^{137}Cs, ^{90}Sr, ^{239,240}Pu in pelagic ecosystems and benthic zones have no toxic influence on the functioning of marine communities. People, feeding mainly on marine products from the Barents Sea, receive the lowest and not dangerous equivalent doses of radiation of order 0.6 μSv / year for population in general and 4 μSv / year for fishermen.

12. The trends in radionuclide concentrations from the beginning of nuclear testing until today and forms the basis for forecasting and modeling, are presented. All scenarios of dissipation of accidental radioactive discharges in the Kola and Abrosimov Bays are elaborated using numerical models, and shows that in case of any scenario of radioactive waste emission there will be no large-scale contamination of the Barents Sea or the Arctic Ocean. These accidents potentially cannot damage the marine biological resources. The high biological productivity, combined with thermohaline circulation, hydrodynamic and lythodynamic factors, neutralize the cumulative impact of radioactive contamination on marine ecosystems and causes the large-scale natural purification capacity of the Ocean.

References

Aagaard K., Carmack E.S. The Arctic Ocean and climate: a perspective, the polar ocean, and their role in shaping the global environment // American Geophysical Union / Ed. O. M. Johannessen, J. E.Overland. 1994. P.5–20.

Aagaard K., Foldvik A., Gammelsrod T., Vinje T. One-year records of current and bottom pressure in the strait between Nordanstlandet and Kvitoya, Svalbard, 1980–1981 // Polar Res. 1983. № 1. P.107–113.

Aagaard K., Carmack E. S., Swift J. N. Termohaline circulation in the Arctic Mediterranean Seas // J. Geophys. Res. 1985. № 90 (C5). P.4833–4846.

Aarkrog A. Environmental studies on Radioecological sensitivity and variability with special emphasis on the fall-out nuclides ^{90}Sr and ^{137}Cs // Report Riso-R-437, Riso National Laboratory. Denmark. 1979. P.31.

Aarkrog A. The radiological impact of the Chernobyl debris compared with that from nuclear weapons fallout // J. Environ. Radioactivity. 1988. № 6. P.151–162.

Aarkrog A. Chernobyl related monitoring and comparison with fallout data // Proceedings from Seminar on the radiological exposure on the population of the European community from radioactivity in North European Waters / Project "Marina", XI/4669/89 – EN Bruxelles: Commission of the European Communities, Directorate-General Environment, Nuclear Safety and Civil Protection. 1989. P.229–249.

Aarkrog A. Radioactivity in Polar Regions – main sources // J. Environ. Radioactivity. 1994. № 25. P.21–35.

Aarkrog A. Inventory of Nuclear Releases in the World, in Proc. NATO Advanced Study Inst. Advanced Course on Radioecolology, Zarechny, Russia, 19–28 June 1995. 1995. 12 p.

Aarkrog A. Doses to the Arctic Population // Materials of 3th International Conference on Environmental Radioactivity in the Arctic. Tromsø, 1997a. P.113–115.

Aarkrog A. AMAP Greenland 1994–1996 // Environmental Project № 356. 1997b. P.211–239.

Aarkrog A., Holm E., Dahlgaard H., Christensen G. Radioactive Tracer Studies in the Surface Waters of the Northern Atlantic Including the Greenland, Norwegian and Barents Sea // Rit. Fiskideldar. 1985. V.9. P.37–42.

Aarkrog A., Boelskifte S., Dahlgaard H., Holm E., Smith J. N., Duniec S., Hallstadius J. Tc-99 and Cs-137 as long Distance Tracers in Arctic Waters // Estuarine and Shelf Science. 1987. V.24. P.637–647.

Abaturov Yu.D., Abaturov A.V., Bykov A.V. Influence of the ionizing irradiation on pine forests in the near zone of the Chernobyl atomic power station. Moscow: Nauka, 1996. 240 p. *Abramov V.A.* On some results of investigations of the water masses of the Kara Sea northern part // Trudy AANII. 1985. V. 400. P. 143-155.

Adlandsvik B., Martinsen E. Model simulation of long-term transport of waters soluble substances // Posters Radioecology. 1996. P.123–124.

Ajbulatov N.A., Sapozhnikov Yu.A., Plishkin A.N., Politova N.V. Cesium-137 in the bottom sediments of the Kara Sea western part. // Doklady RAN. 1994. V. 335. No. 1. P. 98-100.

Aleksakhin R.M. Nuclear energy and biosphere. Moscow: Energoizdat, 1982. 215 p.

Aleksakhin R.M., Naryshkin M.A. Migration of radionuclides in forest biogeocoenoses. Moscow: Nauka, 1977. 124 pp.

Alekseenko V.A. Geochemical peculiarities of the initial period of the biosphere forming. Proceedings of the XI Congress of the Russian Geographical Society. Arkhangelsk, 2000. V.1. P. 117-133

AMAP (Arctic Monitoring and Assessment Programme), Arctic pollution issues: a state of the Arctic environment report, i-188, [AMAP Secretariat, Oslo, Norway]. 1997. 21 p.

Amundsen N. Radiocesium contamination in Norway // Radioactive Protection Dosimetry. 1995. V.62. P.53–57.

Andersen S. S., Livens F. R., Singleton D. L. Radionuclides in grey seals // Mar. Pollut. Bull. 1990. V.21, № 7. P.343–345.

Anisimova N.A., Luppova E.N. To fauna and its distribution of the Crustacea of the Pechora Sea // Modern state and perspectives of the Barents, Kara and Laptev Seas ecosystems researches. Abstr. Int. Conf. Murmansk, 1995. P. 12-14.

Anon. Report of the Arctic Fisheries Nordking Group // ICES. C.M. 1985. № 2. P.1–52.

A Group of Expert by the Commission of the European Communities: The Radioecological Exprosure of the Population of the European Community from the radioactivity in the North European Marine Waters / Project "Marina" Report № EUR 12483. 1990. 48 p.

A survey of artificial radionuclides in the Kara Sea. Results from the Russian-Norwegian 1992 expedition to the Barents and Kara Seas: Materials of joint Russian-Norwegian Expert group for Investigation of Radioactive Contamination in the Northern Areas. 1992. 90 p.

Baklanov A., Bergman R., Segerstahl B. Radioactive sources in the Kola Region: Actual and potential radiological consequences for man // Radiation Safety of the Biosphere. 1996. P.46–51.

Batrakov G.F., Eremeev V.N., Chudinovskikh T.V. Technogenous radioactivity of the Black Sea waters before and after Chernobyl // Diagnosis of the state of ecossytem of the Black Sea and of the land-sea contact zone.Sevastopol, 1997. P. 10-11.

Baxter M. S., Harms I. H., Osvath I., Povinec P. P, Scott E. M. Modelling the potential radiological consequences of radioactive waste dumping in the Kara Sea // Bellona Foundation, First comprehensive radioactivity measurements off the Kola coastline: Bellona Facts, 23 April 1998. 53 p.

Belov N.A., Lapina N.N. Bottom sediments of the Arctic Basin. Leningrad: Gidrometeoizdat, 1961. 53 pp.

Belopolsky l.O. Ecology of the Barents Sea colonial birds. Moscow-Leningrad: Nauka, 1957. 470 pp.

Belopolsky L.O. Composition of feed of the Barents Sea birds // Transactions of the Kaliningrad State University. 1971. Iss.6. P. 41-67.

Berger V.Ya. Adaptations of sea mollusks to the changes of environmental salinity. Leningrad: Nauka, 1986. 216 pp.

Birds of the USSR. Guillemots / Ed. by A.N. Golovkin. Moscow: Nauka, 1990. 205 pp.

Blindheim J., Foyn L., Martinsen E. A., Svendsen E. et al. The sunken nuclear submarine in the Norwegian sea – a potential environmental problem? // Fisken og Havet. 1994. № 7. 46 p.

Bobrov Yu.A. Primary production of the phytoplankton of the White Sea and the Barents Sea coastal zone // Plankton of the east Murman coastal waters. Apatity, 1982. 24 pp.

Bowen V. T., Wong K. M., Noshkin V. E. Plutonium-239 in and over Atlantic Ocean // J. Mar. Res. 1971. V.29, № 1. P.1–10.
Braginsky L.P., Velichko I.M., Merezhko A.I., Pasichny A.P. Buffer properties of river ecosystems in conditions of the complex radioactive and chemical contamination // Gidrobiologichesky Zhurnal. 1994. V.30, No 3. P.78-92.
Brooks J. M., Champ M. A., Baskaran M., Bryant W., Slowey N., Santschi P. N., Kennicutt I. I. Radionuclide contaminants released to the Russian Arctic from land-based and sea sources // Department of defense Arctic Nuclear Waste Assessment Program, FY's 1993–1994. ONP 322-95-5. 1995. 36 p.
Brungot A., Carroll J., Rudjord A., Foyn L. Presentation of the Norwegian national surveillance programme of radioactivity in the marine environment in the period 1996–1998 // Materials of the 4th Inter. Conf. on Environmental Radioactivity in the Arctic. Edinburgh, 1999a. P.189.
Brungot A., Carroll J., Rudjord A., Foyn L., Kolstad A. K., Brown J., Boy B., Hellstrom T. Radioactive contamination in the marine environment. Report № 3 from national surveillance programme // Stralevern Rapport. 1999b. № 6. P.124–125.
Champ M., Brooks J., Wade T. A research and monitoring strategy for assessment of Arctic environmental and human health risks // Background Paper for AMAP Expert Meeting and WG-G. Troms?, Norway.1994. P.1–7.
Chaplygin E.N. Oceanographic characteristics of the border seas (hydrochemical regime) // Soviet Arctic (seas, islands of the Polar Ocean). Moscow: Nauka, 1970. P.153-173.
Chebotina M.Ya., Kulikov N.V. Ecological aspects of investigations of radionuclide migration in the continental basins // Ecologiya. 1998. No 4. P.282-290.
Chelyukanov V.V., Saveljev V.A. On the radiation situation in the area of nuclear weapon test site on the Novaya Zemlya // Meteorologiya i gidrologiya. 1992. No 2. P.65-74.
Christiansen J. S., Bjørnstad H. E. Little evidence of radiocaesium in the White Sea navaga (Pisces: *Eleginus navaga*) // Mar. Pollut. Bull. 1996. V.32. P.689–690.
Chumichev V. B. Sr-90 discharge with main rivers of Russia into the Arctic Ocean during 1961–1990 // Materials the Second Inter. Conf. on Environmental Radioactivity in the Arctic 1995. Oslo. P.79–83.
Chumichev V. B. Long-term investigations of Sr-90 content in the water of the Arctic Ocean Seas // Materials of the Third Inter. Conf. on Environmental Radioactivity in the Arctic. Troms?, 1997. P.189–191.
Cook G. T., MacKenzie A. B., McDonald P., Jones S. P. Remobilization of Sellafield – derived radionuclides and transport from the north-east Irish Sea // J. Environ. Radioactivity. 1997. № 35. P.227–241.
Crene K., Galasso J. L. Arctic Environmental Atlas. 1999. 164 p.
Cross-Border Environmental Problems Emanating from Defence-Related Installations and Activities. Final report. V.1: Radioactive contamination. Rhase I: 1993–1995. Report № 204. North Atlantic Treaty Organization. 1995. 306 p.
Dahlgaard H. Radioecological situation of the Arctic seas // Materials of Inter. Meeting of Actual and potential consequences of dumping of radioactive waste into Arctic Seas. Oslo, 1993. P.24–27.
Dahlgaard H. On Tc-99, Cs-137 and Sr-90 in the Kara Sea // Environmental Radioactivity in the Arctic / Ed. P. Strand, A. Cooke. Osteras. 1995. P.91–95.

Dahlgaard H., Aarkrog A. Radiocaesium transport from the Irish Sea via the North Sea and the Norwegian Coastal Current to East Greenland // Rapp. P.-V. Reun. Cons. Int. Explor. Mer. 186. 1986. P.70–77.

Denisenko S.G. Ecology and resources of scallop in the Barents Sea. Apatity, 1989. 138 pp.

Denisenko S.G., Anisimova N.A., Dahle S., Larsen L.-H. Benthic researches // Results of the International (US - Norwegian - Russian) ecological expedition to the Pechora Sea, Novaya Zemlya, Kolguev, Vajgach and Dolgy Islands. July 1992. Preprint. Apatity: KSC RAS Publishing House, 1992. P.21-24.

Domanov M.M., Kontar E.A., Kosjan R.D., Sapozhnikov Yu.A., Shimkus K.N. Anthropogenic radionuclides in the bottom sediments of the Black Sea north-western part // Okeanologiya. 1996. V.36, No 4. P.550-555.

Dovgusha V.V., Tikhonov M.N. On the problem of the burial of radioactive wastes in the seas // Priroda. 1995. No 1. P.11-20.

Drobysheva S.S., Timofeev S.F. Eufausiides of the Barents Sea. Operations an informational summary. Apatity, 1990. 33 pp.

Dumping of radioactive waste and investigation of Radioactive contamination in the Kara Sea: Results from 3 years of investigation (1992–1994) in the Kara Sea. 1996. 55 p.

Duursma E. K., Carroll J. Environmental compartments: equilibria and assessment of processes between air, water, sediments and biota // Springer. 1995. P.23–26.

Egorov V.N., Polikarpov G.G., Kulebakina L.G., Stokosov N.A., Evtushenko D.B. Model of the large-scale contamination by long-living radionuclides Cs-137, Sr-90 caused by the Chernobyl accident // Vodnye Resursy. 1993. V.20, No 3. P.326-330.

Emelyanov E.M., Kravtsov V.A., Loktionov A.V., Pankratova N.A., Kravchishina M.D. Peculiarities of the hydrochemical regime and the geoecological situation in the Kursh Bay of the Baltic Sea // Geology of seas and oceans. Abstr. Rep. XIII Int. School on Marine Geology. Moscow, 1999. V.1. P.251-254.

Environment and ecosystems of the Franz Josef Land (Archipelago and shelf) / Ed. by G.G. Matishov. Apatity, Kola Sci. Centre Publ., 1993. 262 pp.

Erlenkeuser H., Balzer W. Rapid appearance of Chernobyl Cs-137 in the deep Norwegian Sea sediments // Oceanol. Acta. 1988. V.11, № 1. P.101–106.

Evolution of ecosystems and biogeography of the European Arctic Seas / G.G. Matishov, S.F. Timofeev, S.S. Drobysheva, V.M. Ryzhov. Saint-Petersberg: Nauka, 1994. 222 pp.

Fishek N. S., Nolan C. V. Fowler S. W. Scavenging and retention of metals by zooplankton fecal pellets and marine snow // Deep-Sea Res. 1991. V.38, № 10. P.1261–1275.

Foyn L., Nikitin A. The joint Norwegian-Russian expedition to the dumpsites for radioactive waste in the open Kara Sea, Tsivolki fjord and the Stepovogo fjord. Sept.–Octob. 1993: Report from the expedition on board r/v "V. Buinitskiy", with some preliminary results. 1993. 29 p.

Foyn L., Semenov A. Survey of artificial radionuclides in the Kara Sea. Final results from the Norwegian-Russian 1992 expedition to the Barents and Kara Seas: Draft version. 1993. 68 p.

Foyn L., Svaren I. The Barents Sea, its fisheries and the past and present status of radioactive contamination, and its impact on fisheries // Environmental Radioactivity in the Arctic / Ed. P. Strand, A. Cooke. Osteras. 1995. P.215–220.

Galkin Yu.I. Rate of change of the composition and boundaries of spreading of sea bottom and invertebrate animals // Problems of Quaternary paleoecology and paleogeography of the Barents and White Seas. Abstr. All-Union Conf. Murmansk, 1985. P.33-34.

GEBCO-97: The 1997 Edition of the GEBCO Digital Atlas. C.D.

Gerlach S. Marine pollution. Diagnostics and the Therapy. New York, 1981. 380 p.

Golovkin A.N. Role of birds in the marine ecosystems // Zoology of vertebrates. Results of science and technics. Moscow: VINITI, 1982. P.97-157.

Golovkin A.N., Shirokolobov V.N., Garkavaya G.P. Peculiarities of distribution of biogenous elements in the area of northern Novaya Zemlya bird bazaars // Peculiarities of biological productivity of waters near the Novaya Zemlya northern part. Leningrad; Nauka, 1972. P.46-62.

Golubchikov S. Who is in need of Novaya Zemlya // Energy: economics, technics, ecology. 1993. No 4. P.37-45.

Goncharov V.K., Klementyeva N.V. Forecasting of the spreading of contaminants in the marine environment from the places of their bottom burials // Okeanologiya. 1999. V.39, No 2. P.212-220.

Gordon C., Christensen G. C., Steinnes E. Radionuclides in Minke whale from the Arctic Ocean // Materials of the 4th Inter. Conf. on Environmental Radioactivity in the Arctic. Edinburgh, 1999. P.197–199.

Gromov V.V., Spitsyn V.I. Artificial radionuclides in the sea water. Moscow: Atomizdat, 1975. 224 pp.

Gromov V.V., Moskvin A.I., Sapozhnikov Yu.A. Technogenous radioactivity of the Ocean. Moscow: Energoatomizdat, 1985. 271 pp

Gueguenint P., Hermann J., Kershaw P., Bailly du Bois P., Baron Y. Artificial radioactivity in the English Channel and the North Sea // Radionuclides in the Oceans. 1997. P.23–40.

Gulin S.B., Polikarpov G.G., Aarkrog A., Egorov V.N., Nielsen S., Stokozov N.A. Goechronological investigation of the Cs-137 input to the bottom sediments of the Black Sea north-western shelf, continental slope and deep part // Reports on the National Academy of Scoences of Ukraine. 1997. No 7. P.133-136.

Gulliksen B., Lonne O. V. Distribution, abundance, and ecological importance of marine sumpagic fauna in the Arctic // Rapp. et P.-V. Reun. 1989. V.188. P.133–138.

Gunnerod T. B., Blakan I., Ugedal O., Skogland T. Freshwater and alpine ecosystem response to Chernobyl // Materials of Inter. Conf. Environmental contamination following major Nuclear accident. Vienna, 1990. V.1. P. 56–67.

Gurevich V.I. Cesium-137 distribution in the bottom sediments and benthos of the White, Barents and Kara Seas in summer 1991 // X Int. School on marine Geology. Abstr. Rep. Moscow, 1992. P. 34-36.

Hamilton T. F., Ballestra S., Baxter M. S., Gastaud J., Osvath I., Parsi P., Povinec P. P. Radiometric investigations of Kara Sea sediments and preliminary radiological assessment related to dumping of radioactive wastes in the Arctic Seas // J. Environ. Radioactivity. 1994. № 25. P.113–134.

Hardy J. E., Krey P. W., Volchok H. L. Global inventory and distribution of fallout plutonium // Nature. 1973. V.241. P.444–445.

Harms I. N., Karcher M. J., Dethleff D. Modeling Siberian river runoff – Implications for containant transport in the Arctic Ocean // Marine Systems. 1999. № 9. P.234–253.

Haug T., Nilssen K. T., Oien N., Potelov V. Seasonal distribution of harp seals (*Phoca groenlandica*) in the Barents Sea // Polar Res. 1994. V.13. P.163–172.

Hoibraiten S., Thoresen P. E. The sunken nuclear submarine "Komsomolets" and its effects on the environment // Materials of Inter. Conf. on Environmental Radioactivity in the Arctic. Oslo, 1995. P.350–358.

Holm E., Persson R., Roos P. Radiocesium and plutonium in Atlantic surface waters from 73° N to 72° S // Radionuclides the Study of Marine Processes / Ed. P. J. Kershaw, D. S. Woodhead. London: Elsevier Science, 1991. P.12–22.

Holm E., Persson R., Dahlgaard H. Radiocesium and transuranium elements in the Greenland and Barents Sea // Oceanologica Acta. 1993. V.6, № 4. P.457–461.

Hydrometeorology and hydrochemistry of the seas of USSR. V.1. Barents Sea. Iss.1. Hydrometeorological conditions. / Ed. by F.S. Terziev et al. Leningrad: Gidrometeoizdat, 1990. 280 pp

Hunt G., Kershaw P. L. Remobilization of artificial radionuclides from the sediments of the Irish Sea // J. Radial. Prot. 1990. V.10, № 2. P.147–151.

Ikдheimonen T. K, Ilus E. Monitoring of radioactivity in the vicinity of Finnish nuclear power plants 1991–1992 // STUK-A. 1995. V.121. 24 p.

Ikдheimonen T. K., Ilus E., Sjoblom K.-L. Monitoring of radionuclides of the Vicinities of Finnish Nuclear Power Stations in 1991–1992 // STUK-A. 1995. V.121. 96 p.

Iljenko A.I. Concentration of radionuclides by animals and their influence on the population. Moscow: Nauka, 1974. 173 pp.

Ilus E., Ikдheimonen T. K. Finnish studies on radioactivity in the Baltic Sea in 1987. Finnish centre for radiation and nuclear safety: Supplement 8 to Annual Report 1987 (STUK-A74). 1989. 17 p.

Ilus E., Ikдheimonen T. K. Study of radioactive substances in the Baltic Sea in 1986–1987. Finnish centre for radiation and nuclear safety // STUK-B. 1991. V.69. 23 p.

Ilus E., Sjoblom K.-L., Ikдheimonen T. K. Radioactivity of *Fucus vesiculosus* along the Finnish coast in 1987 // STUK-A. 1988a. V.83. 25 p.

Ilus E., Sjoblom K.-L., Ikдheimonen T. K. Monitoring of radioactivity in the environs of Finnish nuclear power stations in 1987 // Finnish Center for Radiation and Nuclear Safety. Helsinki. 1988b. 86 p.

Ilus E., Sjoblom K.-L., Ikдheimonen T. K. Monitoring of radionuclides in the Baltic Sea in 1989–1990 // STUK-A. 1993. V.103. P.35.

Impact of the radioactive contamination on the land ecosystems in the zone of Chernobyl accident (1986-1996). In 2 volumes. Syktyvkar, 1996. V.1. 204 pp. V.2. 114 pp.

Information on the accident on the Chernobyl power station and its consequences, prepared for IAEA // Atomnaya energiya. 1986. V.61, Iss.5. P. 301-320.

Investigation of the Distribution of Artificial Radionuclides. Data Report "Meteor" – Reise № 71. Federal maritime and Hydrographic Agency. 1997. 45 p.

Ionov V.V., Matishov D.G., Ionov D.V. Possibilities of GIS-technologies in the researches of the shelf marine geosystems // Biogeocoenoses of the West Arctic glacial shelves. Apatity: KSC RAS Publishing House, 1996. P.267-279.

Ivanov G. I. Assessment of Radioactive Contamination in the Pechora Region // Materials of the 4th Inter. Conf. on Environmental Radioactivity in the Arctic. Edinburgh. 1999. P.259–262.

Ivanov G. I., Bordukov Yu. K., Polyak L. V., Krinitski P., Neshcheretov A. V., Krasnyuk A. D., Ellis K. M. Assessment of Radiological Contamination from the Dumpsites in the

Kara Sea // Environmental Radioactivity in the Arctic / Ed. P. Strand, A. Cooke. Osteras. 1995a. P.162–163.

Ivanov G. I., Bordukov Yu. K., Polyak L. V., Rakhmanov A. Yu., Neshcheretov A. V., Shevchenko V. P. Multidisciplinary investigation of the Saint Anna Trough sediments // Environmental Radioactivity in the Arctic / Ed. P. Strand, A. Cooke. Osteras. 1995b. P.254.

Ivanov V.V. Water balance and water resources of the Arctic // Trudy AANII. 1976. V.333. P.4-24.

Izrael Yu.A. Radioactive contamination of hte Earth's surface // Vestnik RAN. 1998. V.68, No 10. P.898-915.

Izrael Yu.A., Tsyban A.V. Anthropogenic ecology of the Ocean. Leningrad: Gidrometeoizdat, 1989. 527 pp.

Izrael Yu.A., Tsyban A.V., Vakulovsky S.M. Radioactive contamination of seas and oceans // Meteorologiya i gidrologiya. 1994. No 10. P.15-23.

Johannessen O. M., Pettersson L. N., Bobylev L. P., Kouraev A. V., Neelov I. A., Volkov V. A., Hasselmann K. Observations and modelling of transport and dilution of radioactive waste and dissolved pollutants in the Kara Sea. NERSC Technical Report № 109. 1996. 65 p.

Josefsson D., Holm E., Persson B. R., Roos P., Smith J. N., Kilius L. Radiocaesium and I-129 along the Russian coast. Preliminary results from the Swedish Russian Tundra Ecology Expedition 1994 // Environmental Radioactivity in the Arctic / Ed. P. Strand, A. Cooke. Osteras. 1995. P.273–275.

JRNEG. Radioactivity in the marine environment radioactive contamination at dumping sites for nuclear waste in the Kara Sea // Joint Russian-Norwegian Expert Group for Investigation of Radioactive Contamination in the Northern Areas. Osteras: Norwegian Radiation Protection Authority, 1993. P.18–22.

Kanisch G., Nagel G. Radioaktivitat in Fischen uns der nord see // Inform. Fischwirt. 1992. V.39, № 1. P.158–169.

Kautsky N. Results of the Radiological North Sea Programme RANOSP 1974–1976 // Deutsche Hydrograph. Ztschr. 1980. V.34, № 4. P.152–157.

Kautsky N. Distribution and content of Cs-137+Cs-134 and Sr-90 in the water of the North Sea during the years 1982–1984 // Deutsche Hydrograph. Ztschr. Hamburg, 1986. V.39. P.139–159.

Kautsky N. Radioactive substance, pollution of the North Sea-an assessment leds. Berlin:Springer-Verlag, 1988. P.390–399.

Kershaw P. J. A review of radioactivity in the Irish Sea: A report prepared for the Marine Pollution Monitoring Management Group. 1992.12 p.

Kershaw P. J. Estimates of fluxes of Cs-137 in northern waters from recent measurements // The Science of the Total Environment. 1997. № 202. P.211–223.

Kershaw P. J., Baxter A. J. The transfer of reprocessing wastes from North-West Europe to the Arctic // Deep-Sea Res. 2. 1995. V.42, № 6. P.1413–1448.

Kershaw P. J., Woodhead D. S. (Ed.) Radionuclides in the study of marine processes. London; New York: Elsevier Applied Science, 1991. 391 p.

Khlebovich V.V. Critical salinity of the biological processes. Leningrad: Nauka, 1974. 230 pp.

Khrustalev Yu.P. Sedimentation patterns in the intercontinental seas of the arid zone. Leningrad: Nauka, 1989. 264 pp.

Kilezhenko V.P., Podymaknin V.N. Accumulation of dissipated elements by some organisms of the Barents Sea southern part // Materials of fishery researches of the Northern Basin. Murmansk: PINRO, 1964. Iss.4. P. 135-141.

Klenova M.V. Marine geology. Moscow-Leningrad: Uchpedgiz, 1948. 493 pp. *Кленова М. В.* Геология моря. М.; Л.: Учпедгиз, 1948. 493 с.

Klenova M.V. Geology of the Barents Sea. Moscow: Academy Publishing House, 1960. 368 pp.

Koblents-Mishke O.I. Photosynthetic primary production // Biollogical resources of the Ocean. Moscow: Nauka, 1985. P.48-62

Kolomiets B.I., Samorodov A.F., Filippov M.P. Radioactive wastes of atomic icebreakers, sources of forming, characteristics, processing and keeping // Materials of 4th Annual Sci. And Tech. Conf. of Nuclear Soc. Nizhny Novgorod, 1993. 12 pp. *Krasnov Yu.V., Matishov G.G., Galactionov K.V., Savinova T.N.* Sea colonial birds of the Murman. Sant-Petersberg: Nauka, 1995. 224 pp.

Krivolutsky D.A. Radioecology of communities of land animals. Moscow: Energoatomizdat, 1983. 87 pp.

Krivolutsky D.A. Dynamics of biodiversion and ecosystems in conditions of radioactive contamination of environment // Bioindication of radioactive contamination. Moscow; Nauka, 1999. P.5-20.

Krosshavn M. Environment pollution and oceanography in the Arctic-Epoca 95: Rapport FFI. 1995. 8 p.

Kulikov N.V., Chebotina M.Ya. Radioecology of freshwater ecosystems. Sverdlovsk: Ural Branch of Academy of Sciences Publishing House, 1988. 129 pp.

Kuptsov V.M., Lisitsyn A.P., Shevchenko V.P., Burenkov V.I. Suspended matter fluxes to ther Laptev Sea bottom sediments // Okeanologiya. 1999. V.39, No 4. P.597-604.

Kuzmenko M.I. Radioecological problems of water basins of Ukraine // Gidrobiologichesky Zhurnal. 1998. V.34, No 6. P. 95-119.

Kuzmenko M.I., Pankov I.V., Volkova E.N., Shirokaya Z.O. Strontium-90 and Cesium-137 content in the hydrobionts of Volga, Danube and Dnieper // Gidrobiologichesky Zhurnal. 1995. V.29. P.53-60.

Kuznetsov A.P. Trophic structure of the Kara Sea bottom fauna // Bottom fauna of the border seas of the USSR. Moscow: Institute of Oceanology, 1976. P.32-60.

Kuznetsov A.P. Ecology of the bottom communities of the Ocean. Trophic structure. Moscow: Nauka, 1980. 244 pp.

Kuznetsov A.P., Dando P., Shmelev I.P., Denisenko S.G., Efimov B.V., Demidov A. M., Shubko V.M. Radionuclides in the bottom fauna of the atomic submarine «Komsomolets» area (the Norwegian Sea) // Proceedings of RAS. Ser. Biol. 1996. No 4. P. 467-471.

Kuznetsov A.P., Shmelev I.P., Denisenko S.G., Efimov B.V., Demidov A. M., Shubko V.M Radionuclides 90Sr and 137Cs in the bottom fauna of the atomic submarine «Komsomolets» area // Benthos of the northern seas of Eurasia. Moscow: VNIRO, 1997a. P.13-18.

*Kuznetsov A.P., Dando P., Shmelev I.P., Denisenko S.G., Efimov B.V., Demidov A. M., Shubko V.M.*Radionuclides in the Kara Sae bottom sediments // Benthos of the northern sea sofEurasia. Moscow: VNIRO, 1997b. P.19-27.

Kuznetsov Yu.V. Radiochemical principles of unbalanced methods of dating of Oceanic sediments: Dr. of Chemistry Thesis. Leningrad, 1971. 21 pp.

Kuznetsov Yu. V. An assessment of the Contribution by the Enisey and Ob into total Radioactive Contamination of the Kara Sea // Environmental Radioactivity in the Arctic / Ed. P. Strand, A. Cooke. Osteras. 1994. P.72–74.

Kuznetsov Yu. V., Nosov N. A., Legin V. K. Dumping of radioactive wastes into the Sea environment: scientific and practical aspects // Materials of Intern. Conf. on Environmental Radioactivity in the Arctic and Antarctic. 1993. P.37–53.

Kuznetsov Yu.V., Rezenkov Yu.A., Legin V.K., Rakov N.A., Zhidkov V.V., Savitsky Yu.V., Tishkov V.P., Pospelov Yu.N., Egorov Yu.M. Of the estimate of the Yenisey River input to the total radioactive contamination of the Kara Sea // Radikhimiya. 1994. V.36, Iss.6. P. 546-552.

Landa E., Reimnitz E., Beals D., Pochkowski J., Rigor I. Transport of Cs-137 and Pu-239,240 by ice rafted debris in the Arctic Ocean // Arctic, accepted. 1999. P.123–127.

Lebedev I.A. On the influence of the deep Atlantic waters on the sea ice and hydrological processes in the Barents Sea northern part // Trudy AANII. 1992. V.430. P.145-156.

Lebedeva N.V. Ecotoxicology and biogeochemistry of the bird geograpfical populations. Moscow: Nauka, 1999. 199 pp.

Leontjev O.K. Ocean bottom. Moscow: Nauka, 1968. 270 pp

Leotjev O.K. Fundamentals of the Ocean physical geography. Moscow: Nayuka, 1974. 288 pp.

Leontjev O.K. On some peculiarities of the feeding of the Arctic Ocean by sedimented matter // Vestnik MGU. Ser. Geogr. 1984. No 1. P. 34-38.

Life and environment of the Polar Seas. Leningrad: Nauka, 1989. 240 pp.

Lisitsyn A.P. Processes of the Oceanic sedimentation. Moscow: Nauka, 1978. 390 pp.

Lisitsyn A.P. Marginal filter of the Ocean // Okeanologiya. 1994. No 5. P.735-747.

Lisitsyn A.P., Shevchenko V.P., Vinogradov M.E., Severina O.V., Vavilova V.V., Mitskevich I.N. Sedimented matter fluxes in the Kara Sea and Ob and Yenisey estuaries // Okeanologiya. 1994. V.34, No 5. P. 748-758.

Lisitsyn A.P., Vinogradov M.E. Global patterns of the life distribution in the Ocean and their reflection in the sediment composition. Formation and distribution of biogenous sediments // Reports of the Academy of Sciences of USSR. Ser. Geol. 1982. No 4. P.5-24.

Lisovsky I. The list of possible radiation accidents in the Ships and coastal objects of the Russian Navy // Materials of the 4th Inter. Conf. on Environmental Radioactivity in the Arctic. Edinburgh, 1999. P.149–152.

Livingston H. D., Kupferman S. L., Bowen V. T., Moore R. M. Vertical profile of artificial radionuclide concentrations in the central Arctic Ocean // Geochimic et Cosmochimica Acta. 1984. № 48. P.2195–2203.

MAFF/SEPA Radioactivity in Food and the Environment. Lowestooft, 1997, 1998.

MacDonald R. W, Bewers J. M. Contaminants in the Arctic marine environ-ment: priorities for protection. // J. Mar. Sci. 1996. V.53. P.537–563.

Marty Yu.Yu., Martinsen G.V. Problems of forming and using of the biological production of the Atlantic Ocean. Moscow: Nauka, 1969. 267 pp

Matishov D.G. Employment of the method of oxygen indication of the downward convection of the sea water in the Spitsbergen archipelago area (by the results of the RV «Polarstern» expedition in June-July, 1991). Preprint. Apatity: Kola Sci. Centre Publ., 1992b. 23 pp.

Matishov D.G. Bioindication of radiation state of the Barents Sea coastal waters and open areas // Arctic seas: bioindication of the environmental status, bioexamination and technology of destruction of pollutants. Apatity: Kola Sci. Centre Publ., 1992c. P.87-97.

Matishov D.G. Radionuclides in the bottom sediments and biota of the shelf and of Barents Sea coasts. Results of radioecological investigations conducted by MMBI in 1991-1992. Preprint. Apatity: Kola Sci. Centre Publ., 1993a.

Matishov D. Radionuclides in the bottom sediments and biota of the shelf and of Barents Sea coasts // Materials of Inter. Conf. on Environmental Radioactivity. Kirkenes, 1993b. P.247–256.

Matishov D.G. Radionuclides and biooceanological phenomena in the ecosystem of the Barents Sea. Cand. of Geogr. Thesis. Saint-Petersberg, 1994. 16 pp.

Matishov D.G. Artificial radionuclides in the environment and biota of the West ArcticSeas // Sci. Session of MMBI KSC RAS young researchers (March 1997). Abstr. Pres. Murmansk, 1997. P.26-27.

Matishov D.G. Levels and patterns of artificial radionuclides accumulation in the East Murman bays and the Azov Sea // Materials of scientific session of MMBI KSC RAS young researchers (April 1998). Abstr. Pres. Murmansk: «MIP-999» Co, 1998. P.56-62.

Matishov D.G. Content of artificial radionuclides in the Azov Sea bottom sediments // Materials of the Conference of MMBI young specialists devoted to the 275th Anniversary of the Russian Academy of Sciences (Murmansk, May 1999). Abstr. Pres. Apatity: Kola Sci. Centre Publ., 1999. P.78-79.

Matishov D., Rissanen K. Investigations of radiation in the Barents Sea Region and Lapland // Joint Workshop of Russian and Finnish Companies and Organizations. Abstr. Pres. Murmansk, 1993. P.10-13

Matishov D.G., Matishov G.G. Anthropogenous radionuclides in the Azov Sae bottom sediments // Dolklady RAN. 1998. V.363, No 5. P.673-677.

Matishov D., Namjatov A.A. Radionuclides in the ecosystems of the Kola Peninsula bays // Third Congress on Radiation Researches. Abstr. Pres. Moscow, 1997. P.313-314.

Matishov D.G., Shaban A.Yu. Some data on the biooceanology of the Barents Sea // Problems of Cainozoic paleoecology and paleogeography of the Arctic Ocean Seas: Abstr. Pres. 3rd All-Union Conf. Apatity: Kola Sci. Centre Publ., 1989. P.115-117.

Matishov D., Shaban A. On structure of water masses in straits of the Franz Josef Land // Franz Josef Land. With Emphasis on Tikhaia Bay, Hooker, Meibell Isl., Norsk Polarinstitute. Meddelseer. 1992. № 120. P.5–10.

Matishov D., Szczypa J. Gamma emitters in the Barents Sea Area // Proceeding of the XX Polar Symposium, Lublin. Poland, Lublin, 1993. P.211–228.

Matishov D.G., Matishov G.G., Szczypa Y., Pavlova L.G. New data on the conent of radionuclides in the Barents Sea and on the coast // Doklady RAN. 1993. V.332. No 1. P.118-119.

Matishov D.G., Matishov G.G., Rissanen K. Contamination of the White Sea bottom sediments by artificial radionuclides // Doklady RAN. 1995a. V.345, No 2. P.256-258.

Matishov D.G., Matishov G.G., Rissanen K., Pavlova L.G., Szczypa Y. Radionuclides in the West Arctic Seas // Proceedings of RAS. Ser. Geogr. 1995b. No 6. P.36-42.

Matishov D., Matishov G., Rissanen K. Peculiarities of radionuclide accumula-tion in benthic and fish of Barents and Kara Sea // Materials of Intern. Conf. on Environmental Radioactivity in the Arctic. Oslo, 1995c. P.233–237.

Matishov D., Rissanen K. Ikдheimonen T. Pu in fish, algae, and sediments in the Barents, Petshora and Kara Seas // Environmental Radioactivity in the Arctic / Ed. P. Strand, A. Cooke. Osteras. 1995b. P.227–232.

Matishov D.G., Matishov G.G., Rissanen K. Radioactive contamination of the Kola Bay of the Barents Sea // Doklady RAN. 1996. V.351. P.571-573.

Matishov D.G., Matishov G.G., Rissanen K. Level of strontium-90 accumulation in the bottom sediments and biota of the Barents and Kara Seas // Doklady RAN. 1997. V.353, No 5. P.700-702.

Matishov G.G. Ocean bottom in the Glacial period. Leningrad, Nauka, 1984. 176 pp.

Matishov G.G. World Ocean and glaciation of the Earth. Moscow: Mysl, 1987. 270 pp.

Matishov G.G. Ecological situation and problems of bioresources protection in seas of Northern Europe (at the example of the Barents Sea). Preprint. Apatity, 1989. 54 pp.

Matishov G.G. Anthropogenous destruction of the ecosystems in the Barents and the Norwegian Seas. Apatity, KSC RAS Publ., 1993. 116 pp.

Matishov G.G., Denisenko S.G. Present state of the bottom fauna of Chernaya Bay Novaya Zemlya Archipelago) // Doklady RAN. 1996. V.346, No 2. P.842-846.

Matishov G.G., Denisov V.V. Ecosystems and biological resources of Russian European Seas at the turn of the 21st century. Murmansk: MMBI, 2000. 118 pp.

Matishov G.G., Drobysheva S.S. General laws of structure and development of marine ecosystems of the European Arctic // Evolution of ecosystems and biogeography of the European Arctic Seas. Saint-Petersberg: Nauka, 1994. P.9-30.

Matishov G., Matishov D. The Investigation of concentration of radionuclides in the sediments in the Barents Sea // Polar Session. Arctic Natural Environment problems. Poland. Lublin, 1992. P.179–190.

Matishov G.G., Matishov D.G. The Arctic. New data. // Ecoinform. 1994. No 9. P.200-202.

Matishov G.G., Matishov D.G. Influence of radioactive contamination on the life activity of marine biota in the polar Arctic // Radioactive wastes, keeping, transportation, processing, influence on man and environment: Abstr. Pres. Saint-Petersberg, 1996. P.15-16.

Matishov G., Matishov D. New data on artificial radionuclides in the Bays of Kola Peninsula // Materials of the Third Inter. Conf. on Environment Radioactivity in the Arctic. Tromsш, 1997. P.34–35.

Matishov G.G., Pavlova L.G. General ecology and paleogeography of the poar oceans. Leningard: Naua, 1990. 224 pp.

Matishov G.G., Matishov D.G., Shaban A.Yu. New data on the role of troughs in the biooceanology of the Franz Josef Land shelf. Preprint. Apatity: Kola Sci. Centre Publ., 1992a. 48 pp.

Matishov G.G., Iljin G.V., Matishov D.G., Makarevich P.R., Timofeev S.F., Denisenko S.G. International US-Norwegian-Russian ecological expedition to the Pechora Sea, Novaya Zemlya, Kolguev, Vajgach and Dolgy Islands (July 1992, RV «Dalnie Zelentsy»). Preprint. Apatity: Kola Sci. Centre Publ., 1992b. 42 pp.

Matishov G.G., Matishov D.G., Podobedov V.E., Szczypa E.,Soletsky Ya., Yanush V. Radionuclides on the Kola Peninsula, Novaya Zemlya, Franz Josef Land and in the Barents Sea. Preprint. Apatity: Kola Sci. Centre Publ., 1992c. 67 pp.

Matishov G.G., Iljin G.V., Matishov D.G. Summer oceanological situation in the Pechora Sea (Materials of expedition in the 67 cruise of RV «Dalnie Zelentsy», July 1992). Preprint. Apatity: Kola Sci. Centre Publ., 1993a. 50 pp.

Matishov G.G., Matishov D.G., Szczypa E. Cesium-137 in bottom sediments, biota of the Barents Sea shelf and coast // Contamination of the seas along the CIS coast, mainly the Arctic. Abstr. Pres. Part 1. Arkhangelsk, 1993b. P.52-54

Matishov G., Matishov D., Szczypa E. Cs-137 in the bottom sediments and biota of the shelf and Barents coast. Pollution of the seas around the coasts of the GIS with special emphasis on the Arctic // Materials of Inter. Conf. Archangelsk, 1993c. P.52–54.

Matishov G.G., Matishov D.G., Nazimov V.V. Levels and main pathways of the radionuclides in the Barents and Kara Seas. The map of scale 1:4704075. Rovaniemi, Finland. 1994a.

Matishov G.G., Matishov D.G., Szczypa E., Pavlova L.G. Radioactive contamination of the environment and biota on the Novaya Zemlya because of nuclear tests // Doklady RAN. 1994b. V.337, No 6. P.824-826

Matishov G.G., Matishov D.G., Szczypa E., Rissanen K. Radionuclides in the ecosystem of the Barents and the Kara Seas region. Apatity: Kola Sci. Centre Publ., 1994c. 237 pp.

Matishov G.G., Matishov D.G., Rissanen K. Peculiarities of the radionuclide accumulation in benthic organisms and fishes of the Barents and Kara Seas // Doklady RAN. 1995a. V.342, No 1. P.134-135.

Matishov G.G., Matishov D.G. Rissanen K., Rachor E. Cs-137 levels in the Laptev Sea bottom sediments. Doklady RAN. 1995b. V.334, No 6. P.810-811.

Matishov G., Denisov V., Matishov D. Factors of environmental risk connected with oil and gas extraction in Barents and Kara Seas // Ecological Summit-96. Copenhagen. 1996a. 31 p.

Matishov G.G., Galaktionov K.V., Timofeev S.F., Pavlova L.G., Matishov D.G. Tasks of marine biology and the nature protection in russian Arctic // Materials of VII Congress of Russian Hydrobiological Society. Kazan, 1996b. P.5.

Matishov G.G., Iljin G.V., Matishov D.G. General patterns of oceanological regime and oceanological conditions of the ice-free period // Ecosystems, biological resources and anthropogenic contamination of the Pechora Sea. Apatity: Kola Sci. Centre Publ., 1996c. P.25-40.

Matishov G., Matishov D., Namjatov A. Radioactive Contamination of sediments in the Kola Fiord near RTP "Atomflot", the base of the nuclear fleet // Reducing Wastes from Decommissioned Nuclear Submarines in the Russian Northwest. Proceedings the NATO Advanced Research Workshop. 1996d. P.85–101.

Matishov G.G., Matishov D.G., Rissanen K. Cesium-137 in the Barents Sea birds and seals // Doklady RAN. 1996e. V.351, No 6. P. 844-846

Matishov G.G., Matishov D.G., Rissanen K. Radioactive contamination of grounds and bottom organisms of the West Arctic shelf // Biogeocoenoses of the West Arctic glacial shelves. Apatity: Kola Sci. Centre Publ., 1996f. P.251-266.

Matishov G.G., Pavlova L.G., Matishov D.G. Role of hydrochemical and radiochemical phenomena in the biological productivity of the Barents and Kara Seas // X Int. Conf. on the Fishery Oceanology. Abstr. Pres. Moscow, 1996g. P.42.

Matishov G., Matishov D., Namjatov A. Distinctions in conditions of Cs-137 accumulation by bottom sediments in the West Arctic and Southern Russia Seas // Materials of the

Third Inter. Conf. on Environmental Radioactivity in the Arctic. Tromsш, 1997a. P.235–236.

Matishov G.G., Matishov D.G., Namjatov A.A. Levels and conditions of accumulation of artificial radionuclides in the Kola and Motovsky Bays // Doklady RAN. 1997b. V.357, No 6. P.812-814.

Matishov G.G., Matishov D.G., Namjatov A.A. Levels and main pathways of the artificial radionuclide transfer in the West Arctic Seas // Third Congress on radiation researches. Abstr. Pres. Moscow, 1997c. P.313-314.

Matishov G., Matishov D., Namjatov A. Modern level of the radioactive contamination in the coast water of the Barents and White Seas // Materials of the Third Inter. Conf. on Environmental Radioactivity in the Arctic. Tromsш, 1997d. P.239–241.

Matishov G., Matishov D., Namjatov A. Risk assessment from Real and Poten-tial Sources of Radioactivity Contamination in the Kola Bay of the Barents Sea // Materials of Inter. Conf. "Bordomer 97" – Coastal Environment Management and Conservation. 1997e. T.2. P.92–101.

Matishov G.G., Matishov D.G., Namjatov A.A., Zuev A.N., Kirillova E.E. Radionuclides in the bay ecosystem // Kola Bay: oceanography, biology, ecosystems, pollutants // Apatity: Kola Sci. Centre Publ., 1997f.

Matishov G.G., Makarevich P.R., Matishov D.G., Larionov V.V., Lubin P.A., Mityaev M.V., Kondakov A.A., Iljin G.V. Complex ecological investigations of the Azov Sea (by the results of the MMBI expedition on the RV «Geophysik», September 1997). Preprint: Apatity, Kola Sci. Centre Publ., 1998a. 64 pp.

Matishov G.G., Matishov D.G., Namjatov A.A. Artificial radionuclides in the Kara Sea bottom sediments // Biology and oceanography of the Kara and the Barents Seas (on the Northern Sea Route). Apatity: Kola Sci. Centre Publ., 1998b. P.422-430.

Matishov G.G., Matishov D.G., Namjatov A.A., Kukina N.A., Mityaev M.V. Peculiarities of cesium-137 accumulation by the different types of bottom sediments in the Barents and Kara Seas coastal waters // Litologiya i poleznye iskopaemye. 1998c. No 5. P.540-543.

Matishov G.G., Volkov V.A., Denisov V.V. Modern views of the structure of warm Atlantic waters circulation in the Barents Sea northern part // Doklady RAN. 1998d. V.362, No 4. P.553-556.

Matishov G.G., Matishov D.G., Namjatov A.A. Content of artificial radionuclides in the Azov Sea bottom sediments // Modern development of estuarine ecosystems on the example of the Azov Sea. Apatity: Kola Sci. Centre publ., 1999a. P.249-271.

Matishov G.G., Matishov D.G., Namjatov A.A. Content of strontium-90 and plutonium-239,240 in the Azov Sea bottom sediments // Doklady RAN. 1999b. V.368, No 4. P.552-554.

Matishov G., Matishov D., Namjatov A. Distinction in conditions of Cs-137 accumulation by bottom sediments in the West Arctic and Southern Russia Seas // Materials of the 4th Inter. Conf. on Environmental Radioactivity in the Arctic. Edinburgh, 1999c. P.239–241.

Matishov G.G., Matishov D.G., Namjatov A.A. Estimate of the contamination of the Kola Bay marine ecosystem elements at the permanent and instantional discharge of radioactive wastes // Modern technologies and forecasting in the polar oceanology and biology. Apatity: Kola Sci. Centre Publ., 1999d. P.395-401.

Matishov G., Matishov D., Namjatov A. Modern level of the content of Cs-137 in fish and seaweed of the Barents sea // Materials of the 4th Inter. Conf. on Environ-mental Radioactivity in the Arctic. Edinburgh, 1999e. P.242–243.

Matishov G.G., Matishov D.G., Namyatov A.A., Berestovsky E.G., Moiseev D.V., Bronnikov I.V. Biooceanographical indications of the cooling in the West Arctic // Doklady RAN. 1999f. V.368, No 2. P.254-258.

Matishov G., Matishov D., Namjatov A., Carroll J., Dahle S. Antropogenic radionuclides in Kola and Motovsky Bays of the Barents Sea, Russia // J. Environ. Radioactivity. № 43. 1999g. P.77–78.

Matishov G., Matishov D., Namjatov A., Carroll J., Dahle S. Human Health Impacts from Antropogenic radionuclides in Kola and Motovsky Bays of the Barents Sea // Materials of the 4th Inter. Conf. on Environmental Radioactivity in the Arctic. Edinburgh, 1999h. P.99–100.

Matishov G., Matishov D., Namjatov A., Carroll J., Dahle S. Modern level of the radioactivity contamination in the coastal water of the Barents and White Seas // Materials of the 4th Inter. Conf. on Environmental Radioactivity in the Arctic. Edinburgh, 1999i. P.329–241.

Matishov G., Matishov D., Namjatov A., Carroll J., Dahle S. Discharges of nuclear waste into the Kola Bay and their impact on human radiological doses // J. Environ. Radioactivity. 2000a. V.48. P.5–21.

Matishov G.G., Matishov D.G., Namjatov A.A., Carroll J., Dahle S. Geochronological investigations of the artificial radionuclides input into bottom sediments of the coastal waters of the Barents Sea // Long-term changes of the Arctic marine ecosystems. Climate, marine periglacial, bioproductivity. Abstr. Pres. Murmansk, 2000b. P.96-99.

Matishov G.G., Matishov D.G., Namjatov A.A., Risik N.S., Bufetova M.V. Radioactive contamination of environment // Regularities of oceanological and biological processes in the Azov Sea. Apatity: Kola Sci. Centre Publ., 2000c.

Meese D., Cooper L., Larsen I. L., Tucker W., Reimnitz E., Grabmeier J. Radionuclides in sea ice // Proceedings of Arctic Nuclear Waste Assessment Program Workshop. Woods Hole, MA. 1995. 14 p.

Melnikov I.A. Ecosystem of the Arctic sea ice. Moscow, 1989. 191 pp.

Merezhko A.I., Pasichny A.P., Velichko I.M., Gusynskaya S.L., Yakubovsky K.B., Sherstyuk V.V., Pasichnaya E.A., Savitskaya L.L. Production of hydrobionts in conditions of radioactive contamination of rivers // Gidrobiologichesky Zhurnal. 1996. V.32, No 5. P.22-26.

Methods of the control of contamination of aquatic objects (MVI.01-7/96). /Ed. by A.I. Nikitin. 1996.

Miretskiy G. I., Lebedev G. B., Komarov E. I., Popov A. O., Ramzaev P. V. New radioecological data for the Russian Federation (from Alaska to Norway) // Environmental Radioactivity in the Arctic and Antarctic / Ed. P. Strand, A. Cooke. Osteras, 1993. P.269–272.

Miretskiy G. I., Lebedev G. B., Komarov E. I., Popov A. O., Ramzaev P. V. Radioecology situation and health status of native population in the far east of Russian Arctic // Environmental Radioactivity in the Arctic / Ed. P. Strand, A. Cooke. Osteras. 1995. P.303–306.

Moiseenko T.I., Kudryavtseva L.P. Моисеенко Т. И., Кудрявцева Л. П. Ecotoxicological estimate of the technogenous hydrogeochemical anomalies (on example of the Kola Peninsula) // Geokhimiya. 1999. No 10. P.1112-1130.

Moiseev A.A. Ramzaev P.V. Cesium-137 in the biosphere. Moscow: Atomizdat, 1975. 182 pp.

Namjatov A.A. Influence of the dumps of «Atomflot» enterprise on the ecosystem of the Kola Bay // Modern state and perspectives of the Barents, Kara and Laptev Seas ecosystems researches. Abstr. Pres. Murmansk, 1995. P.67-68.

Namjatov A.A. Radiation contamination of the Kola and Motovsky bays of the Barents Sea. Cand. Geogr. Sci. Thesis. Saint-Petersberg, 1998. 16 pp.

Nansen Centennial Arctic Programme. Plan for a scientific expedition 1994–1996 / Ed. T. O. Vorren. Oslo, 1992. 25 p.

Nesvetova G.I., Chepurnov A.N. Seasonal variability of the Barents Sea primary production // Problems of the fishery oceanology of the Northern Basin. Murmansk, 1989. P.139-154.

Nies H., Nielsen S. P. Radioactivity in the Baltic Sea // Radionuclides in the Oceans, Inputs and Inventories. Coordinators: Guequeniat, Germain and Metivier, Les editions de Physique, Les Ulis, France. 1996. P.219–231.

Nies H., Bahe C., Harms I. H. Transport and Dispersion of Artificial Radio-activity in the Arctic Ocean – Model Studies and Observations // Radioprotection-Colloques. 1997. V.32. P.46–49.

Nies H., Harms I. H., Bahe C., Karcher M. J., Dethleff D., Kuhlmann G., Oberhuber J. O., Backhaus E., Kleine E., Loewe P., Matishov D., Stepanov A., Vasiliev O. F. Anthropogenic Radioactivity in the Nordic Seas and the Arctic Ocean – Results of a Joint Project // German J. Hydrography. 1998. V.50, № 4. P.313–343.

Nikiforov E.G., Shpajkher A.O. Patterns of the formation of large-scale variations of the Arctic Ocean hydrological regime. Saint-Petersberg: Gidrometeoizdat, 1980. 270 pp.

Nikitin A., Foyn L. A joint Norwegian-Russian project "Monitoring of the radioactive contamination in the sea environment in the area of discharge of wastes from the liquid radioactive waste treatment plant at "Atomflot": Results of the background survey, 1998. Arctic Radioecology-Aquatic. P.20–22.

Nikitin A.I., Katrich I.Yu.,Kabanov A.I.,Chumichev V.B., Smagin V.M. Radioactiv e contamination of the Arctic Ocean by results of observations in 1985-1987 // Atomic energy. 1991. V.71, Iss.2. P.169-172.

Nikitin A., Chumichev V. B., Vakulovsky S. M. Results of field investigation of the sea water radioactive contamination during some operations on low-level liquid radioactive waste dumping into the Barents Sea in 1984–1990 // Materials of the 4th Inter. Conf. on Environmental Radioactivity in the Arctic. Edinburgh, 1999. P.179– 180.

Nikitin A., Bovkun L. A., Katrich I. Yu., Chumichev V. B., Valetova N. K. Field investigation of marine environment radioactive contamination in the seas of Eastern Russian Arctic during the year 1993 // Materials of the Third Inter. Conf. on Environmental Radioactivity in the Arctic. Troms?, 1997. P.41–42.

Nikitin A., Chumichev V. B., Valetova N. K. et al. Joint Russian-Norwegian Collaboration on Radioactive Contamination from dumped nuclear Waste in the Kara Sea – 3 years with Expeditions to the Dumping Sites: Goals and Investigation // Environmental Radioactivity in the Arctic / Ed. P. Strand, A. Cooke. Osteras. 1995. P.125–128.

Nilsen T., Kudrik I., Nikitin A. The Russian Northern Fleet; Potential risk of radioactice contamination of the region // Bellona rapport. 1996. V.2. 168 pp.

Novitsky M. A., Vakulovsky S. M., Nikitin A. I., Tereb L. A. Evaluation of the Kara Sea radioactive contamination as a consequence of potential accidents at the Mayak PA // Materials of the 4th Inter. Conf. on Environmental Radioactivity in the Arctic. Edinburgh, 1999. P.36–38.

Novitsky V.P. Constant currents of the Barents Sea northern part // Trudy GOIN. 1961. Iss. 64. P.3-32.

Paluszkiewicz T., Hibler L. F., Richmond M. C., Bradley D. J. An assessment of the flux of radionuclide contamination through the Ob and Yenisei rivers and Estuaries to the Kara Sea // Environmental Radioactivity in the Arctic / Ed. P. Strand, A. Cooke. Osteras. 1995. P.38–51.

Panteleev G. P., Livingston H. D., Sayles F. L., Medkova O. N. Deposition of plutonium isotopes and Cs-137 in sediments of the Ob Delta from the beginning of the nuclear age // Environmental Radioactivity in the Arctic / Ed. P. Strand, A. Cook. 1995. P.57–64.

Pavlidis Yu.A., Shcherbakov F.A., Shevchenko F.Ya. Clayey minerals of the bottom sediments of the Cuba and White Sea shelves: geology and climate - comparisons // Oceanologiya. 1995. V.35, No 1. P.121-127.

Pavlik M., Zigmunt J., Matishov D., Szczypa J. Gamma emiters in the Tshoshska and Pechora Bays // Polar Session. Poland. Lublin, 1996. P.245–253.

Pavlyutin A.P., Babitskaya V.A. Higher water vegetation in the lake, contaminated by radionuclides, composition, distribution, stocks and Cs-137 accumulation // Gidrobiologichesky Zhurnal. 1996. V.32, No 4. P.79-86.

Pavlov V. K., Stanovoy V. V., Nikitin A. I. Possible causes of radioactive contamination in the Laptev Sea // Land-Ocean System in the Siberian Arctic: Dynamics and History. Berlin: Springer-Verlag, 1999. P.65–72.

Pavshtiks E.A. Seasonal changes in the plankton and feeding migrations of herring // Trudy PINRRO. 1956. V.9. P.93-124.

Peredelsky A.A. Fundamentals and tasks of the radioecology // Zhurnal obshchej biologii. 1957. V.18. P.17-30.

Petrov Yu.M., Fedorov A.F. Strontium-90 content in the bones of Greenland seal and crested seal // Materials of fishery researches of the Northern Basin. Murmansk, 1967. Iss. 10. P.195-197.

Pfirman S. L., Kogeler J. W., Rigor I. Potential for rapid transport of contaminants from the Kara Sea // The Science of the Total Environment. 1997. V.202. P.185–198.

*Polikarpov G.G.*Radioecology of marine organisms. Moscow: Atomizdat, 1964a. 254 pp.

Polikarpov G.G. Some biological aspects of radioactive contamination of seas and oceans // Radioactive contamination of seas and oceans. Moscow: Nauka, 1964b. P.98-125.

Polikarpov G.G. Radioecology of marine organisms. Moscow: Atomizdat, 1971. 275 pp.

Polikarpov G.G., Aarkrog A. Problems of Eurasian radioecology. Sources of radioactive contamination of environment in the former USSR // Radiobiologiya. 1993. V.33, Iss.1. P.15-24.

Polikarpov G.G., Egorov V.N. Marine dynamic radiochemoecology. Moscow: Energoatomizdat, 1986. 176 pp.

Potanin V.A., Korotkov C.V., Ershtadt T.A. On the problem of the Barents Sea water circulation // Problemy Arktiki i Antarktiki. 1989. Iss.64. P.110-117.

Pozolotina V.N. Adaptative processes of plants in conditions of radiation impact // Ecologiya. 1996. No 2. P.11-116.

Radioactive contamination at dumping sites for nuclear waste in the Kara Sea // Materials of joint Russian-Norwegian Expert group for Investigation of Radioactive Contamination in the Northern Areas. 1994. 122 p.

Radioecological consequences of the Chernobyl accident / Ed. by I.I. Kryshev. Moscow, IAE Publ., 1991. 199 pp.

Radiosensitivity and postradiation recovery of plants / Ed. by N.V. Kulikov. Sverdlovsk: Ural Branch of RAS, 1989. 80 pp. *Rahola T., Tillander M., Lilukka E., Jaakola T., Miretsky G. I., Doudarev A. A., Theodorovich O. A., Popov A. O.* Intercomparison of Finnish and Russian whole-body counters used for the determination of Cs-137 body burdens in reindeer-herding populations // Environmental Radioactivity in the Arctic / Ed. P. Strand, A. Cooke. Osteras. 1995. P.294–298.

Rissanen K., Rahola T. Radiocesium in lichens and reindeers after Chernobyl accident // Rangefer. Finnish Center for Radiation and Nuclear Safety. 1992. № 3. P.55–61.

Rissanen K., Matishov G., Matishov D. Radioactivity level in Barents, Petshora, Kara, Laptev and White Seas // Environmental Radioactivity in the Arctic / Ed. P. Strand, A. Cooke. Osteras. 1995. P.208–214.

Rissanen K., Matishov G., Matishov D., Ikəheimonen T., Tarja K. Radionuclide concentrations in sediments, soil and plant samples from the archipelago of Franz Joseph Land, an area affected by the Chernobyl fallout // The Science of the Total Environment. 1996. P.325–327.

Rissanen K., Ikəheimonen T., Matishov D., Matishov G. Radioactivity levels in fish, benthic fauna, seals and sea birds collected in the Northwest Arctic of Russia // Radioprotection-Colloques. 1997a. V.32. P.C2-323–C2-331.

Rissanen K., Ikəheimonen T., Nielsen S., Matishov D., Matishov G. Gammanuc-lide and Pu concentrations in the White Sea // Materials of the 3th Inter. Conf. on Environmental Radioactivity in the Arctic. Tromsш, 1997b. P.222–224.

Rissanen K., Matishov G., Matishov D., Ikəheimonen T. Radioactivity levels in Kola bay // Radiation Protection Dosimetry. 1998. V.75, № 1–4. P.223–228.

Rissanen K., Ikəheimonen T., Matishov D., Matishov G. Radionuclide concentrations in sediments, soil and plant samples from the archipelago of Franz Joseph Land, an area affected by the Chernobyl fallout // Materials of the 4th International Conference on Environmental Radioactivity in the Arctic. Edinburgh, 1999a. P.325–327.

Rissanen K., Ikəheimonen T., Pempkowiak J., Matishov D., Matishov G. Cs-137, Sr-90, Pu-239,240 and selected metal concentrations in organs of Greenland seal pups in the White Sea area // Materials of the 4th International Conference on Environmental Radioactivity in the Arctic. Edinburgh. 1999b. P.186–188.

Roos P., Holm E., Persson R. B. R., Aarkrog A., Nielsen S. P. Environmental Radioactivity in the Antarctic Peninsula Area: Materials of Inter. Conf. on Environ-mental Radioactivity in the Arctic and Antarctic. Kirkenes, 1993. 183 p.

Rudels B. On the mass balance of the Polar Ocean, with special emphasis on the Fram Strait // Norsk Polarinstitutt Skrifter. 1987. V.188. P.53.

Ryabinin A.I., Belyavskaya V.B., Dolotova I.S. Technogenous radionuclides and strontium in the Azov Sea in 1987-1988 // Trudy GOIN. 1992. Iss.205. P.96-105.

Salbu B., Nikitin A. I., Strand P. et al. Radioactive contamination from dumped nuclear waste in the Kara Sea-results from the joint Russian-Norwegian expeditions in 1992–1994 // The Science of the Total Environment 1997. V.202. P.185–198.

Saetre R., Foyn L., Klungsoyr I., Loeng H. The Barents Sea Ecosystem-Man and Barents Sea Ecosystem. Groningen. 1992. 30 p.

Saetre R. The sunken Nuclear Submarine in the Norwegian Sea – A Potential pollution problem? Bergen: Inst. Mar. Res., 1994. P.28–34.

Safyanov G.A. Estuaries. Moscow: Nauka, 1987. 189 pp.

Sapozhnikov Yu.A., Plishkin A.N. Cesium-137 in the bottom sediments of the Kara Sea western part // Doklady RAN. 1994. V.335, No 1. P.98-100.

Savinov V.M. Estimation of the Barents Sea primary production by the utilization of the solar energy // Theoretical approaches to the study of ecosystems of the Arctic and Subarctic seas. Apatity: Kola Sci. Centre Publ., 1992. P.18-22.

Savinova T.N., Gabrielssen G. Chlorinated hydrocarbons and heavy metals in the Barents Sea birds. Preprint. Apatity: Kola Sci. Centre Publ., 1994. 36 pp.

Savchuk O.P., Matishov D.G. On the positing of problems of the Barents Sea ecosystem modeling. Preprint. Apatity: Kola Sci. Centre Publ., 1990. 46 pp.

Schlosser P., Swift J. H., Lewis D., Pfirman S. L. The role of the large-scale Arctic Ocean circulation in the transport of contaminants // Deep-Sea Res. 1995. V.42, № 6. P.1341–1367.

Sediment K_dS and Concentration Factors for Radionuclides in the Marine Environment: Technical Report Series. № 247. Vienna: IAEA, 1985. 74 p.

Seliverstov V.S., Korytov V.G., Shulga A.D. Estimation of the abundance and biomass of the mature part of the Barents Sea cappelin population by the Allan method // Trudy PINRO. 1977. Iss. 38. P.106-110.

Shvedov V.P., Patin S.A. Radioactivity of seas and oceans. Moscow: Atomizdat, 1968. 288 pp.

Sickel M. A., Selnas T. D., Christensen G. C., Strand P. Radioactivity in the Marine Environment // Rep. from the National surveillance programm Strallvern Rapport. 1995. P.23–24.

Sivintsev Yu.V., Kiknadze O.E. Radioecological danger of the ship nuclear reactors sunken in the Arctic // Atomnaya energiya. 1995. V.79, Iss.3. P.204-211.

Smith J. N., Ellis K. M., Kilius L. R. I-129, Cs-137 tracer measurements in the Arctic Ocean // Deep-Sea Res. 1998. № 6. P.959–984.

Smith J. N., Ellis K. M., Naes K., Matishov G. Radioactivity levels in Barents Sea sediments of Novaya Zemlya // Materials of Inter. Conf. on Environmental Radioactivity in the Arctic and Antarctic. Kirkenes, 1993. P.17–25.

Smith J. N, Ellis K. M, Aarkrog A., Dahlgaard H., Holm E. Sediment mixing and burial of the ^{239,240}Pu pulse from the 1968 Thule, Greenland nuclear weapons accident // J. Environ. Radioactivity. 1994. № 25. P.135–159.

Smith J. N., Ellis K. M., Naes K., Matishov D., Dahle S. Sedimentation and mixing rates of radionuclides in Barents Sea sediments off Novaya Zemlya // Deep-Sea Res. 1995a. № 6. P.1471–1493.

Smith J. N., Ellis K. M., Forman S., Polyak L., Ivanov G., Matishov D., Kilius L., Dahle S. Radionuclide sources in the Barents and Kara Seas // Environmental Radioactivity in the Arctic / Ed. P. Strand, A. Cooke. Osteras. 1995b. P.179–185.

Sokolov V.E., Arsenjev V.A. Mammals of Russia and adjoining regions. Whalebone whales. Moscow: Nauka, 1994. 200 pp.

Solberg T., Hellstrom Y., Eikelmann I. M. H. Radiocaesium in marine fish, shrimps and scallops in coastal waters of Northern Norway and in the Barents Sea – measured over a period of two years // Environmental Radioactivity in the Arctic / Ed. P. Strand, A. Cooke. Osteras. 1995. P.221–226.

Standards of radiation safety. NRB-76/87. Moscow: Energoatomizdat, 1988. 23 pp.

Stepanets O. V., Borisov A. P., Solovyeva G. Yu. Distribution of anthropogenic radionuclides in the estuaries of Ob and rivers Adjacent Kara Sea // Berichte zur Polarforschung. 1999. V.300. P.132–140.

Stepanov A.V., Tsvetkov O.S., Tishkov V.P., Belyaev B.N., Domkin V.D., Ivanova L.M., Osokina A.A., Plekhov V.S., Bobylev K.L., Iohannessen O. Isotope composition of plutonium in the bottom sediments of the Norwegian and Greenland Seas and identification of contamination sources // Atomnaya energiya. 1999. Iss.87, No 4. P.286-294.

Strand P. Radioactivity, AMAP, Assessment Report: Arctic Pollution Issues, Oslo. 1998. P.526–552.

Strand P., Nielsen S. P. Doses to man from dumping of radioactive waste in the Arctic Seas // Environmental Radioactivity in the Arctic / Ed. P. Strand, A. Cooke. Osteras. 1995. P.156–158.

Strand P., Rudijord A. L., Nikitin A. A survey of Artificial Radionuclides in the Kara Sea: Results from the Norwegian-Russian Expedition to the Barents Sea and Kara Sea. 1992. 15 p.

Strand P., Rudjord A., Salbu B., Christensen G., Foyn L., Lind B., Bjornstad H., Bjerk T. Survey of artificial radionuclides in the Kara Sea // Materials Intern. Conf. on Environmental Radioactivity in the Arctic and Antarctic. Kirkenes, 1993. P.53–55.

Strand P., Sickel M., Aarkrog A. Radioactive Contamination of the Arctic Marine environment // Materials of the Third Inter. Conf. on Environment Radioacti-vity in the Arctic. Tromsш, 1997. P.102–130.

Strand P., Sickel M., Aarkrog A., Bewers J. M., Tsaturov Y., Magnusson S. Radioactive Contamination of the Arctic Marine environment // Radionuclides in the Ocean / Ed. P. Gueguenïat et al. 1996. P.95–119.

Strand P., Nikitin A, Rudjord A. L., Salbu B., Christensen G., Foyn L., Kryshev I. I., Chumichev V. B., Dahlgaard H., Holm E. Survey of Artificial Radionuclides in the Barents Sea and the Kara Sea // J. Environ. Radioactivity. 1994. № 25. P.99–112.

Study on Environmental Radioactivity in Finland // STUK. Helsinki. 1983. 45 p.

Sykes L.R., Dewis D.M. On the capacity of the Soviet strategic weapons // V mire nauki. 1987. No 3. P.21-32.

Szczypa J. Monitoring of radioactive contamination of the shores of the Kola Peninsula, Novaya Zemlya and Franz Josef Land // Disturbance and Recovery in Arctic Lands / Ed. R. M. M. Crawford. Environment, Kluwer Academic Publishers. 1997. V.25. P.391–407.

Szczypa J., Matishov G. G., Matishov D. G., Janusz W., Solecki J. The investigations of concentration of radionuclides in the sediments in the Barents Sea // Polar session. Arctic natural environment problems. Lublin, 1992. P.179–191.

Tande K. E. Ecological investigation of the zooplankton community of Bals-fjorden, northern Norway: population structure and breeding biology of the chaeto-gnath Sagitta elegans verrill // J. Exp. Mar. Biol. and Ecol.1989. V.68. P.13–24.

Tantsyura A.I. On the Barents Sea currrents // Trudy PINRO. 1959. Iss.11. P.35-53.

Tantsyura A.I. On the seasonal shanges of the Barents Sea currents // Trudy PINRO. 1973. Iss. 34. P.108-113.

Timoshchuk V.I. Strontium in the waters of the Azov and Caspian Seas // Marine radioecology / Ed. by G.G. Polikarpov. Kiew: Naukova dumka, 1970. P.151-158.

Timofeev S.F. Ecology of onthogenesis of Euphausia crustaceans (Crustacea, Euphausiacea) of the northern seas. Saint-Petersberg: Nauka, 1996. 156 pp.

Timofeev S.F., Shirokolobova O.V. Trophic structure of the Barents Sea pelagic zone in the area of the Atlantic water mass influence // Influence of the Gulf Stream on the life in the pelagic zone of Arctic Seas. Apatity, Kola Sci. Centre Publ., 1995. P.104-115.

Timoshchuk V.I., Sokolova I.A. Radioecological investigations of Danube and adjoining part of the Black Sea // Marine radioecology / Ed. by G.G. Polikarpov. Kiew: Naukova dumka, 1970. P.142-150.

Trapeznikov A.V., Pozolotina V.N., Yushkov P.I., Trapeznikova V.N., Chebotina M.Ya., Aarkrog A., Nielsen S.P. Investigation of radioactive contamination of the Ob-Irtysh river system by discharges of the integrated plant «Mayak» // Third congress on the radiation researches. Moscow, 1997. P.335-336.

Treshnikov A.F. Actual problems of the polar oceanology // Ocean researches / Ed. by A.F. Treshnikov et al. Moscow: Nauka, 1984. P.38-53.

UNSCEAR Ionizing Radiation: Sources and biological effects. United Nations Scientific Committee on the effects of atomic Radiation 1982: Report to the General Assembly. 1982.

Uspenskaya E.Yu., Kalyakin V.N., Khakhin G.V. Radiation contamination on the Novaya Zemlya Islands // Vestnik MGU. Ser.5. Geogr. 1999. No 6. P.46-50.

Uspensky S.M. Bird bazaars of the Novaya Zemlya. Moscow: Nauka, 1956. 178 pp. [*Успенский С. М.* Птичьи базары Новой Земли. М.: Наука, 1956. 178 с.]

Vakulovsky S.M., Nikitin A.I., Chumichev V.B. On the contamination of Arctic seas by the radioactive wastes of West European radiochemical plants // Atomnya energya. 1985. V.58, Iss.6. P.445-449

Vakulovsky S.M., Nikitin A.I., Chumichev V.B. Contamination of the White Sea by the radioactive wastes of West European counries // Atomnaya energya. 1988. V.65, Iss.1. P.66-67

Vakulovsky S.M., Kryshev I.I.,, Nikitin A.I., Savitsky Yu.V., Malyshev S.V., Tertyshnik E.G. Estimate of the influence of the Krasnojarsk mine and chemical complex on the radioecological state of Yenisey River // Proceedings of high school. Nuclear energetics. 1994. No 2-3. P. 32-35.

Vakulovsky S., Nikitin A., Chumichev V., Katrich I. Yu. Radioactive Contamination of Russian Arctic Seas from results of observation during the years 1960–1962 // Environmental Radioactivity in the Arctic and Antarctic / Ed. P. Strand, E. Holm. 1993. P.177–179.

Vakulovsky S., Nikitin A., Chumichev V., Katrich I. Yu., Voitsekhovich O. A., Medinets V. I., Pisarev V. V., Bovkun L. A., Khersonsky E. S. Cs-137 and Sr-90 Contamination of Water Bodies in the Areas Affected from the Chernobyl Nuclear Power Plant Accident: An Overview // J. of Environmental Radioactivity. 1994b. № 24. P.103–122.

Vartanov R., Hollister C. Nuclear legacy of the cold war // Mar. Pol. 1997. V.21, № 1. P.1–15.

Vernadsky V.I. Biosphere. Leningrad Sci. and Tech. Chemical Publishing House, 1926. 146 pp.

Vernadsky V.I. On the concentration of radium by living organisms // Doklady AN SSSR. 1929. No 2. P. 33-34.

Vise V.Yu. Ice drift from the kara to the greenland Seas // Problemy Arktiki. 1937. No 1. P. 103-116.

Vinogradov A.P. Introduction to the Ocean geochemistry. Moscow: Nauka, 1967. 215 pp.

Vozzhinskaya V.B., Kuzin V.S., Shmelev I.P., Pestrikov V.V. Radionuclide content in laminaria algae of the Arctic coastal ecosystems // Doklady RAN. 1994. V.336, No 2. P.269-272.

Volkov V.A., Lukin V.V. Water masses and water stratification of the Arctic basin // Trudy AANII. 1977. V.398. P.30-39.

Yablokov A.V., Karasev V.K., Rumyantsev V.M., Kokeev M.E., Petrov O.I., Lystsov V.N., Emeljanenkov A.F., Rubtsov P.M. Facts and problems connected with the burial of radioactive wastes in the seas surrounding the territiry of Russian Federation // Materials of the report of Governmental Commission on the issues connected with the burial of radioactive wastes. Moscow: Administration of President of RF, 1993. 108 pp.

Yudakhin F.N., Kiselev G.P., Kiseleva I.M. Technogenous radioactive isotopes in the West Arctic Seas in the period 1972-1996. Proc. Rus. Geogr. Soc. 1998. V.130, Iss.5. P.13-21.

Zakharov V.D. Arctic ices and modern natural processes. Leningrad: Gidrometeoizdat, 1981. 180 pp.

Zenkevich L.A. Biology of the Seas of USSR. Moscow-Leningrad: Academy Publishing House, 1963. 739 pp.

The New Springer Global Website
Be the first to know
Benefit from new practice-driven features.
Search all books and journals – now faster and easier than ever before.
Enjoy big savings through online sales.
springeronline.com – the innovative website with you in focus.
springeronline.com
The interactive website for all Springer books and journals
Springer
010048x

Printing: Mercedes-Druck, Berlin
Binding: Stein+Lehmann, Berlin

GPSR Compliance

The European Union's (EU) General Product Safety Regulation (GPSR) is a set of rules that requires consumer products to be safe and our obligations to ensure this.

If you have any concerns about our products, you can contact us on ProductSafety@springernature.com

In case Publisher is established outside the EU, the EU authorized representative is:

Springer Nature Customer Service Center GmbH
Europaplatz 3
69115 Heidelberg, Germany

Batch number: 10620056

Printed by Printforce, the Netherlands